Sylvia Schein is Senior Lecturer in History at the University of Haifa.

Fideles Crucis

FIDELES CRUCIS

The Papacy, the West, and the
Recovery of the Holy Land
1274–1314

SYLVIA SCHEIN

CLARENDON PRESS · OXFORD
1991

Oxford University Press, Walton Street, Oxford OX2 6DP
Oxford New York Toronto
Delhi Bombay Calcutta Madras Karachi
Petaling Jaya Singapore Hong Kong Tokyo
Nairobi Dar es Salaam Cape Town
Melbourne Auckland
and associated companies in
Berlin Ibadan

Oxford is a trade mark of Oxford University Press

Published in the United States
by Oxford University Press, New York

© Sylvia Schein, 1991

All rights reserved. No part of this publication may be reproduced,
stored in a retrieval system, or transmitted, in any form or by any means,
electronic, mechanical, photocopying, recording, or otherwise, without
the prior permission of Oxford University Press

British Library Cataloguing in Publication Data
Schein, Sylvia
Fideles crucis: the papacy, the West and the recovery of
the Holy land, 1274–1314.
1. Crusades
I. Title
909.7
ISBN 0-19-822165-7

Library of Congress Cataloging in Publication Data
Schein, Sylvia.
Fideles crucis: the Papacy, the West, and the recovery of the
Holy Land, 1274–1314/Sylvia Schein.
p.   cm.
Revision of the author's thesis (Ph. D.)—University of Cambridge,
1980.
Includes bibliographical references and index.
1. Crusades—Later, 13th, 14th, and 15th centuries.   2. Papacy—
History—To 1309.   I. Title.
D171.S34 1991
909'.2—dc20   90-39512
ISBN 0-19-822165-7

Typeset by Rowland Phototypesetting Ltd,
Bury St Edmunds, Suffolk
Printed and bound in
Great Britain by Biddles Ltd,
Guildford & King's Lynn

In Memoriam
Raymond Charles Smail
1914–1986

PREFACE

This book is a revised and enlarged version of a dissertation which was accepted for the degree of Ph.D. in Cambridge in 1980. The subject is due to the inspiration of the late Dr R. C. Smail of Sidney Sussex College, Cambridge, under whose supervision the dissertation was written. The late Professor J. Prawer of the Hebrew University of Jerusalem supervised it during the research period in Israel. I owe them both more than I can say.

The completion of the study was made possible by the awards granted by the British Council, the Memorial Foundation for Jewish Studies, the Hebrew University Friends in London, the Alexander von Humboldt-Stiftung, and Haifa University. The research was conducted while I was in residence in Lucy Cavendish College, Cambridge, in the Cambridge University Library, the British Museum, the Bibliothèque Nationale, Paris, and the libraries of Haifa University, the Hebrew University of Jerusalem, and the Monumenta Germaniae Historica in Munich. I would like to express my thanks to the offices of all these institutions for their assistance. The final version of this study was written during my stay at the hospitable Institute for Advanced Studies of the Hebrew University in Jerusalem and at the Monumenta Germaniae Historica in Munich whose staff provided me with most patient and skilful assistance.

During the period of research I was happy to benefit, in addition to my supervisors, from the advice of a number of colleagues and scholars. My particular thanks go to Professor J. Riley-Smith, the late Professor L. Butler, Professor B. Z. Kedar, Professor J. N. Hillgarth, Dr D. S. H. Abulafia, Professor H. E. Mayer, Professor A. Grabois, Dr M. Barber, and Dr C. J. Tyerman. They read the typescript and their comments and suggestions were a source of encouragement and help.

S.S.

CONTENTS

ABBREVIATIONS	ix
INTRODUCTION	1
1. *1274–1276* After St Louis: The Beginning of a New Era	15
2. *1276–1291* Between *Crux Cismarina* and *Crux Transmarina*	51
3. *1291–1292* The Loss of the Holy Land and the First Attempts at its Recovery	74
4. *1291–1292* The Loss of the Holy Land in Public Opinion	112
5. *1292–1305* The Years of Transition	140
6. *1305–1308* In Search of a Project	181
7. *1309* The Papal–Hospitaller *passagium particulare* and the 'Crusade of the Poor'	219
8. *1310–1314* The Crusade at the Council of Vienne	239
CONCLUSION	258
APPENDIX 1: List of *de recuperatione Terrae Sanctae* Treatises 1274–1314	269
APPENDIX 2: The Monetary System and Prices	271
BIBLIOGRAPHY	273
INDEX	299

ABBREVIATIONS

AA	*Acta Aragonensia*, ed. H. Finke (Leipzig and Berlin, 1908–22)
AE	*Annales Ecclesiastici ab anno 1198 usque ad annum 1565*, ed. O. Raynaldus (Lucca, 1738–59)
AHR	*American Historical Review*
AOL	*Archives de l'Orient latin*
BEC	*Bibliothèque de l'École des Chartes*
BIHR	*Bulletin of the Institute of Historical Research*
CCCM	*Corpus Christianorum: Continuatio Medievalis* (Turnhout, 1951–)
DHGE	*Dictionnaire d'histoire et de geographie eccesiastique* (Paris, 1912–)
EHR	*English Historical Review*
Golubovich	*Biblioteca bio-bibliografica della Terra Santa e dell'Oriente francescano*, ed. G. Golubovich (Quaracchi, 1906–27)
HL	*Histoire littéraire de France* (Paris, 1933–)
Mansi	*Sacrorum conciliorum nova et amplissima collectio*, ed. G. D. Mansi (Florence and Venice, 1759–98)
MGH	*Monumenta Germaniae Historica*
MGH SRG	*Monumenta Germaniae Historica: Scriptores rerum Germanicarum in usum scolarum* (Hanover, 1839–)
MGH SS	*Monumenta Germaniae Historica: Scriptores*, ed. G. H. Pertz *et al.* (Hanover, Weimar, Berlin, Stuttgart, Cologne, 1826–1934)
MIÖG	*Mittheilungen des Österreichischen Instituts für Geschichtsforschung*
PL	*Patrologiae cursus completus: Series Latina*, ed. J.-P. Migne (Paris, 1844–64)
Potthast	*Regesta Pontificum Romanorum inde ab anno post Christum natum 1198 ad annum 1304*, ed. A. Potthast (Berlin, 1874–5)
RHC Hist. arm.	*Recueil des historiens des croisades: Historiens arméniens* (Paris, 1869–1906)
RHC Hist. occ.	*Recueil des historiens des croisades: Historiens occidentaux* (Paris, 1844–95)

Abbreviations

RHGF	*Recueil des historiens des Gaules et de la France*, ed. M. Bouquet *et al.* (Paris, 1737–1904)
RHSE	*Revue historique du Sud-Est Européen*
RIS	*Rerum Italicarum Scriptores*, ed. L. A. Muratori (Milan, 1723–38)
RIS NS	*Rerum Italicarum Scriptores* NS, ed. G. Carducci *et al.* (Città di Castello and Bologna, 1900–)
Röhricht, *Regesta*	*Regesta regni Hierosolimitani 1097–1291*, ed. R. Röhricht (Innsbruck, 1893); *Additamentum* (1904)
ROL	*Revue de l'Orient latin*
RS	(Rolls Series) *Rerum Britannicarum Medii Aevi Scriptores* (London, 1858–96)
Setton's *History*	*A History of the Crusades*, ed. K. M. Setton *et al.* (Madison, 1969–)
TRHS	*Transactions of the Royal Historical Society*
ZDPV	*Zeitschrift des Deutschen Palästina-Vereins*

INTRODUCTION

The fall of Acre (28 May 1291), the last major bastion of Christian Outremer, brought to an end the two-hundred-year-long Christian domination of Syria and Palestine. Yet the contemporaries of the event hardly perceived it as final and irreversible. The loss of Acre—and of the Holy Land—was regarded as a sad but ephemeral episode, a temporary setback, a transient chastisement to propitiate the Divinity. It was generally believed and expected that the situation would soon be reversed, the Holy Land recovered, and a new kingdom created, a kingdom in which lessons of the past would be applied and which would therefore follow a new course of history.

Nevertheless, the traumatic experience could not have passed without any immediate reaction. The long drawn-out decline of Outremer had culminated in the fall of Acre, its wealthy and densely populated capital, one of the most famous cities in Christendom. The news made an impact throughout western Europe and at all levels of society. Although not everyone was affected, the event was dramatic enough to cause an outcry and to stir public conscience; its initial force left its impression on all future plans and attempts to launch a crusade.

How did the loss of the Holy Land affect the ideology of the crusades and the crusading movement? What reactions can be identified among the governments and social groups in western Europe? How great was the impact on the policy of that main pillar of the movement, the papacy, in terms of planning and attempting to organize a new crusade? These problems, surprisingly enough, are still an almost uncharted territory. One of the reasons why these problems have been so little studied seems to have been a division of effort among the scholars who might have done so. Some have been historians of the Latin Kingdom and of the crusades in the twelfth and thirteenth centuries, while others have concerned themselves with the 'crusades of the later middle ages'. For the first group, which includes some illustrious names,[1] not only the Latin Kingdom but, to quote J. F. Michaud, 'the glorious epoch of the crusades ended in 1291'.[2] They treated developments in the crusading field after that date, therefore, as a kind of epilogue, which they did not closely study. As for those historians, fewer in number, who took as their subject the crusades after 1291, they simply assumed that the final loss of the Holy Land resulted in a strong European reaction, which in its turn became the departure-point of future planning. Taking the events of 1291 as the starting-point of their studies, they regarded the loss of the Holy Land not only as an end of a chapter in crusading history and the beginning of a new one, but also as a definite break in the ideology of the crusades and the thinking which lay behind them.[3] A cataclysmic epilogue for the historians of the 'classical age' of the crusades and a prologue for the historians of the 'crusades in the later middle ages', the reaction itself was somehow lost between general assumptions and the duly re-corded and chronologically arranged summaries of plans for the recovery of the Holy Land. The result is an amorphous picture which rests on sources restricted both in number and in type. The reaction, or rather the variety of reactions and their evolution, their ramifications, and mutual influences have never been satisfactorily described, let alone explained.

Of the scholars mentioned, only R. Röhricht, that most erudite and productive of all historians of the crusades and the crusaders in the East, consulted anything like the full range of the available documentation, as he demonstrates in his detailed account of the last days of the Latin Kingdom and of the crusading movement at the end of the thirteenth and the beginning of the fourteenth centuries.[4] But Röhricht, true to his

[1] See bibliography for works by F. Wilken, J. F. Michaud, R. Röhricht, L. Bréhier, R. Grousset, P. Alphandéry and A. Dupront, S. Runciman, A. Waas, H. E. Mayer, J. Prawer, and E. Stickel.

[2] J. F. Michaud, *Histoire des croisades* (Paris, 1824–9), iii. 341 ff.

[3] J. Delaville le Roulx, *France en Orient au XIVe siècle: Expeditions du maréchal Boucicaut* (Paris, 1885–6); N. Iorga, *Philippe de Mézières 1327–1405 et la croisade au XIVe siècle* (Paris, 1896); F. Heidelberger, *Kreuzzugsversuche um die Wende des 13. Jahrhunderts* (Leipzig and Berlin, 1911); L. Thier, *Kreuzzugsbemühungen unter Papst Clemens V 1305–1314* (Düsseldorf, 1973); A. S. Atiya, *Crusade in the Late Middle Ages* (London, 1938).

[4] R. Röhricht, 'Die Eroberung Akkas durch die Muslimer im Jahre 1291', *Forschungen zur Deutschen Geschichte*, 20 (1879); id., 'Untergang des Königreiches Jerusalem', *MIÖG* 15 (1894); id., 'Études sur les derniers temps du royaume de Jérusalem', *AOL* 1 (1881), 2 (1884).

way of writing, listed rather than analysed his enormous source-material. He seldom went beyond historical annalistic narratives proper. And yet the collection of sources listed by Röhricht (though it is not exhaustive) clearly indicated that the reaction to the loss of the Holy Land was neither socially restricted, nor was the crusade in the years that followed ignored. However, Röhricht's painstakingly collected source-material was overlooked by the historians who might have used it, and research remained in many areas at the stage in which Röhricht left it. The most neglected of all is the period immediately preceding the loss of the Holy Land, the years between the death of Pope Gregory X (1276) and the fall of Acre (1291). There are some references in the monograph of M. H. Laurent on Innocent V, R. Sternfeld on Nicholas III, and N. Schöpp on Hadrian V as well as in the studies of M. Purcell, K. M. Setton, and N. Housley on papal crusading policy.[5] P. A. Throop's well-known analysis of criticism of the crusade, which is mainly concerned with the pontificate of Gregory X (1271–6), deals briefly with the subject within a chapter on 'The Failure of Gregory X's Plans for the Crusade'. Throop's study will always remain valuable because of the immense volume of source-material on which it is based; but as the author himself declares, its main theme is 'the decline of the crusades from the viewpoint of an increasingly hostile public opinion in Europe'.[6] The inevitable result of such a prejudiced starting-point is a prejudiced approach to the entire subject-matter of the study. The author tends to find criticism and opposition to the crusade and its ideology even when it is simply not there. He describes the 'Opus tripartitum' of Humbert of Romans, for example, as a work that 'makes one realize to the

---

[5] M. H. Laurent, *Le Bienheureux Innocent V (Pierre de Tarentaisse) et son temps* (Vatican, 1947), 228–55; R. Sternfeld, *Kardinal Johann Gaetan Orsini (Papst Nikolaus III) 1244–1277* (Berlin, 1905; repr. Vaduz, 1965), *passim*; N. Schöpp, *Papst Hadrian V (Kardinal Ottobuono Fieschi)* (Heidelberg, 1916), *passim*; M. Purcell, *Papal Crusading Policy: The Chief Instruments of Papal Policy and Crusade to the Holy Land from the Final Loss of Jerusalem to the Fall of Acre 1244–1291* (Studies in the History of Christian Thought, 11; Leiden, 1975), 3–22; K. M. Setton, *The Papacy and the Levant 1204–1571*, i (Phil., 1976), 123–47; N. J. Housley, *Italian Crusades: The Papal–Angevin Alliance and the Crusades against Christian Lay Powers 1254–1343* (Oxford, 1982), *passim*.

[6] P. A. Throop, *Criticism of the Crusade: A Study of Public Opinion and Crusade Propaganda* (Amsterdam, 1940), p. viii.

full the difficulty of inspiring zeal for the Crusade in a Christendom profoundly discouraged, skeptical, and disgusted'.[7] His general conclusion that 'the evidence of contemporaries throws a great deal of light on the decay of the crusading movement, a decay that had progressed so far that a strong and able pope . . . labored in vain for the recovery of the Holy Sepulchre',[8] seems unjustified. Whereas Throop tended to exaggerate the extent and the amount of the opposition to crusading, the recent J. E. Siberry's *Criticism of Crusading 1095–1274*,[9] intended to replace Throop's study, underestimates it too easily. The critics are dismissed as either the enemies of the papacy or as influenced by personal interests or simply prejudiced. The study, quite unconvincingly, denies the critics any meaningful influence upon public opinion. As a result a detailed and unprejudiced study of the years 1274–91 is still needed.

The more recently explored subject, the 'crusades in the later middle ages', had so far attracted relatively few scholars. Their output has never approached the avalanche of studies of the crusades in their 'classical age'. The activities of the Société de l'Orient latin, inspired by P. Riant,[10] and the studies of J. Delaville le Roulx and N. Iorga[11] considerably modified the view that the crusader movement coincided with the Latin domination of Syria and Palestine, but all those, as A. S. Atiya has pointed out, are important primarily as pioneering works. Yet, without minimizing the scholarly achievement of A. S. Atiya himself, whose *opus magnum*, *The Crusade in the Later Middle Ages*, is still the standard study of the subject (and the only one in English), it hardly replaced older studies and above all those of Delaville le Roulx. Atiya was in fact very much indebted to Delaville le Roulx's *La France en Orient au XIVe siècle*. The sources Atiya used—and here he followed Delaville le Roulx, if not quantitatively at least qualitatively—were restricted to crusade plans and petitions, missionary treatises, and *itineraria*.[12]

---

[7] Throop, *Criticism of the Crusade*, p. 183.   [8] Ibid. 284.
[9] J. E. Siberry, *Criticism of Crusading 1095–1274* (Oxford, 1985), e.g. pp. 151, 187, 213, 217–20.   [10] Namely the *Archives de l'Orient latin* and the *Revue de l'Orient latin*.
[11] Quoted above, n. 3.
[12] Atiya, *Crusade*, pp. v–vi. For Atiya's opinions, see also, ibid. 3–26; id., 'Crusades: Old Ideas and New Conceptions', *Cahiers d'histoire mondiale*, 2 (1954), 469–75; F. Pall 'Croisades en Orient au Bas Moyen Âge: Observations critiques sur l'ouvrage de M. Atiya', *RHSE*, 19 (1942), 521–83.

Introduction

The picture presented by the above-mentioned studies was that of a movement of which the main feature was a flood of new and often far-fetched plans for a new crusade on the one hand and a series of sterile attempts to launch it on the other. Far from being a universal movement, the crusade appears as a papal monopoly compounded by a sort of hobby, bordering on obsession, of isolated groups of naïve individuals, or, in particular cases, a dream of certain megalomaniac rulers.[13]

The studies of Delaville le Roulx and Atiya were complemented by those of F. Heidelberger and L. Thier on the pontificate of Clement V (1305–14).[14] Both are monographs and both restricted the subject-matter to papal policy. Dealing mainly with attempts to launch crusades, both followed Delaville le Roulx as far as crusade-planning was concerned. The sources they used were almost exclusively limited to papal registers and to the collection of documents edited by H. Finke.[15] Their particular contribution was to provide a chronological framework for the attempted crusades; their weakness, even in the limited scope of their studies, in the almost complete disregard of the ideological and socio-political background.

Recently the Holy Land crusade has been discussed within a broader framework. The studies of J. Riley-Smith and N. Housley[16] have firmly established that, as the crusade was by definition an expedition proclaimed by the papacy, in which the participants took a crusading vow and consequently were granted certain privileges, including the indulgence bestowed on crusaders going to the Holy Land, it is impossible to define the crusade in terms of its geographical objective.[17] Housley, in his recent book on the *Italian Crusades* attempted to show that

---

[13] For such comments on the crusade after 1291 see E. Barker, *Crusades* (New York, 1923; repr. 1971), 90–6; F. M. Powicke, *The Thirteenth Century 1216–1307* (Oxford, 1953), 83; Throop, *Criticism of the Crusade*, pp. 262–91; see also e.g. S. Runciman, *History of the Crusades* (Cambridge, 1955), iii. 427–68; H. E. Mayer, *Crusades*, trans. J. Gillingham (Oxford, 1972); J. Prawer, *World of the Crusaders* (Jerusalem and London, 1972), 147–52.

[14] Quoted above, n. 3.

[15] *Papsttum und Untergang des Templerordens*, ed. H. Finke (Münster, 1907); *AA*.

[16] J. Riley-Smith, *What were the Crusades?* (London, 1977), 11–33; Housley, *Italian Crusades*, pp. 4–7.

[17] For an attempt to define the crusade precisely in such geographical terms see Purcell, *Papal Crusading Policy*, pp. 3–22.

the popes of the period treated the crusades against Christian lay powers on an equal footing with those against the Moslem enemy in the East.[18] The idea is a tempting one for any 'Guelf' historian to adopt; but was this attitude of the popes more than just a legal position aimed at the defence of a policy which was bound to be criticized both inside the Church and outside it? Was the papacy indeed treating the crusades against Christian powers in Europe in the same way as those against the infidels in the Holy Land? If this was so, why the link the papacy continually attempted to establish between the various European crusades and the welfare of the Holy Land?[19] It seems that though the crusades against Christian lay powers often took priority over the crusade to the Holy Land, the most sublime form of crusading was still, as Housley does not admit but clearly demonstrates, the crusades to the Holy Land and not those against Christians in Europe.[20]

Moreover, Housley is unconvincing in his analysis of public attitudes to the Italian crusades. He claims that on the whole the crusades against Christian lay powers were regarded like those against Moslems and that public opinion was less overwhelmingly hostile to the use of crusades against Christian rulers than historians have hitherto assumed. Criticism when expressed arose from specific causes not directly linked to the crusades in question. For example, the Christians of Outremer were critical of crusades within Europe because of their belief that they prevented the dispatch of aid to the Holy Land.[21] It seems that Housley is correct in pointing out that the crusades against Christian powers have been over-criticized by historians and have been wrongly described as the misuse of a spiritual weapon for political ends.[22] However, Housley himself takes too seriously the task of *defensor diaboli*. He is not only too readily inclined to dismiss the criticism of the crusade against Christian powers made by modern historians but, more important still, to minimize the objections made in the Middle Ages. His study, exhaustive as it is regarding papal policy and its instruments, lacks a systematic survey of public opinion in

[18] Housley, *Italian Crusades*, esp. pp. 62–70.
[19] Such linkage is clearly demonstrated; see ibid. 35–70 and *passim*.
[20] Ibid., esp. pp. 62–110.
[21] Ibid. 75–9. [22] Ibid. 3–8.

Introduction 7

the thirteenth and fourteenth centuries. Therefore his conclusive assertion that 'most contemporaries did not view the Italian crusades with indignation',[23] should still be regarded as debatable. So is the entire issue of the interaction between the Italian and the Holy Land crusades in terms both of papal policy and its instruments, as well as of public opinion. Once the view is accepted that the papacy, in practice, treated the crusades against the lay powers as equal to those against the Saracens in the East, the subject should be treated *in toto*. Because of the interaction between papal policies in respect of crusades against its Christian opponents in Italy and those against Moslems in the Holy Land, both must be discussed together. If either is considered without the other, the only result will be an incomplete, and perhaps obscure, picture.

There are two recent studies which deal with the crusade to the Holy Land within the framework of papal policy. One is K. M. Setton's monumental *The Papacy and the Levant 1204–1571*. The first volume, dealing with the thirteenth and fourteenth centuries focuses on papal policy towards Byzantium and thus can be hardly considered, as claimed by the author, 'a history of the later crusades'.[24] Papal policy towards the Holy Land is dealt with as a side-issue, in respect of which the author is often content to rely on secondary material. So, for example, his evaluation of the attitude of Europeans to the crusade at the time of the Second Council of Lyons (1274) is seriously biased towards P. A. Throop, *Criticism of the Crusade*.[25] The result is the unconvincing conclusion that 'neither the chivalry of Europe nor the clergy was notably anxious for self-sacrifice in the later thirteenth century... For some time to come the religious motive seemed spent...'[26] On the whole Professor Setton is content, it seems, to adopt every view which minimizes the importance of the crusade both in the mentality of contemporaries and in papal policy. Thus, when he refers, for example, to the succession of Philip IV to the throne of France, Professor

---

[23] Ibid. 252. For similar critical comments regarding this subject see R. S. Smail's review of Housley in *Journal of Theological Studies*, 35 (1984), 259–61.
[24] Setton, *The Papacy and the Levant*, vol. i, p. vii.
[25] Ibid. 106–22; Throop, *Criticism of the Crusade*, esp. pp. 282–91.
[26] Ibid. 109.

Setton quotes J. R. Strayer's statement that: '1285 marks . . . the end of the crusade as a regular and reliable instrument of papal policy . . . deprived of the steady support of the French king the pope was in a poor position to combat the rising tide of secularism and indifference'.[27] This rather careless statement is certainly incorrect. First of all, the crusade, as it persisted after 1285, was anything but a powerful instrument of papal policy. Secondly, it is true perhaps that Philip was far less inclined than his father to support some of the policies of popes like Honorius IV and Boniface VIII but subsequently he lent steady support to Benedict XI and Clement V.

The second study of papal policy is Maureen Purcell's book on the period 1244–91. Its title is, however, misleading. It is a monograph which deals almost entirely two or the chief instruments of papal crusading policy, with vows and indulgences and their administration. There is no attempt to examine papal crusading policy in relation either to the West or Outremer nor to examine papal crusading politics.[28]

The limited type of sources used in some studies—crusading literature (Delaville le Roulx and Atiya), papal and ecclesiastical documentation (Heidelberger and Thier) and limited subject-matter (Housley, Setton, and Purcell)—has left a deep mark on the current state of research. There is actually no concept of a crusading movement after the disastrous crusade to Tunis of Louis IX, but rather a discussion of isolated events precariously strung together by an elusive goal and a fluid, non-defined ideology. A *vue d'ensemble* therefore is still missing. A broader and more serious outlook, which could legitimately furnish a comprehensive framework is absent; it appears as if plans and acts following the loss of the Holy Land were newly sprung events, a creation *ex nihilo*. Studied by historians whose departure-point was the débâcle of 1291, the crusades in the years to follow were treated without any reference or relation to the past. This could only result in a biased view of the phenomenon as a whole. A good example of such an approach is

---

[27] Setton, *The Papacy and the Levant*, vol. i, p. 146; J. R. Strayer, 'Crusade against Aragon', *Speculum*, 28 (1953), 102–3. It is worth pointing out that Strayer thought it necessary to devote only two pages to the theme of the king's attitude to crusade in his 450-page study *Reign of Philip the Fair* (Princeton, NJ, 1980), see pp. 296–7.
[28] Purcell, *Papal Crusading Policy, passim.*

Introduction 9

the recent study by Thier. Whatever its intrinsic merits, it treats the crusades without any reference to the wider European scene; in the best of cases the European background is presented as something given and static, and not, as it really was, in a process of change. When referred to, it is in the framework of what seems to be the most irrelevant, though certainly most favourite subject, namely the interpretation of the repeated failures of attempts to launch a crusade. The long list of causes put forward consists in fact of the symptoms and characteristics of a new age, late medieval Europe.[29] An exception to this kind of approach is the treatment of the crusade around 1300 by J. N. Hillgarth in his study of Ramon Lull.[30] Though the crusade is marginal to the main subject-matter, here a serious attempt is made to insert and place the crusade in the framework of the everchanging European scene.

This critical review of the existing literature indicates that a different approach to the subject is desirable. In the first place a more extensive range of source-material needs to be consulted. In the past, entire categories of sources have been completely neglected and others read out of context. The categories of sources to be brought into consideration go beyond both crusading literature *stricto sensu* and papal and diplomatic official documents. They include decrees of Church councils, documents of the military and monastic orders, royal and princely correspondence, both secular and ecclesiastical, chronicles and annals, pseudo-prophetic treatises, and even purely literary works. Such material, coming often from unexpected quarters, is surprisingly rich.

The analysis of these sources shows that the subject-matter cannot be treated as if it was anchored to the mood prevalent in Europe in the period immediately following the loss of the Holy Land. It is necessary to go back to the earlier period between the last major pan-European crusade (1270) and the fall of Acre in 1291. Only thus can the difference in the moods before and after 1291 be established and the impact of the loss of the Holy

---

[29] L. Thier, *Kreuzzugsbemühungen*, *passim*. See also e.g. P. Alphandéry and A. Dupront, *Chrétienté et l'idée de croisade* (Paris, 1954–9), ii. 274–5; Heidelberger, *Kreuzzugsversuche*, pp. 8–9; E. Stickel, *Fall von Akkon: Untersuchungen zum Abklingen des Kreuzzugsgedankens am Ende des 13. Jahrhunderts* (Berne and Frankfurt-on-Main, 1975), 96–216. Stickel follows Throop, *Criticism of the Crusade* almost word for word.

[30] J. N. Hillgarth, *Ramon Lull and Lullism in Fourteenth-Century France* (Oxford, 1971).

Land estimated and put into relief. This analysis also shows that the alleged dichotomy between the crusades of the classical age of the twelfth and thirteenth centuries, and those of the later Middle Ages, is an imaginary and artificial construction created by historians who never looked back to the crusade-planning of the late thirteenth century. As both Riley-Smith and Richard have recently implied, it is 1270 and not 1291 that marks an end of well-defined period in the history of the crusading movement.[31] Many of the ideas considered by historians to be new in the period after 1291, and to be typical and representative of that period, were actually already in the air at the time of the Second Council of Lyons (1274) and perhaps even earlier. Prevalent at least a quarter of a century before the loss of Acre, they became only more accentuated and better defined in the decades before and after 1300. The evidence shows that the loss of the Holy Land marked neither an end nor a fresh beginning. Its effect was that of a catalyst, a crystallizing of attitudes and policies which were already current.[32] This prompted the formation of a question: Was it the loss of the Holy Land which brought about the change of attitudes which is visible between 1291 and 1314? The wider range of sources which have been used show that the change may have been brought about by other conditions also; the importance of such conditions had therefore to be evaluated. Thus the scope of this study was markedly broadened. The crusading movement underwent more than the single impact of the loss of the Holy Land. The subject cannot therefore be discussed within the framework of the history of the Latin establishments in the East only, nor can it be squarely placed under the heading of that literature which is devoted to the theme *de recuperatione Terrae Sanctae*. Anchored in both, it transcends their scope. One needs to take into consideration contemporary developments in western Christendom, which might have left their mark on Europe's attitude to the crusade. The problem has to be viewed in its European setting. It can be demonstrated that this approach is

---

[31] Riley-Smith, *What were the Crusades?*, pp. 50–1; J. Richard, *Saint Louis* (Paris, 1983), 574.
[32] This is implied by A. Luttrell, 'The Crusade in the Fourteenth Century', in J. R. Hale *et al.* (eds.), *Europe in the Late Middle Ages* (London, 1965), 126–7; Purcell, *Papal Crusading Policy*, p. 6.

vindicated by the sources. Attitudes to the crusade, and attempts to plan and organize such projects, proved to be more a part and parcel of European developments or a reflection of European problems than the result of well-thought-out ideas of popes or of other individuals concerned with the Holy Land. Account had to be taken of political, economic, and intellectual changes current in Europe in the years on either side of 1300. Papal policies, power politics of European secular rulers, economic enterprises and interests, expansive tendencies of governments, events in the Levant, Hospitaller expeditions, *Pastoraux* movements, activities of individual crusade-planners when fully investigated, fill many a gap and become linked and interrelated so as to present a far more coherent picture of the crusading movement.

In this context it was tempting to treat the subject as a study of public opinion, of which so many are made nowadays. However, the reconstruction of the mentality of a medieval society is always difficult, and in this instance the sources raise special problems. The information which they provide about different social groups is markedly uneven, and so is that concerning any one group within a particular period of time. Sometimes it is possible to discern, on a particular occasion, the expression of a commonly held attitude or opinion within a social group, but usually we are left in the dark about the thoughts and emotions of a majority which is too large and too silent. In those circumstances it seems better to refer to those attitudes and opinions which can be identified, rather than to public opinion in general. It is sometimes possible to link them with particular groups, and when there is evidence about a number of groups, it may be possible to speak in terms of a prevailing mood or attitude.

The period investigated in this study had been limited to that between the Second Council of Lyons (1274) and the aftermath of the Council of Vienne (1311–12), namely 1314. The *terminus a quo* needs some qualification. The Second Council of Lyons —the peak of the crusading activities that characterized the pontificate of Pope Gregory X (1272–6)—reflects the beginning of a new age in the history of the crusades. It marks a new trend in crusade-strategy as well as a new attitude to the concept of the crusade. These new views expressed at this period became more pronounced immediately after the loss of the Holy Land. It was, however, only some four decades later at the Council of Vienne that those views were fully defined. At Vienne the heads of Europe took official and final decisions on a crusade, summing up in more than one sense crusader thinking since the Second Council of Lyons. Moreover, during the second session of the council (3 April 1312) a new general crusade was officially proclaimed for the first time since the loss of the Holy Land; however, this general crusade as seen and planned at Vienne was a far cry from that envisaged at the Second Council of Lyons. Thus the Council of Vienne and its aftermath—the taking of the Cross by Philip IV of France, the most powerful of European monarchs (5 June 1313)—marks, as the end of a well-defined period in European attitudes to the crusade and the Holy Land, the *terminus ad quem* of this study.

The two opening chapters deal with the attitude of Europeans to the crusade in the twenty years preceding the fall of Acre. Chapter 1 describes the current views and attitudes during the pontificate of Pope Gregory X and analyses the success of this great pope's project for a crusade. Chapter 2 is devoted to one of the most neglected periods in the history of the crusading movement, the years 1276–91, and seeks to explain why the West then almost entirely abandoned the Latin Kingdom to its fate, thus indirectly causing its fall. Chapters 3 and 4 describe the immediate reaction to the news of the loss of the Holy Land. Plans and preliminary moves for the recovery of the Holy Land which immediately followed its loss formed, as will be shown, only a part of Europe's immediate reaction. The deep stir caused by the loss found its expression in an avalanche of accusations, justifications, explanations, and interpretations which originated in almost all sections of European society. Chapter 5 covers the years 1292–1305. This period, which begins with the papal interregnum following the death of Nicholas IV (14 April 1292) and ends with the elevation of Clement V (5 June 1305), is a kind of intermission in the series of papal attempts to launch a crusade. It is, however, most important as a transitional period between the immediate reaction and the resumption of active crusade-making under Clement V. Chapters 6 and 7 deal with one of the most intensive periods of crusade-planning and making. The general expectations, current since around 1295, and intensified by the events of 1300, seemed to become more tangible after 1305. Propitious political conditions and vested State interests made Philip IV of France the *spiritus movens* of the crusading movement. The envisaged double sponsorship of the crusade by the two heads of Christendom—the pope and the king of France—made the realization of a general crusade more feasible than ever. The overall picture of the period was that of tremendous agitation, incessant diplomatic activities, and efforts to adapt the new concept of the crusade, and especially its strategy, to the circumstances of the Avignon Papacy, and a divided, particularist, and centrifugal Europe. This became clearly evident at the Council of Vienne, the subject-matter of the last chapter of this study. As for crusade-planning, the new general crusade officially proclaimed at Vienne was actually meant to be the final stage of the western thrust against Islam by the means of the maritime blockade. Thus the *passagium generale*, as envisaged by the Council of Vienne, was a far cry from that envisaged by the Second Council of Lyons some forty years earlier.

I

1274–1276
AFTER ST LOUIS:
THE BEGINNING OF A NEW ERA

During the two decades between the crusade of St Louis to Tunis in 1270 and the final loss of the Holy Land in 1291, the only serious attempt in the West to organize a general crusade to the East was made during the pontificate of Gregory X (1272–6). During those four years the organization of a general crusade for the aid of the Holy Land was at the centre of European politics.

The New Strategic Approach to Crusade

Such a crusade, however, never departed to the Levant, and this failure to organize a major international expedition has been seen by some historians as the consequence of a growth both in criticism of the crusading movement and in opposition to its ideology.[1] A number of recent studies, however, have assigned the failure to launch a large expedition to disenchantment with this particular kind of crusade rather than with the crusade as such.[2] In the 1260s there was a loss of confidence in the traditional form of major crusade. Face with a series of defeats at the hands of Baybars and of territorial losses to him (Caesarea, Haifa, Arsuf, Safed, Toron, Belfort, Chastel Neuf,

[1] P. A. Throop, *Criticism of the Crusade: A Study of Public Opinion and Crusade Propaganda* (Amsterdam, 1940), 15–68; id., 'Criticism of Papal Crusade Policy in Old French and Provençal', *Speculum*, 13 (1938), 379–412; J. Prawer, *Histoire du royaume latin de Jérusalem* (Paris, 1969), ii. 382–9; E. Stickel, *Fall von Akkon: Untersuchungen zum Abklingen des Kreuzzugsgedankens am Ende des 13. Jahrhunderts* (Berne and Frankfurt-on-Main, 1975), 96–252; see also G. B. Flahiff, '*Deus non vult*: A Critic of the Third Crusade', *Medieval Studies*, 9 (1947), 162–88.
[2] J. Riley-Smith, 'Note on Confraternities in the Latin Kingdom of Jerusalem', *BIHR* 44 (1971), 301–8; id., *What were the Crusades?* (London, 1977), 50–1, 62–5; B. Z. Kedar, 'The Passenger List of a Crusader Ship 1250: Towards the History of the Popular Element on the Seventh Crusade', *Studi Medievali*, 13 (1972), 267–79.

Beaufort, Jaffa, and the principality of Antioch), contemporaries came to understand that the traditional crusade was no longer the answer to the urgent needs of the crusader state, and that what it needed above all was a permanent mercenary army, disciplined and dependable. Hence crusade-planning came to reflect a new strategic approach of which the main feature was the creation of permanent garrisons in the Holy Land and the launching of small manageable expeditions which would succeed one another in the East and thus form, to use an expression coined by P. A. Throop, a 'perpetual crusade'. This was described during the pontificate of Boniface VIII by the phrase *passagium particulare*.[3]

This term is less well known than its famous counterpart, the *passagium generale*. Though it was not coined until 1301, and did not become widely used until the pontificate of Clement V (1305-14), it denotes a form of expedition already current in the 1170s (the crusade of Stephen of Blois (1171-2), the German crusade (1172), and that of Philip of Flanders (1177-80)) which was to become even more so during the thirteenth century. Whereas the term 'general crusade' stands for a large international expedition, the term *passagium particulare* refers to one which was small and often locally or nationally organized. The difference between these two types of crusade is therefore quantitative rather than qualitative. Both were considered as crusades *stricto sensu*, namely as expeditions proclaimed by the papacy whose participants took a crusading vow and consequently gained certain privileges, including the indulgences bestowed on crusaders going to the Holy Land.[4] As far as their composition was concerned, there was no marked difference. It is true that the advocates of the strategy of the *passagium particulare* made it increasingly dependent on mercenaries and on the professional warfare of the military orders. Thus Gilbert of Tournai did not hesitate, on the eve of the Second Council of Lyons (1274), to recommend the abolition of personal crusader

---

[3] Boniface VIII, *Registres de Boniface VIII*, ed. G. Digard *et al.* (Paris, 1884-1939), nos. 4380-2; Throop, *Criticism of the Crusade*, pp. 184-213, esp. pp. 197-8.

[4] Nicholas IV, *Registres de Nicholas IV*, ed. E. Langlois (Paris, 1886-91), nos. 6850-1; Clement V, *Regestum Clementis Papae V*, ed. monks of the Order of St Benedict (Rome, 1885-92), no. 2988; L. Thier, *Kreuzzugsbemühungen unter Papst Clement V 1305-1314* (Düsseldorf, 1973), 96 n. 74; Prawer, *Histoire*, ii. 548; see below, chs. 3, 4, 6, 7.

services altogether. However, mercenaries were regularly employed in the thirteenth century, both in general crusades and in the armies of the kingdom of Jerusalem, in which both native Turcopoles and Europeans were employed. On the other hand, *passagium particulare* was not always, even during the fourteenth century, composed of mercenaries or the military orders only, but contained genuine voluntary crusaders as well.[5]

The change that occurred in favour of small crusades manned mainly by professional soldiers meant, theoretically at least, a deliberate attempt to exclude from crusading the unruly popular element which had lent earlier expeditions their enthusiasm, and to replace them with matter-of-fact professionals. Such characteristics have led some historians not only to see a general decline in the crusading movement, but also to believe that the crusade became, as M. Mollat puts it, 'l'affaire des barons', confined to the upper classes of the day. The employment of mercenaries was often interpreted by those scholars as the result of the shrinking of crusader manpower rather than that of a change in strategic thinking.[6] It has been recently shown, however, that the zeal of the popular element, though it diminished in the thirteenth century, did not disappear; the crusading idea did not lose its traditionally wide appeal. The popular element was easily incorporated into the traditional general crusade, as happened during the crusade of Louis IX to Egypt, but this was hardly possible in the new type of crusading expedition, the professional *passagium particulare* which dominated the twenty years preceding the fall of Acre. And yet, it should be remembered that it was the Italian popular element and the burgesses who answered Nicholas IV's appeal to aid the Latin Kingdom after the fall of Tripoli (1289); the fall of Acre resulted in a strong popular pro-crusade fervour in Europe, and Clement V's proclamation of the papal–

---

[5] Gilbert of Tournai, 'Collectio de Scandalis Ecclesiae', ed. A. Stroick, *Archivum Franciscanum Historicum*, xxiv (1931), 40; Throop, *Criticism of the Crusade*, pp. 101–3; R. C. Smail, *Crusading Warfare 1097–1193* (Cambridge, 1956), 93–9; J. Richard, 'Account of the Battle of Hattin referring to the Frankish Mercenaries in Oriental Moslem States', *Speculum*, 27 (1952), 168–77.

[6] M. Mollat, 'Problèmes navals de l'histoire des croisades', *Cahiers de civilisation médiévale*, 10 (1967), 350, *contra* Kedar, 'Passenger List', p. 267.

Hospitaller *passagium particulare* in 1308 attracted popular masses and town burgesses. Thus not even during the fourteenth century did not appeal of the crusade become confined to the upper classes of society.[7] Moreover, the policy of discouraging the poor from crusading was hardly new. Already Pope Urban II had tried to persuade non-combatants not to join the First Crusade. During the course of the twelfth century measures were taken in England, France, and Germany to prevent those unable to bear arms or to support themselves from going to the East. Thus, for example, Emperor Frederick I decreed in 1189 that those who could not finance themselves during the years on a crusade should not be allowed to depart. Following the Third Crusade, the criticism of the participation of the non-combatants in crusades led popes to encourage the redemption or even absolution of vows due to 'want of means or weakness of body'.[8]

The strategy of the *passagium particulare* gained more and more support from the 1260s onwards. By 1267 William II of Agen, patriarch of Jerusalem (1262–1270) had already insisted on the recruitment and maintenance of permanent troops in the Holy Land. It was his view that the Templars should assure a corps of professional archers and crossbowmen whereas the king of Sicily should maintain a fleet of six armed galleys in the waters of Latin Syria. Moreover, he declared, the pope should ban the departure of non-combatants not only from crusading but from pilgrimage to the Holy Land as they would be only a burden for the kingdom.[9] A similar demand, it seems, was also voiced at the Second Council of Lyons (1274) and it certainly appeared in the treatises submitted to Gregory X on its eve. At the time of the council, the strategy of small expeditions had already the support of Gregory X, James I of Aragon, the masters general of the Temple and the Hospital, Erart de

---

[7] Riley-Smith, 'Note', pp. 301–8; Kedar, *Passenger List*, pp. 267–79; see below, chs. 3, 4, 7.

[8] M. Purcell, *Papal Crusading Policy: The Chief Instruments of Papal Policy and Crusade to the Holy Land from the Final Loss of Jerusalem to the Fall of Acre* (Studies in the History of Christian Thought, 11; Leiden, 1975), 118–32; J. E. Siberry, *Criticism of Crusading 1095–1274* (Cambridge, 1956), 25–8.

[9] G. Servois, 'Emprunts de Saint Louis en Palestine et en Afrique: Appendice', *BEC* 19 (1858), 291–3; see also Röhricht, *Regesta*, no. 1347; M. L. de Mas-Latrie, *Histoire de l'île de Chypre sous la règne des princes de la maison des Lusignan* (Paris, 1852–61), ii. 71–2.

Valeri the representative of the king of France, as well as of King Philip III of France himself. It was also supported by Gilbert of Tournai and Humbert of Romans; by the 1290s it was adopted by Edward I of England, Charles II of Anjou, and James I of Aragon.[10]

This dispatch of small expeditions and the maintaining of permanent garrisons in the Latin Kingdom was effectively practised during the second half of the thirteenth century. Louis IX left behind him in the Holy Land (1254), under the command of Geoffrey of Sergines, a contingent of 100 knights with a treasury to employ additional crossbowmen and sergeants; this he financed until his death (1270).[11] In 1273 Louis IX's successor Philip III sent to the Latin Kingdom a contingent of twenty-five knights and 100 crossbowmen under the command of Oliver of Termes. In 1275 the king of France dispatched to the East, under William of Roussillon, a contingent of forty knights, sixty mounted sergeants, and 400 crossbowmen (the latter were in the pay of the Church). Gregory X sent three small contingents which he regarded as precursors of a general crusade. One of these, of 500 knights and footsoldiers, financed with the aid of Philip III and of Charles I of Anjou, departed under Thomas Agni of Lentino, patriarch of Jerusalem (1272–7) in October 1272. The two other expeditions were financed with the aid of Philip III. They sailed in the following year; one, which numbered 400 crossbowmen, sailed under Giles of Sanci, and the other of 300 crossbowmen sailed under Pierre of Aminnes.[12]

---

[10] Throop, *Criticism of the Crusade*, pp. 100–4, 197–9, 228–34, 253; Prawer, *Histoire*, ii. 393–4; below, pp. 85–6, 107–111.

[11] J. R. Strayer, 'Crusades of Louis IX', in *Setton's History*, ii. 508. The French treasury estimated this to cost the king an average of 4,000 *livres tournois* a year between 1254 and 1270. See also W. C. Jordan, *Louis IX and the Challenge of the Crusade: A Study in Rulership* (Princeton, NJ, 1979), 78 and n. 93; Riley-Smith, *What were the Crusades?*, pp. 65–70; Kedar, 'Passenger List', pp. 277–8.

[12] 'Estoire d'Eracles empereur et la conqueste de la Terre d'Outremer', *RHC Hist. occ.* ii. 462–3, 464, 467; Marino Sanudo Torsello, 'Liber Secretorum Fidelium Crucis', in *Gesta Dei per Francos*, ed. J. Bongars (Hannau, 1611; repr. Jerusalem, 1973), ii. 225; Potthast, no. 20534; R. Röhricht, *Geschichte des Königsreiches Jerusalem* (Innsbruck, 1898), 966, 968; Throop, *Criticism of the Crusade*, pp. 229, 273–4.

Gregory X and the Crusade

Though the strategy of *passagium particulare* gained considerable support from the 1260s onward, Pope Gregory X (1272–6) made the organization of the traditional *passagium generale*—the aim of which he defined in his *Constitutiones pro Zelo Fidei* (1274) as *subsidium Terrae Sanctae*[13]—the leitmotiv of his pontificate.

Tedaldo Visconti of Piacenza, archdeacon of Liège had been an ardent advocate of the crusade and took the Cross in St Paul's in London (1267), when he served as an aid to the papal legate to England (1265–9). Involved with King Louis IX's preparations for a crusade, he was on his way to join the saintly king in Tunis when he learnt of his death (1270). Instead he went to the Holy Land where he received the news of his election to the Holy See (1 September 1271).[14] In their notification of Tedaldo's election, the cardinals expressed their hope that the future pope would do all in his power to save the Holy Land whose suffering he had beheld personally. This, they stated, was one of the principal reasons for electing him a pope as during his stay in the Holy Land he had become acquainted with its needs.[15]

Leaving Acre for Europe in early November 1271, the future pope Gregory X delivered a sermon in the church of the Holy Cross on the text 'Let my tongue cleave to the roof of my mouth if I remember thee not, Jerusalem.' (Ps. 136: 6.)[16] Gregory X indeed truly lived up to the text of his sermon. Until his death he preserved a vivid recollection of Jerusalem and worked for its recovery. His genuine devotion to the cause of the Holy Land became the basis of this whole policy. This was clearly seen by his contemporaries such as Salimbene, and one of his anonymous biographers remarked: '[He was] consumed with zeal for the Holy Land'.[17] Indeed Gregory X's obsession with

[13] See in Purcell, *Papal Crusading Policy*, pp. 196–9.
[14] Throop, *Criticism of the Crusade*, pp. 12–13.
[15] 'Estoire', pp. 471–2; *AE* ad an. 1271, no. 15; Throop, *Criticism of the Crusade*, p. 13; see also 'Cronica S. Petri Erfordensis Moderna', *MGH SS* xxx. 407; P. V. Laurent, 'La Croisade et la question d'Orient sous le pontificat de Grégoire X', *RHSE* 22 (1945), 107 and n. 1.
[16] Marino Sanudo Torsello, 'Liber', p. 225; 'Memoria Terrae Sanctae', in 'Deux projets de croisade en Terre Sainte', ed. C. Kohler, *ROL* 10 (1903/4), 435–6.
[17] Salimbene, 'Chronica', *MGH SS* xxxii. 494–5; Throop, *Criticism of the Crusade*, p. 14; P. V. Laurent, 'Croisade', pp. 105–6, 122–3.

sending a crusade there is unequalled in any pope save Innocent III.[18]

From the time of his elevation to the papal throne Gregory worked on two levels. First of all, to ensure immediately the safety and protection of the crusader kingdom, he sponsored the dispatch of small expeditions to the East.[19] At the same time he made preparations for a general council whose primary aim would be, as the pope emphatically stated, to consider means of recovering the Holy Land,[20] though the council also had on its agenda two additional issues: the reform of the Catholic Church and its reunion with the Orthodox Church. The bull of summons to the council, the *Salvator noster in* of 31 March 1272, contained a violent *excitatorium* in which the wretched condition and the devastation of the Holy Land was described in such a way as to justify the classic insult of the Saracens: 'Ubi est Deus eorum' (Ps. 115: 2). It would be the duty of this council, the pope wrote, to find means of extirpating vice, ending the schism, and regaining the Holy Land; suggestions should also be offered as to how to keep the Holy Land once it was regained and restored. During the two years necessary for the preparation of the council, industrious and effective study should be applied to the provision of necessary aid to the Holy Land, and assistance should be given to those sent on its behalf.[21]

Some of the copies of the *Salvator noster in* contained a request for written advice concerning the planned crusade, as well as to how to keep the Holy Land once it is recovered. Such copies were sent to, for example, the archbishop and province of Sens, the archbishop and province of Tours, and the patriarch of Jerusalem.[22] Gregory X, who in requesting such advice, was thus following the example set by Pope Innocent III on the eve of the Fourth Lateran Council (1215), demanded such information from the clergy. Additionally, Gregory sought more professional information. Writing on the eve of the meeting of the council to King Philip III of France (28 August 1273) the

---

[18] Riley-Smith, *What were the Crusades?*, pp. 19–20.
[19] See above, n. 12.
[20] Gregory X, *Registres de Grégoire X*, ed. J. Guirand *et al.* (Paris, 1892–1906), no. 336; Potthast, no. 20754; *AE* ad an. 1273, no. 35; Throop, *Criticism of the Crusade*, pp. 16–17.
[21] Gregory X, *Registres*, no. 160; Throop, *Criticism of the Crusade*, pp. 17–19.
[22] Gregory X, *Registres*, nos. 160, 657; Mansi, xxiv. 39; Throop, *Criticism of the Crusade*, p. 19.

pope requested him to send to the East skilled military men to observe and report the conditions in the area so that the plans for an expedition, which was to aid the Holy Land until the general crusade would take place, would have a solid foundation. Secular rulers on whose advice Gregory X particularly counted were invited to attend the council personally rather than submit memoirs in writing. The pope preferred to have them at Lyons and then to be able to discuss their plans 'intimately'.[23]

The Crusade Memoirs of the Second Council of Lyons: Crusade versus Mission

One of the results of Pope Gregory's demand for written advice was a number of memoirs composed as working papers to be used at conciliar discussions. They dealt with the three subjects on the council's agenda: the reform of the Catholic Church, its reunion with the Orthodox Church, and the crusade. The memoirs were all composed by churchmen and mainly by mendicants. Humbert of Romans and William of Tripoli were both Dominicans; Gilbert of Tournai, a Franciscan; and Bruno was the bishop of Olmütz. This perhaps explains the fact that their memoirs dealt little with the strategy of a crusade, but mainly with subjects like its ideology, preaching, and financing. This feature of the memoirs submitted before the Fathers of Lyons II is especially striking in comparison with those composed after 1291 (but including that of Fidenzio of Padua, who wrote before that year) which are almost exclusively concerned with strategy and tactics, and even with detailed plans to be followed during a crusade.[24] It is true that some of the memoirs of c.1274 contain practical advice regarding the strategy of a crusade, but this advice is extremely meagre. Both Humbert of Romans and Gilbert of Tournai supported the strategy of small expeditions, of the *passagium particulare* or 'perpetual crusade'. Moreover, Humbert of Romans stressed the importance of Europe's sea-power in the struggle against the Saracens

[23] Gregory X, *Registres*, no. 336; *AE* ad an. 1273, no. 35; Throop, *Criticism of the Crusade*, pp. 19–20.
[24] See below, chs. 3, 6, 8.

claiming that 'quia virtus eorum in mari, in comparatione ad virtutem nostrorum, pro nihil reputatur'.[25] Still, as far as advice on crusade-strategy is concerned, Pope Gregory had to seek it in other quarters, in the advice provided during the council by secular rulers of their nuncios as well as by professionals like the masters of the two main military orders of the Holy Land, the Temple and the Hospital.[26]

Another feature of the memoirs is that they often deal with subjects other than the crusade for the aid of the Holy Land. This evolved, it seems, from the very fact that Pope Gregory asked for advice not only on the crusade but also on the two other matters to be debated at the council at Lyons, namely, the reform of the Catholic Church and its union with the Orthodox Church. As Europeans and churchmen the authors of the memoirs were as interested in these subjects, and especially in the reform of the Church, as they were in the crusade; Gilbert of Tournai, for example, regarded the reform of the Church as a necessary preliminary for the organization of a successful crusade.[27] Bruno of Olmütz saw the northern crusade as such a preliminary. His treatise is indeed a plea for a crusade against the pagan Slavs. He argues that the crusade should begin at home where there are innumerable pagans to be brought under the Cross. This should be preceded by the pacification of Europe as well as by the choice, at the council, of a new emperor to settle European affairs. On the whole Bruno of Olmütz's plan was biased in favour of King Ottokar of Bohemia who, as long as Pope Gregory did not acknowledge the election of Rudolph of Habsburg, had hopes for gaining the imperial throne for himself. Bruno of Olmütz tried to direct the papal crusading policy toward northern Europe so that the King of Bohemia would be designated as 'defender of the faith'.[28]

The memoir of the Franciscan Gilbert of Tournai dealt even less with the crusade to the Holy Land than did that of Bruno of Olmütz. The so-called 'Collectio de Scandalis Ecclesiae' is

---

[25] Humbert of Romans, 'Opus tripartitum', ed. E. Brown, in *Fasciculus rerum expetendarum et fugiendarum* (London, 1690), 203–4; Gilbert of Tournai, 'Collectio', p. 40, and below, pp. 32–3.
[26] See below, pp. 32–8.
[27] Gilbert of Tournai, 'Collectio', pp. 39–40.
[28] Throop, *Criticism of the Crusade*, pp. 105–14.

mainly a collection of critical remarks about the morals of both churchmen and laity. The author sees the failure of the West to recover the Holy Land as the result of its sins—*nostris peccatis exigentibus* was a traditional and much-used phrase—and he proceeds to describe some of those sins in detail. His harshest criticism is directed against the clergy. Papal legates are blamed for their inordinate love of fine horses and the clergy at large for their reluctance to make any financial contribution to the crusade. The author also fiercely condemns the practice of the redemption of crusader vows and the abuses of this practice by clergy, as well as the practice of giving the Cross to the weak, old, and unfit; another condemned practice is that of forcing criminals to take the Cross as a means of penance, a practice which had already been strongly criticized by Jacques de Vitry at the beginning of the century. Gilbert of Tournai's charges against the laity are addressed to the nobility who are blamed for their selfish quarrels which hinder the Holy War and for their reluctance to sacrifice their lives and resources for the cause of the crusade.[29]

The positive advice of the author of the 'Collectio' is far more modest than his description of the abuses hindering a successful crusade. He recommends the Church to devote itself to prayers. The recommendation of a 'general contribution' to be ordered for the financing of the crusade implies that the author denounced the practice of crusader tithes.[30] More meaningful and important, however, is Gilbert of Tournai's military advice: 'Let a standing force of mercenaries [*stipendarii*] be assembled to replace the constantly changing soldiers in the Holy Land and let these mercenaries carry on the war of the Lord and the affairs of the crusade.'[31] Gilbert of Tournai expressed thus an opinion shared by others at this period, the total negation of the strategy of the 'general crusade'. He suggests a new type of crusade; a perpetual crusade carried on by the device of the *passagium particulare*. This, he argued, should be realized by mercenaries, by professional soldiers. Thus the advice as

---

[29] Gilbert of Tournai, 'Collectio', pp. 36–52; Jacques de Vitry, *Historia Orientalis: Iacobi de Vitriaco Libri Duo* (Douai, 1577), 162–3; Throop, *Criticism of the Crusade*, pp. 69–100.
[30] Gilbert of Tournai, 'Collectio', p. 40; Throop, *Criticism of the Crusade*, pp. 100–1.
[31] Gilbert of Tournai, 'Collectio', p. 40.

formulated by Gilbert of Tournai meant the absolute negation of crusading *stricto sensu*, as it replaced the *crucesignati* with professionals. As such, Gilbert of Tournai's advice was unique, as even the most 'progressive' of crusade-planners before and following the loss of the Holy Land in 1291 never suggested a complete abolition of the institution of *crucesignati* but only their partial replacement by mercenaries.[32]

Another attack on the *passagium generale* as a means for the recovery of the Holy Land, but from an entirely different point of view was that of William of Tripoli in his 'Tractatus de Statu Saracenorum'. This Dominican, who composed his memoir in 1273 in Acre, believed in the imminent fall of Islam and the conversion of the Moslems to Christianity.[33] Hardly an opponent of crusading or a pacifist,[34] William of Tripoli therefore insisted that there was no need for a crusade.[35] He was thus hardly one of those opponents of the crusade listed by Humbert of Romans in his 'Opus tripartitum', whose opposition was based on the conviction that it was not in accordance with the Christian religion to spill the blood of the infidels.[36] Nevertheless, such opponents of the crusade, the supporters of peaceful missions to the infidels were quite numerous at the time of the Second Council of Lyons. One of the supporters of such missions was the Franciscan scientist Roger Bacon (d. 1292) who questioned, like the sixth kind of men who condemned the crusade described by Humbert of Romans,[37] the point of a war against the Saracens which did not bring them to conversion but rather stirred them up against the Christian faith. He asserted that, due to the wars, the Saracens were made more

[32] See below, chs. 3, 5, 6. For a different opinion see Throop, *Criticism of the Crusade*, pp. 100–2, who insists that this conviction of Gilbert of Tournai was shared by many others.

[33] William of Tripoli, 'Tractatus de Statu Saracenorum', ed. H. Prutz, in Prutz, *Kulturgeschichte der Kreuzzuge* (Berlin, 1883), 589.

[34] N. Daniel, *Islam and the West: The Making of an Image* (Edinburgh, 1960), 122, and B. Z. Kedar, *Crusade and Mission: The Interplay between Two European Approaches towards The Muslims in Medieval Times* (Princeton, NJ, 1984), 180–3, *contra* Throop, *Criticism of the Crusade*, pp. 115–46, esp. pp. 120–2, and E. R. Daniel, *The Franciscan Concept of Mission in the High Middle Ages* (Ky., 1975), 12.

[35] William of Tripoli, 'Tractatus', p. 589.

[36] Humbert of Romans, 'Opus tripartitum', pp. 191–2. For English trans. of this text see J. and L. Riley-Smith, *Crusades: Idea and Reality 1095–1274* (London, 1981), 104–6.

[37] Humbert of Romans, 'Opus tripartitum', p. 196.

antagonistic to Christianity, so that it became increasingly difficult to convert them in many parts of the world, especially in the crusader East. Faith, in his opinion, 'does not enter this world by arms but through preaching'. He was, however, aware of the fact that such peaceful preaching would not always ensure their conversion of the infidels. Therefore, he argued that those among them who would not respond to preaching should be forced into conversion by the use of arms.[38]

Yet another supporter of peaceful missions to the infidels at that time was Ramon Lull, the great Catalan missionary and philosopher. Lull, who was probably familiar with the writings of Roger Bacon, argued the greatest obstacle to the conversion of the world was the very existence of Islam. Therefore, he concentrated in his contest for conversion of the world on the Saracens. His philosophy of conversion advocated the means of peaceful persuasion, of love and intellect, with a special emphasis on the necessity for free, and not forced, conversion.[39] However, in the next twenty years Ramon Lull's opinions began to change and after the disaster of 1291 he became an ardent supporter of the crusade as an instrument of mission. Although the crusade was to be a legitimate means for the conversion of the infidels and the recovery of the Holy Land, he regarded the mission as a better one.[40]

Both Roger Bacon and Ramon Lull insisted upon advancing the knowledge of foreign languages and especially Greek, Arabic, Hebrew, and Syriac as essential for missionary purposes. Roger Bacon advocated the study of Arabic, Greek, and Hebrew while Ramon Lull insisted upon Arabic, Mongol, and Greek. Bacon, as a scientist, saw the knowledge of such languages as particularly important in enriching the sciences in the West, but he also stressed the importance of preaching to the unbelievers in their own languages. This, as a matter of fact, had already been recognized much earlier. The General Chap-

---

[38] Roger Bacon, *Opus Majus*, ed. J. H. Bridges (Oxford, 1897–1900), ii. 164, 221, 391, iii. 120–5; id., 'Opus Tertium', in *Opera Inedita* ed. J. Brewer, (RS, London, 1857), 116–17; Throop, *Criticism of the Crusade*, pp. 132–3, presents Bacon as a pacifist opponent of the crusade. For a more balanced description of Bacon's concept of mission see E. R. Daniel, *Franciscan Concept*, pp. 55–6, and Kedar, *Crusade and Mission*, pp. 177–80.
[39] J. N. Hillgarth, *Ramon Lull and Lullism in Fourteenth-Century France* (Oxford, 1971), 12–27.   [40] See Kedar, *Crusade and Mission*, pp. 189–95; below, ch. 3.

ter of the Dominican Order recommended in 1236 the study of local languages in the provinces populated by infidels. In 1248 Innocent IV gave ten grants to the University of Paris for young students learning Arabic and other oriental languages to study theology so that they would be able to serve as missionaries in Outremer. Humbert of Romans exhorted twelve Dominicans, in a *littera encyclica* of 1255, to study Arabic, Hebrew, Greek or any other of the foreign languages in the countries in which those languages were spoken; he claimed also that the *amor soli matalis* and the absence of linguistically qualified missionaries were the worst enemies of successful missions. This was probably influenced by the practice already at work in Spain where, under the instigation of Ramon of Peñáfort, friars were living among Saracens, studying Arabic. It is known that in 1259, for example, the General Chapter of the Dominican Order sent friars willing to study Arabic to the convent of Barcelona; in 1281 ten friars went to study Arabic in Valencia and nine to Barcelona to study Hebrew.[41] Ramon Lull, one of the most ardent supporters of the study of foreign languages as an instrument for missions, founded in 1276 the convent of Miramar at Deya in Majorca, where thirteen Franciscans could study the languages of the infidels.[42] In the years 1287–9 Lull tried to persuade Philip IV of France, as well as the University of Paris, to establish a similar college in Paris where missionaries could be trained. In 1287 he also appealed to Rome with the same end in view. His constant efforts for the foundation of missionary colleges were, however, not crowned with success until 1312.[43]

It therefore follows that even though, on the whole, the Second Council of Lyons was the golden age of peaceful missions to the infidels, only a few of their protagonists rejected the crusade altogether. Moreover, even those who opposed the crusade *per se*, were, like Ramon Lull, to modify their views after

---

[41] E. R. Daniel, *Franciscan Concept*, p. 10; J. Richard, 'Enseignement des langues orientales en Occident au Moyen Âge', *Revue des études islamiques*, 44 (1970), 149–64; R. Weiss, 'England and the Decree of the Council of Vienne on the Teaching of Greek, Arabic, Hebrew and Syrian', *Bibliothèque d'Humanisme et de Renaissance*, 14 (1952), 1–3.

[42] For Miramar see Ramon Lull, 'Vitae Coaetanea', ed. H. Harada, *CCM* xxxiv (1980), 282. The foundation was confirmed by Pope John XXI on 16 Feb. 1276: see E. R. Daniel, *Franciscan Concept*, pp. 66–79.

[43] Hillgarth, *Ramon Lull*, pp. 49–50; below, ch. 8.

the disaster of 1291. From then on, the belief in a peaceful conversion of the infidels, as opposed to a forced one, as a means of recovering the Holy Land, would be in decline.[44]

It is perhaps paradoxical that the man whose memoir is perhaps the best source of information regarding the various attitudes to mission on the eve of the Second Council of Lyons was Humbert of Romans, its fierce opponent as far as the missions to the Saracens was concerned. From among the authors of the memoirs submitted to Pope Gregory it was only he, it seems, who shared the pope's zeal and even obsession with crusading. A master general of the Dominican Order (1254–63) and an adviser on crusading to popes and kings, among them Louis IX of France, and the author of two treatises on preaching, one general and the other on crusade-preaching ('De praedicatione Crucis', composed c.1266–8),[45] he was certainly well qualified on the question of organization of a crusade. Also the idea of mission was not foreign to him; his *littera encyclica* of 1255 reminded the friars that under his mastership he hoped to see the Greeks reunited with the Latin Church and Christianity preached to the Jews, Moslems, and other pagans. He pinpointed the scarcity of linguistically qualified missionaries and appealed to those friars who had learnt Arabic, Hebrew, Greek, and other languages to become missionaries.[46] Later, however, Humbert of Romans evidently changed his opinion on the màtter of the conversion of the Saracens through the means of peaceful missions. In his 'Opus tripartitum' composed between the years 1272–4,[47] he argued that warfare was the only way of dealing with the Saracens in the East, though he still supported missionary activities in Aragon and Castile.[48]

The 'Opus tripartitum', written for Pope Gregory X, is

[44] See Kedar, *Crusade and Mission*, pp. 199–203. For a different opinion see Throop, *Criticism of the Crusade*, pp. 124–6.

[45] As this treatise exists only as an incunabulum all the references are to chapter numbers only. For the description of this treatise see A. Lecoy de la Marche, 'Prédication de la croisade au treizième siècle', *Revue des questions historiques*, 48 (1890), 14–24; E. R. Brett, *Humbert of Romans: His Life and Views of Thirteenth Century Society* (Toronto, 1984), 167–75.

[46] Throop, *Criticism of the Crusade*, p. 168; E. R. Daniel, *Franciscan Concept*, p. 10.

[47] For these dates see Throop, *Criticism of the Crusade*, p. 148 n. 3.

[48] Humbert of Romans, 'Opus tripartitum', pp. 187–90; Kedar, *Crusade and Mission*, pp. 155–6.

divided into three books, according to the subjects on which the pope sought advice. Consequently the first part of the treatise is devoted to the question of the crusade, the second deals with the union of the Churches, and the third with the reform of the Catholic Church. The first part is the most detailed treatment of the crusade on the eve of the Second Council of Lyons, far more detailed than that found in the other memoirs submitted to Gregory X.

Humbert of Romans begins his discourse on the crusade with replies to the current objections to crusading in the East. He counts seven kinds of such objections. First, that it is not in accordance with the Christian religion to shed blood, even that of infidels; second, that even if Saracens are not spared, it is necessary to spare Christians, and in the crusades countless Christians die. Third, that in a war between Christians and Saracens in Outremer the first have no chance of victory because of the great size of the Saracen armies, and the disadvantages of fighting in a strange land with an unfamiliar climate. The fourth of the objections was that although Christendom has the duty of defending itself against the Saracens' attacks, it does not have the same duty to attack them when they leave Christians in peace. The fifth argument was that if Christendom attempts to rid the world of the Saracens, why not treat all other non-Christians in this way? The sixth condemnation of the crusade to the East was on the ground that the crusades not only failed to bring the Saracens to conversion, but actually stirred them up against the Christian faith. Moreover, 'when we conquer and kill them [i.e. the Saracens], we send them to Hell, which seems to be against the law of charity. Also when we gain their lands we do not occupy them as colonists, because our countrymen do not want to stay in those parts, and so there seem to be no spiritual, corporal or temporal fruits from this stort of attack.' The seventh and last objection disclosed by Humbert of Romans was 'that it does not appear to be God's will that Christians should proceed against Saracens in this way [i.e. in the way of crusade] because of the misfortunes which God has allowed and is allowing to happen to the Christians engaged in this business'.[49]

---

[49] Humbert of Romans, 'Opus tripartitum', pp. 191–8. For English translation see J. and L. Riley-Smith, *Crusades*, pp. 103–17.

Humbert of Romans's reply to these objections is without doubt one of the most detailed pieces of apologetics for crusading ever written. He argues, answering the first objection, that the crusade is in complete accordance with orthodox theology and therefore should be used against Saracens, 'extremely wicked men and particular enemies of Christendom'. Referring to the second condemnation he declares that the Saracens would cause a far greater number of Christian deaths if they invaded the West, as they doubtless would do unless attacked in the East. Besides, he continues, 'the aim of Christianity is not to fill the earth but to fill heaven! What did it matter if the number of Christians was lessened by deaths suffered for God? By this kind of death people make their way to heaven who perhaps would never reach it by another road.' Answering the fourth condemnation Humbert of Romans argues that the Saracens, if left in peace, would already have overwhelmed the whole of Christendom. Moreover, the lands they held were Christian before the time of Mohammed. They seized them from the Christians without a 'just cause', so when the Christians invaded their lands they do not invade other people's territories but rather intend to recover their own land. Arguing against the fifth condemnation Humbert of Romans replies that the Jews would be converted anyway and besides, unlike the Saracens, they are in the power of Christians and can do nothing to harm the faithful. One can also always force Jews to listen to Christian preaching, which is impossible with the Saracens. As for idolaters, some of them, like the Prussians, do not usually attack Christians and it is to be hoped that, like the Poles, Saxons, and Bohemians before them, they will be converted to the Christian faith. Others, like the Mongols—and this is a strange argument in view of the terror caused in western Europe by Mongol attacks earlier in the thirteenth century—live too far away to attack Christendom. Besides there was hope that the idolaters could be converted. In his opinion the *oblatores* raising the sixth kind of condemnation of crusading were wrong in claiming that crusade hinders the conversion of the Saracens to Christianity, as there was more hope of converting the Saracens if they became subjects of the Christians: 'When our men had strongholds ... the Saracens themselves and many Greeks and others of that kind remained willingly under our dominion and cultivated the lands and were our tributaries; and the same is true of the Greeks in Achaea. And in this situation there was more hope that those Saracens who were under our rule would be converted more quickly than their fellow Saracens!' Faced with the seventh condemnation, the most serious of all and perhaps the most widespread, that crusading was obviously against the will of God, Humbert of Romans declared that misfortunes do not indicate that those who suffered them displeased God. On the contrary, those who suffer please God most! It is clear from the examples to be found in the Old Testament that in this world those who do good suffer more calamities than those who do evil. Sometimes, and here Humbert of Romans turns once again to the traditional phrase *nostris peccatis exigentibus*, these misfortunes happen to our men on account of our sins. 'For our sins have deserved whatever we suffer, as the judgement of Judith says: "Let us esteem those punishments to be what our sins deserve!"' (Judith 8: 27).[50] Following Bernard of Clairvaux he claimed that warfare is the only way to deal with the Saracen aggression as the Saracens are the worst of all the persecutors of the Church and endanger it; in the Holy Land far more Christians are converted to Islam than Moslems to Christianity! As there is no hope of converting the Saracens, they must be subjected to military force.[51]

Humbert of Romans perceives the crusade as the common task of the whole of Christendom and no Christian should refrain from aiding this task.[52] Moreover, it is the action of the Church against the Saracens and especially on behalf of the Christian religion, because the Saracens intend to banish Christianity from the world.[53] Thus, according to Humbert of Romans, there is nothing more important to Christendom than the crusade against the Saracens as the latter 'intend to banish Christianity from the world'. The crusade should be organized by the pope as he alone has authority over all Christians and so

---

[50] Humbert of Romans, 'Opus tripartitum', pp. 191–8. For a detailed account of Humbert of Romans's arguments see also Throop, *Criticism of the Crusade*, pp. 162–83.

[51] Humbert of Romans, 'Opus tripartitum', pp. 187–9. For Bernard of Clairvaux see his 'De Laude Novae Miliciae', *PL* 182, cols. 924–7.

[52] Humbert of Romans, 'De praedicatione crucis', incunabulum of Bibliothèque Mazarine, XV cent., no. 259 (= 12360), ch. 3.

[53] Id., 'Opus tripartitum', p. 202.

can move them to undertake this task. He alone has the power to force the clergy to give their assistance and moreover he alone can grant plenary indulgences and other privileges to promote the *negotium*.[54] Nevertheless, as it is the common enterprise of the whole of Christendom, all believers should aid it. There are three ways to do so: by 'corporal aid' that is by crusading; by 'real aid', by which is meant financial help, and by 'oral aid', i.e. by prayers. In Humbert of Romans's opinion: 'The last one is good, the second better, but the first one is the best.'[55] Particularly important is the aid to be given by the clergy as it is needed to preach the crusade. As it was his opinion that the clergy was then indifferent to the cause of the crusade, Humbert of Romans devotes four chapters to convince the clergy that their duty was to support the pope in his determination to recover the Holy Land. He argues that the clergy should be more enthusiastic for the crusade than all others, as they represent God on earth and thus they should see to his affairs. Moreover, the clergy, as the preachers of the faith, should defend their teaching. As pastors of Christ they should be diligent against his adversaries, the Saracens, who are wolves devouring the faithful.[56]

Humbert of Romans insists upon three pre-conditions necessary for the promotion of the crusade. First of all, as the crusade was the enterprise of all Christendom, common prayers should be said and the wrath of God placated by fasts, acts of charity, processions, and other suitable ceremonies. Secondly, all plans for the crusade should be carefully considered and no hasty decisions taken; competent men should be consulted and they should consult old histories in order to apply the lessons of the past to the present. Thirdly, as the crusade was the most important enterprise of Christendom, the best possible counsel was needed for its success. The advice of inexperienced, uninterested, and evil men should be rejected in favour of that of experienced, zealous men, inspired by the spirit of God.[57]

The crusade itself appears in Humbert of Romans's 'Opus tripartitum' as a sort of perpetual *passagium particulare*, hardly the conventional general crusade. He claims that the West

[54] Id., 'Opus tripartitum', pp. 186, 202–3.
[55] Id., 'De praedicatione', ch. 3.
[56] Id., 'Opus tripartitum', pp. 200–4. [57] Ibid. 204.

should permanently keep in Outremer an army of such size and force that it could always stand against the Saracens. It should be made up not of mercenaries, but of strong men inspired by zeal for the faith rather than by the lure of wages. A permanent army of this sort should be substituted for transient warriors and for those who have been driven from Europe because of crimes. Christians should realize, Humbert of Romans argues, that their previous attempts to aid the Holy Land failed because of the smallness of their forces, long intervals in warfare, weak fighters, the evil ways of many crusaders, and their lack of zeal. He is well aware that the upkeep of such a large army posed a financial problem. He therefore drew attention to the excess wealth available among Christians and especially that of the Church. The clergy alone, he declared, without making excessive sacrifices, were rich enough to finance a very large army.

What would be the harm if the Church sold its superfluous ornaments in each diocese, ornaments which waste with age? ... if in the cathedral churches and other rich chapters ... the number of offices should be reduced and the return of the prebends applied to the crusade? ... There are other resources, too many, indeed, to be estimated, both among the clergy and the laity. These can best be discovered by wordly wisdom. Thus perpetual returns can be obtained from movable goods and assigned to the *negotium* in perpetuity.[58]

The treatise reveals also Humbert of Romans's particular interest in the preaching of a crusade, a topic on which he had already written two treatises. He indeed outlined a propaganda campaign to be followed before the council meets. First of all, all *excitatoria* which could move believers to aid the war against the Saracens should be put down in writing, in brief and appropriate form, so they can be used to convince suitable men to take the Cross. There are many, Humbert of Romans claimed, not only laymen but even clergy, who had never heard about either Mohammed or the Saracens! He suggested that the pope should send messengers, accompanied by nobles who had taken the Cross and were zealous for the crusade, to secular rulers with letters setting forth reasons for the conquest of the

[58] Ibid. 205.

Holy Land, so as to gain the support of kings and princes for the crusade. As another means of gaining support, especially that of bishops, Humbert advised that provincial councils of the clergy should be summoned. In such assemblies the crusade could be preached, and those taking part could prepare themselves on the subject in advance of the general council. Yet another propaganda device was aimed at the nobility. Aware of the importance of gaining their support, Humbert suggested that there should have impressed upon them the examples to be found in the Holy Scripture of kings and great men fighting against the enemies of faith. The nobility should also be reminded that the crusade permitted them to do penance in a way appropriate to their station in life, as 'what finer penance could be found for those not suited to support abstinence and ascetic afflictions than that of serving God with arms?'[59] In his 'De praedicatione Crucis' he offers crusade-peachers seven reasons that should inspire the faithful, and above all the nobility, to take the crusader vow. First he lists divine love, commenting that the Holy Land is the place on the earth which is the dearest to Christ, but now it is defiled by the Saracens. Second is love of Christian law in contrast to the law of Mohammed; the faithful should take up the Cross to protect the true law and destroy the false. Third is brotherly love, namely Christians in the East suffer from the Moslems and their appeals for aid deserve an answer. The fourth reason is love for the Holy Land; this is the land of Christian heritage where the great figures of the Old and New Testament, including Christ, lived. There is not one Moslem who had not visited the sepulchre of Mohammed; how much more should all Christians desire to visit the places where Christ walked. Another reason, the fifth one, is that crusaders fighting against Christ's enemies save not only the Church but achieve personal salvation: 'who is there who wishes to show himself a soldier of Christ, who is there who wishes to accumulate merit, who is there who wishes to reach heaven quickly and with little trouble? Let him come forth and receive the sign of the Cross.' The sixth reason calls upon the warrior class to imitate the heroism of such knightly predecessors as Charlemagne, God-

[59] Id., 'Opus tripartitum', pp. 205–6.

frey of Bouillon, and Richard the Lion Heart. Humbert concludes his list with the seventh reason, namely the spiritual and material advantage offered by the Church to *crucesignati*.[60]

The memoirs of Humbert of Romans, as well as those of Gilbert of Tournai and William of Tripoli, seem to reflect some doubts as to the compatibility of the crusading idea with the teaching of Christian orthodoxy, as well as growing criticism of crusader indulgences and the spiritual value or merits of the crusade. Moreover, they reveal the conviction that the crusades, far from being a divinely inspired movement, were actually against the will of Providence.[61] Memoirs like the 'Opus tripartitum' (*c.*1272–4) of Humbert of Romans have suggested to modern scholars not only an escalation in the criticism of the crusading movement and a growing opposition to its ideology, but also the shrinking of its human reservoir; the memoirs suggest that, by the time of Gregory X (1271–6), both clergy and nobility, the dominant strata of European society, were increasingly withdrawing from the crusading movement. Before accepting this thesis it is only legitimate to ask whether the problem has been correctly stated. To put it differently: was there a real or an imagined decline in the crusading movement following the crusade of Louis IX to Tunis? The 'Opus tripartitum', it should be remembered, like other treatises which report anti-crusade sentiments, as for example the 'Collectio de Scandalis Ecclesiae' of Gilbert of Tournai and the 'Tractatus de Statu Saracenorum' of William of Tripoli, were all submitted to Gregory X following the latter's request for advice regarding means of aiding the Holy Land, advice he intended to present at the council. Among the measures included in the treatises were those to be taken to combat anti-crusade sentiments. Such sentiments were therefore very fully stated, so that the false impression is given that all contemporaries were convinced of the extent and the gravity of the abuses of the crusading movement. Moreover, independent criticism, such as that brought together in the studies of Throop, J. Prawer, and E. Stickel, reveals, or at least can be interpreted as revealing, no more criticisms of the crusade than those current in the early

---

[60] Id., 'De praedicatione', chs. 11–17; Brett, *Humbert of Romans*, pp. 169–71.
[61] See above, pp. 22–9, and also Humbert of Romans, 'Opus tripartitum', pp. 191–8.

thirteenth century.⁶² The failure of Louis IX's crusades certainly threw its dark and sinister shadow on the European's attitude to crusading in the East, but this shadow was, it seems, less sinister than has often been supposed. This is reflected by the crusade-planning of the Second Council of Lyons.

At the Second Council of Lyons

The council, attended by some 300 archbishops and bishops, one monarch (King James I of Aragon), and royal envoys from England, France, Germany, Sicily, Cyprus, and Byzantium, opened on 7 May 1274 with the pope's sermon on the text used by Innocent III while opening the Fourth Lateran Council (1215): 'Desiderio desideravi hoc pascha manducare vobiscum' (Luke 22: 15). This was primarily intended to raise enthusiasm for the crusade among the clergy as well as to persuade them to make sacrifices for its sake.⁶³ At the same session Pope Gregory publicly declared that if he lived and circumstances permitted, he would go on the crusade with its armies.⁶⁴

The issue of the Holy Land was introduced by the pope, perhaps following the advice of Humbert of Romans, as a matter of concern to all Catholics, and was the major topic of the second session of the council, which opened on 18 May 1274 with a sermon preached by Cardinal Bonaventura on the text: 'Exsurge, Jerusalem, sta in excelso et circumspice ad Orientem . . .' (Baruch 5: 5). A good part of this session was given to the promulgation of measures for the aid of the Holy Land, the *Constitutiones pro Zelo Fidei*.⁶⁵ It seems that the pope devoted

---

⁶² Throop, *Criticism of the Crusade*, pp. 69–183, 283–91; id., 'Criticism of Papal Crusade Policy', pp. 379–412; Prawer, *Histoire*, ii. 375–95; Stickel, *Fall von Akkon*, pp. 96–216. See also A. S. Atiya, *Crusade in the Later Middle Ages* (London, 1938), 3–10; S. Runciman, 'Decline of the Crusading Ideal', *Sewanee Review*, 79 (1971), 498–513; Purcell, *Papal Crusading Policy*, esp. pp. 183–6, *contra* Siberry, *Criticism of Crusading*, pp. 190–220.

⁶³ K. M. Setton, *The Papacy and the Levant 1204–1571*, i (Phil., 1976), 112–13 and n. 34.

⁶⁴ Gregory X, *Registres*, no. 1041 (dated 31 July 1274); Salimbene, 'Chronica', p. 494.

⁶⁵ For the *Constitutiones pro Zelo Fidei* see Finke, *Konzilienstudien zur Geschichte des 13. Jahrhunderts* (Münster, 1891), 113–17, and a reprint in Purcell, *Papal Crusading Policy*, app. A, pp. 196–9; see also Throop, *Criticism of the Crusade*, pp. 236–61; Purcell, *Papal Crusading Policy*, pp. 21–31.

most of the interval between the first and the second sessions to the preparation of this document. Between 7 and 18 May the pope and his cardinals called on each archbishop present to exact promises of financial aid from his clergy, both regular and secular, so as to lay the financial basis for the Holy War in the form of a tenth for six years. Moreover, Pope Gregory used these ten days of intermission in the council's sessions to hear the advice of the laity. However, the absence from the council of the leading secular rulers in the West must have caused the pope bitter disappointment. He had hoped for the attendance both of Philip III of France and Edward I of England, a former crusader whose advice the pope considered invaluable. The royalty at the Second Council of Lyons were therefore hardly those who might have ensured the success of a crusade. Maria of Antioch was present merely in order to assert her claim to the crown of the Latin Kingdom against Hugh III of Cyprus, whereas the other nobles present, Baldwin II the former emperor of the Latin Empire of Constantinople and the counts of Nuremburg and Statin, were hardly the sort to advise the pope on the subject of the crusade. The only ruler of importance who was present and could be of use to the pope on the issue of the crusade was the old King of Aragon James I. Consequently, professional advice from the laity could be obtained mainly from the representative of the king of France, Erart de Valeri, a former friend of Louis IX who stood high in the counsels of Philip III, as well as from the members of the delegation from Outremer, William of Beaujeu the master general of the Temple, Hugh Revel the master general of the Hospital, and John of Grailly the seneschal of the Latin Kingdom.[66]

The advice of the laity, among them experts like the masters general of the two most important military orders of the crusader kingdom, was more practical than that of the memoirs, with the exception of that of Humbert of Romans. Still, the advice given was hardly that which the pope wanted to hear, as on the whole it was directed against his plan for a *passagium generale*. James I of Aragon favoured the dispatch to the kingdom of a force of 500 knights and 2,000 foot-soldiers to

---

[66] 'Estoire', pp. 464, 472–3; C. J. Hefele and H. Leclercq, *Histoire des conciles*, vi (Paris, 1914–15), 169–71; Throop, *Criticism of the Crusade*, pp. 214–28.

garrison the castles and other strategic places for the period of two years till the arrival of the armies of the general crusade. He also offered to join the pope with a contingent of 1000 knights, if the latter fulfilled his promise to accompany the crusade. The master general of the Temple William of Beaujeu insisted upon a garrison of 250 to 300 knights and 500 foot-soldiers. He pointed out the weakness of the Mamluks' navy as an advantageous factor in the war against them. He also observed that a crusade required careful consideration in the matters of arms, food, and suitable men. The Holy Land, he said, was insufficiently provided with those necessities and therefore could not supply them to the crusader armies. It must be assumed that, by his reference to 'suitable men', the master general meant that the armies of the crusade should consist of professional soldiers only. The advice given by Erart de Valeri, the representative of the king of France at the council, was even more sceptical with regard to a general crusade. According to the *Chronicle of James I of Aragon* Erart declared that: 'This is a mighty affair. Great forces have passed thither long ago on various occasions. I will tell you what this is like; it is like the little dog barking at the big one, who takes no heed of him. Kings and great men have passed beyond sea and yet have been unable to recover and hold the land'. Erart thus agreed with William of Beaujeu that a general crusade was altogether futile, and that the best that could be done for the sake of the Holy Land would be to send a force to serve as its garrison and to protect the remaining crusader strongholds.[67]

The *Constitutiones pro Zelo Fidei*, though aiming at a general crusade, reflect some influence of the experts' opinions. However, as they envisaged a general crusade they ignored their reservations against the plan.[68] The document opens with a description of the miserable state of the Holy Land and the ineffable 'atrocities of the Saracens' which the pope had seen with his own eyes. What follows is the declaration of the pope that he had made his decisions concerning the crusade accord-

---

[67] James I of Aragon, *Chronicle of James I of Aragon*, trans. J. Forster and P. de Gayngos (London, 1883), ii. 645–50; Throop, *Criticism of the Crusade*, pp. 228–33; Mayer, *Crusades*, tr. J. Gillingham (Oxford, 1972), pp. 282–3 and n. 143.

[68] For a different opinion on this point see Throop, *Criticism of the Crusade*, pp. 236–54.

ing to the advice he had received at Lyons, referring probably to both the memoirs and the various consultations held at the council. Nevertheless, it is obvious that his *Constitutiones* were also profoundly influenced by the measures taken in regard to the Holy Land by previous councils, the Fourth Lateran Council (1215) and the First Council of Lyons (1245).[69] Thus, the measures decided upon, as they appear in the *Constitutiones*, reflect to a large extent a pope's and a council's adherence to both the traditional means and instruments of organizing a crusade as well as the traditional model of a crusade, that is, the *passagium generale*.

First of all, the pope announced that there were to be no exceptions from the payment of a six years' tithe to be levied on ecclesiastical incomes and the financial imposts. Secular authorities were required to collect a sum equivalent to one *denier tournois* or one sterling a year from every Christian within their jurisdictions. All Christendom was divided into twenty-six districts administered by collectors and sub-collectors. Following Gregory IX, the pope ordered all blasphemers to be fined and those fines dedicated to the cause of the Holy Land. The faithful were also to be exhorted, as during the time of Innocent IV, to remember the Holy Land in their wills. Following the example of Innocent III, empty chests with three locks were to be placed in churches all over Christendom and the clergy were to exhort believers to put money into them for remission of their sins.[70] Again, according to the example set by the Fourth Lateran Council and the First Council of Lyons, and this time verbatim, excommunication was pronounced on pirates and anyone aiding them; no contract of sale or purchase should be made with them under the pain of anathema. Those who were supplying the Saracens with war-materials, such as arms, iron, or timber for the construction of vessels, and those who served them as pilots or gave them counsel or aid of any sort and especially in the Holy Land, were to be deprived of all their goods, which, when confiscated, were to be devoted to the

---

[69] See the *Constitutiones pro Zelo Fidei* as well as the decrees of 1215 and 1245 in Purcell, *Papal Crusading Policy*, app. A, pp. 187–99.

[70] *Constitutiones pro Zelo Fidei*, in Purcell, *Papal Crusading Policy*, app. A, pp. 196–8. The declaration of 'no exceptions' was never kept. See Gregory X, *Registres*, nos. 384, 399, 409, 1048, 1056, 1069.

cause of the crusade. Furthermore, they were to be slaves of those who arrested them. This sentence, Pope Gregory X ordered, was to be read in all sea-ports on every Sunday and feast day. Those who confessed their guilt were not to be permitted to re-enter the Church unless they had given up all their gains from their illegal commerce. Moreover, Christians were forbidden to send vessels to the Saracens' lands for the period of six years; this measure, Gregory pointed out, would increase the number of ships at the disposal of a crusade and rob the Saracens of considerable income.[71]

The measures concerning trading with the Saracens were directed at the maritime powers of Europe and especially the most important among them, namely Genoa and Venice and their communities in Outremer. Both, it should be remembered, had taken an active part in the crusading movement since the First Crusade. A Venetian chronicler, the so-called 'Monk of Lido', boasts in his account of the First Crusade, written some twenty years after the event, that the Venetians use their ships as others, namely knights, use their horses. And indeed because no European power except for the maritime communes possessed fleets in the twelfth and thirteenth centuries, those of Venice and Genoa became indispensable for the crusading movement to the East. Actually, no crusade, unless it was to go by land, could depart without the co-operation of at least one of the most important of the maritime powers, Venice or Genoa. Not only the crusading movement but also the crusading states in Outremer, which also had no fleets of their own, were dependent for their existance on a constant influx of men and merchandise from the West, transported by the ships of the maritime cities of Italy. Consequently during the twelfth and the thirteenth centuries the maritime republics were granted and enjoyed extensive fiscal exemptions, and a great measure of judicial autonomy in the port-cities of the kingdom of Jerusalem where they had acquired quarters of their own. Venice, for example, held in the thirteenth century one third of Tyre and a large quarter in Acre where also other maritime powers, namely Genoa, Pisa, Marseilles, Montpellier, and

---

[71] *Constitutiones pro Zelo Fidei*, in Purcell, *Papal Crusading Policy*, app. A, p. 198. For other measures taken by Gregory to stop the trade with the Saracens see Gregory X, *Registres*, nos. 351–3, 355, 821, 1088; Throop, *Criticism of the Crusade*, pp. 245–6.

Barcelona, had their own quarters. Though by the 1270s the communes were the richest element in the port-cities of the crusading kingdom (Tripoli, Tyre, and Acre), due to their autonomy and privileges they did not contribute to either its economy or its defence. On the contrary. First of all they used their quarters in the port-cities of the kingdom as bases for their illicit trade with the Moslems and even made Acre into one of the main harbours from which prohibited merchandises, that is war-materials, originating in the West, had been exported to Egypt. The trade with Egypt was simply too profitable to be abandoned. Turning now as before a deaf ear to papal prohibitions, merchants preferred to be fined for infringing them and even to be excommunicated than to give up the lucrative trade with Egypt. Secondly there were the frequent collisions among the communes inside Acre. These conflicts and especially the 'War of St Sabas' between Genoa and Venice (1256–61), won the communes the unenviable reputation of imperilling by their selfish behaviour the very existence of the kingdom. This accusation together with two others, already raised against the communes by Jacques de Vitry, namely their devotion to profit rather than to the welfare of the crusading kingdom and their friendly attitude to the Saracens was repeated time and again by Europeans after 1291 when reflecting on the loss of the Holy Land.[72]

Other traditional measures announced by the *Constitutiones* included a ban on all tournaments until after the crusade had taken place. Such a prohibition had been issued earlier both by Innocent III and Innocent IV.[73] This was intended to ensure that the nobility preserved their energy for the Holy War, as well as to keep them from illegal warfare. The pope proclaimed that there should be no war in Christendom for a period of six years and that the prelates were to use the weapons of excommunication and interdict to force truces upon the quarrelling factions.[74] The document ends with the usual promise of

[72] Mayer, *Crusades*, trans. J. Gillingham, pp. 172–6; S. Schein, 'From "Milites Christi" to "Mali Christiani": The Italian Communes in Western Historical Literature', in B. Z. Kedar and G. Airaldi (eds.), *Comuni italiani nel regno crociato di Gerusaleme* (Genoa, 1986), 681–9; below, ch. 3.

[73] On this measure see Throop, *Criticism of the Crusade*, p. 249.

[74] *Constitutiones pro Zelo Fidei*, in Purcell, *Papal Crusading Policy*, app. A, p. 198; for the decrees of 1215 and 1245 see Purcell, *Papal Crusading Policy*, app. A, p. 194.

plenary indulgence for all who would take part in the crusade or would redeem their vows by monetary contributions to the crusade, according to their means, intended to finance the participation of one or more warriors. The pope promised the same indulgence to those who gave aid and counsel on behalf of the crusade.[75] In the latter case Gregory X was less generous than his predecessors Urban IV and Clement IV, who granted a plenary indulgence to all preachers and collectors who had worked for one year; Gregory X required three years' service for such an indulgence. As to commutation of vows, Gregory's policy was on the whole directed against commutation of crusading vows away from the Holy Land. On the other hand he was liberal in his attitude towards redemption of vows and was the only pope to allow women to redeem them specifically on the grounds of 'fragilitas sexus'.[76]

Pope Gregory also laboured during the council to win recruits and allies among the secular rules of Europe and even outside its borders. He recognized the claims of Rudolph of Habsburg to the Roman Empire upon the latter's promise to take the Cross.[77] The issue of the crusade was also at stake in the discussions with the Greeks, as Pope Gregory counted on Greek co-operation in the planned *passagium generale*. The union of the Churches, proclaimed at the fourth session of the council held on 6 July 1274, secured their support of the project. According to an embassy headed by George Metochites, archdeacon of Hagia Sophia, dispatched in the spring of 1275 to the pope, the Emperor Michael VIII Palaeologus was willing to allow the crusader armies free passage through his territories if they chose the land-route. He hoped that the armies might help him against the infidels in Asia Minor and to re-establish Byzantine authority there. The emperor would then assist the papacy in the recovery of the Holy Land, and his help would include the provision of food, money, and soldiers. Moreover, if some of his enemies in the West were neutralized, and here the Emperor had in mind the aspirations of Charles I of Anjou to

[75] *Constitutiones pro Zelo Fidei*, in Purcell, *Papal Crusading Policies*, app. A, p. 199.
[76] Gregory X, *Registres*, nos. 337, 497, 539, 569; Throop, *Criticism of the Crusade*, pp. 252–4; Purcell, *Papal Crusading Policy*, pp. 114, 120–1.
[77] 'Annales Placentini Gibellini', *MGH SS* xviii. 559; *MGH: Legum Sectio IV, Constitutiones*, ed. J. Schwalm (Hanover, 1904–6), vol. iii, nos. 14–16, pp. 16–20; nos. 34–5, pp. 32–4; P. V. Laurent, 'Croisade', pp. 125–7.

the Byzantine throne, he would have the crusade preached in his empire. The Greek emperor, it should be remembered, was largely motivated in his support of the union by the necessity of restraining Charles from his plan of conquest.[78] His offer of 1275 had the same aim and it was indeed generous. However, if his plan was realized it would mean a diversion of crusading effort from the Holy Land to Asia Minor. Gregory X, whose whole crusader policy was aimed at the recovery of Jerusalem, was hardly the man to agree to such a project. Moreover, it is doubtful if Gregory X planned to choose the land-route for his crusade.[79] His efforts to secure for his project the navies of Europe rather point out that he had the sea-route on his mind. Sicily, Genoa, Marseilles, Pisa, Venice, and Aragon were all requested more than once to furnish galleys for the coming expedition.[80]

Another potential ally was the Mongol il-khan of Persia, Abaga, whose delegation appeared in Lyons on 4 July 1274. This was not the first Mongol delegation at the papal Curia; an earlier embassy had arrived in Europe in 1262. In a letter of 1262 addressed to King Louis IX of France written not long after the defeat of the Mongols in the battle of Ain-Jalud (September 1260), the il-khan Hulagu, a grandson of Genghis Khan, tried to persuade the West to undertake a joint expedition against the Mamluks of Egypt. The il-khan suggested that a Latin naval force should take up a defensive position along the coast, so that when the Mongols launched their offensive, the enemy would have no way of retreat and so would be finally crushed. He promised that the city of Jerusalem once captured by his army would be handed over to the papacy. King Louis IX, it seems, rejected the offer.[81] A few years later, nevertheless,

---

[78] 'Annales Placentini Gibellini', p. 560; Throop, *Criticism of the Crusade*, pp. 256–7; D. Geanakoplos, *Emperor Michael Palaeologus and the West 1258–1282* (Cambridge, Mass., 1959), 285–9; Setton, *Papacy*, i. 116–22. For Charles I's aspirations to the throne of Byzantium see Geanakoplos, *Emperor Michael*, pp. 189–228, and below, ch. 2. For Emperor Michael VIII Palaeologus's attitude to the union see Geanakoplos, 'Michael VIII Palaeologus and the Union of Lyons 1274', *Harvard Theological Review*, 46 (1953), 79–89.   [79] For a different opinion on this subject see Setton, *Papacy*, i. 121.
[80] Gregory X, *Registres*, nos. 336, 343, 356; Throop, *Criticism of the Crusade*, pp. 246–8; above, n. 71.
[81] P. Meyvaert, 'Unknown Letter of Hulagu, Il-Khan of Persia, to King Louis IX of France', *Viator*, 11 (1980), 245–59; J. Richard, 'Une ambassade mongole à Paris en 1262', *Journal des savants* (1979), 295–303.

Hulagu's successor Abaga sent another embassy to Europe. In a letter written in 1268 to Pope Clement IV he repeated the proposal for a joint expedition against the Mamluks. This time the forces of Byzantium were also included, as Abaga became a son-in-law to the emperor by marrying the latter's natural daughter Maria. This offer, like the previous one, was never realized, as the forces of the crusade of Louis IX landed in 1270 not on the shores of the Holy Land but on those of Tunis. When Prince Edward of England reached Acre in May 1271 he contacted Abaga; the il-khan, however, was at that time engaged in operations in the East against the Chagatai Mongols.[82]

At Lyons the Mongol embassy once again proposed a joint expedition and repeated the promise that the Holy Land, including Jerusalem, once recovered, would be handed over to Christendom. In his reply to the il-khan's letter Pope Gregory expressed his wish that Abaga and his people would recognize the truth and embrace the Christian faith. As for the crusade, the pope wrote that before Christian armies appeared in the East a papal mission would inform the il-khan and provide the Mongols with further means of achieving salvation.[83]

Addressing the assembly at the last session of the council, the sixth general session held on 17 July 1274, Pope Gregory X stated with satisfaction that all the three reasons for convoking the council had been successfully solved.[84] And indeed in the summer of 1274 the prospects of his project for a crusade seemed excellent.

The Preparations for the Crusade of 1280 (1274–1276)

It is apparently futile to speculate as to the chances of success of the general crusade planned by Gregory X to depart in 1280,[85]

---

[82] J. A. Boyle, 'Il-Khans of Persia and the Prince of Europe', *Central Asiatic Journal*, 20 (1976), 30–1; Schein, '*Gesta Dei per Mongolos 1300*: The Genesis of a Non-Event', *EHR* 95 (1979), 808–9; Meyvaert, 'Unknown Letter', pp. 250–1.

[83] Gregory X, *Registres*, no. 577; Setton, *Papacy*, i. 114–16.

[84] Mansi, xxiv. 68; Hefele and Leclercq, *Histoire des conciles*, vol. vi, pt. 1, p. 181.

[85] According to Throop, *Criticism of the Crusade*, p. 272, the crusade was to depart on 2 Apr. 1277. This, however, is not verified by the sources. It seems that Pope Gregory had kept the date a secret to surprise the enemy and intended to send it in six years' time from the proclamation of the crusade as his predecessors did, i.e. in 1280.

since the plan was brought to an end by the pope's death on 10 January 1276. However, the progress which he had made in the organization of the crusade in the two years following the council proves that, even in the second half of the thirteenth century, an energetic pope who was willing to sacrifice at least a part of his European interests for the sake of Outremer could successfully pacify Europe and unite its conflicting forces. By 1276 the anti-Moslem league formed by Gregory X with the crusade in view included the principal sovereigns of Europe: Rudolph of Habsburg, Philip III of France, James I of Aragon, Charles I of Anjou and his son Charles of Salerno, the later Charles II of Anjou—all of whom took the Cross. The future emperor Rudolph of Habsburg did so in exchange for the pope's promise to crown him in Rome. The ceremony took place in Lausanne on 18 October 1275 when Rudolf took the Cross from the hands of Pope Gregory along with his wife and many German nobles.[86]

Philip III of France, and many others in that kingdom, had taken vows to go on crusade in 1270, but had never done so. On 19 September 1271 they were ordered by the pope to honour that undertaking. In the following year, as the result of further urging by Gregory, the king ceremoniously took the Cross from the hands of Cardinal Simon of Brie, and his example was followed by his queen, his brothers, and many of the French nobility.[87] Charles I of Anjou took the Cross on 13 October 1275,[88] and was immediately granted the tithes authorized at the Council of Lyons which were due from Sicily, Provence, and Foucalquier, on condition that he took part in the general crusade. If he could not go in person his son Charles of Salerno, who seems to have taken the Cross in company with his father, was to replace him.[89] Edward I of England also promised in 1275 that he would go on crusade. It was arranged that, when he had taken his vows and was ready to depart, he would have at his disposal the tithes granted at the council in respect of

---

[86] 'Annales Basilenses', *MGH SS* xvii. 198; Bernard Gui, 'Flores Chronicorum', *RHGF* xxi. 703.

[87] Bernard Gui, 'Flores', p. 703; Gregory X, *Registres*, nos. 494, 496, 498, 539; Throop, *Criticism of the Crusade*, pp. 269–70.

[88] For this date see Gregory X, *Registres*, no. 636; P. V. Laurent, 'Croisade', p. 128 and n. 1.

[89] Gregory X, *Registres*, no. 636. For Charles II of Anjou see below, ch. 3.

England, Wales, and Ireland and, if its king agreed, of Scotland as well.[90] Edward did not take the Cross, however, until 1287–8.[91]

The pope also worked for the pacification of Europe, insisted upon by the authors of the memoirs as a *sine qua non* of a successful crusade. Philip III of France was reconciled with Rudolph of Habsburg; the latter was recognized as emperor by both his rivals for the title, Ottokar of Bohemia and Alfonso X of Castile, as well as by Charles I of Anjou. In Italy the pope managed to secure a truce of two years between the everwarring Genoa and Venice. Of the rulers of Italy, the most likely to threaten the papal crusade plan was Charles I of Anjou. Following his victory over Conradin at the battle of Tagliacozzo (23 August 1268) Charles, now secure on his Sicilian throne, resumed his plan for an attack on the Byzantine Empire. The election of Gregory X to the papal throne meant a defeat for the Angevin party as he was hardly prepared to support its plan for the recovery of the Latin Empire. Both the crusading policy of Gregory X as well as his Byzantine policy, which was aimed at the union of the Churches, were constraints upon Charles's plans. Pope Gregory firmly refused to link his plan for the recovery of the Holy Land with Charles's plan for the recovery of Constantinople. Charles, under papal pressure, had no alternative but to make peace with the Greeks for the time being, and to participate in the papal crusade plan.[92] Thus by 1276 the planned general crusade could count not only on the political and military aid of the Byzantine emperor, but also on that of Charles I of Anjou.[93]

There were two other urgent subjects, each a necessary part of the organization of the crusade, with which the pope had to deal. One was to ensure its sound financial basis by securing the payment of the tenth granted at the Second Council of Lyons. The other was to arouse the enthusiasm of the faithful for the crusade, and to induce them to take the Cross. The Lyons' tenth met with opposition even before it was officially pro-

---

[90] Gregory X, *Registres*, no. 945 (dated 14 Nov. 1275).
[91] Edward I took the Cross between 3 Apr. 1287 and 22 Feb. 1288. For this date see W. E. Lunt, 'Papal Taxation in the Reign of Edward I', *EHR* 30 (1915), 411–12 and n. 114.
[92] Geanakoplos, *Emperor Michael*, pp. 237–45.
[93] See ibid. 277–90 and n. 78 above.

claimed, during the informal consultations the pope held with the prelates after the first session of the council (7–18 May 1274). Richard Mepham, the dean of Lincoln, presented the pope with a memorandum pleading for the exemption of the English clergy on the ground that their property had been devastated by civil war and burdened by recent taxation.[94] After the council the clergy's resistance to the new taxation became even stronger. In France there were objections not only to the tithe itself but also to the behaviour of its collectors. In England the clergy protested that the collectors were overvaluing their property and pleaded for the valuation prepared by the bishop of Norwich twenty years before. The clergy of Scotland asked at the Council of Perth (1275) to lower the tithe but their request was refused. In Germany the opposition was so strong that archbishop of Magdeburg went as far as to hold a council where it was decided that the collection of the tithe should be forbidden in that province and those who payed it would be excommunicated![95]

Besides the opposition to the amount of the taxation and the methods of collection, exemptions were being demanded. Though Pope Gregory explicitly stated in his *Constitutiones pro Zelo Fidei* that no exceptions would be allowed, the pressure was probably so strong that concessions had to be made, and especially to the military orders. During the years 1274–5 the orders of St James, Calatrava, the Temple, and St John were granted their exemptions from the Lyons' tithe.[96] The Cistercians, one of the wealthiest of the European monastic orders, forced the pope into accepting £100,000 in return for an exemption.[97] Exemptions were also granted to certain orders and hospitals caring for the poor and sick, and, at the discretion of the collector, to poverty-stricken churches and monasteries.[98] Another criticized device for financing the crusade announced at the council were the money-chests to be placed in all churches to receive the offerings of the faithful. Already

---

[94] Lunt, 'Papal Taxation', pp. 401–2; Throop, *Criticism of the Crusade*, pp. 239–40.
[95] Mansi, xxiv. 154; Lunt, 'Papal Taxation', pp. 402–5; Throop, *Criticism of the Crusade*, pp. 237–41.
[96] Gregory X, *Registres*, nos. 384, 409, 1056, 1069.
[97] Ibid., no. 399.
[98] See e.g. ibid., nos. 630, 465, 571.

employed by Innocent III, this device always aroused resentment and it is known that it was not carried out promptly after 1274.[99]

The resentment and opposition of the clergy to taxation caused the pope to undertake strong measures to win support for the enterprise and its financial backing. On 17 September 1274 he issued the bull *Si mentes fidelium*, an ardent *excitatorium* aimed at arousing the fervour of the clergy. Crusader indulgences were to be granted to those contributing to the expenses of crusaders in the Holy Land. Preaching of the Cross was ordered to take place everywhere, even in churches under interdict. Following Innocent III, Gregory X ordered indulgences of one year and forty days to be granted to anyone who would come to listen to a crusade-sermon and confess his sins.[100]

In the preaching of the crusade Pope Gregory depended much upon the mendicants, the Franciscans and the Dominicans. The preachers were—and here it seems that the pope had once again followed the advice of Humbert of Romans —given the old French *Pèlerinage de Charlemagne* to read, so as to arouse their fervour for the crusade.[101] Those who worked to promote the undertaking, like preachers and collectors, were granted a plenary indulgence for a period of three years' service. Gregory X had thus meaningfully raised the period of service, as in 1262 Urban IV granted the same indulgence for a year's service, while Clement IV reduced it to six months.[102]

The preaching, as it is revealed by the faculties granted to Simon of Brie, cardinal-priest of St Cecilia and papal legate in France,[103] was intensively pursued from August 1274 and continued after Pope Gregory's death (10 January 1276).[104] Yet contemporaries were well aware of the disastrous effect of this pope's death on the fate of both the crusade and the Holy Land. The anonymous continuator of William of Tyre wrote

---

[99] See above n. 70 and Throop, *Criticism of the Crusade*, p. 242.

[100] Gregory X, *Registres*, no. 569. See also ibid., nos. 497, 502, 522, 544; Throop, *Criticism of the Crusade*, pp. 262–3.

[101] Throop, *Criticism of the Crusade*, pp. 263–6. For Humbert of Romans, see above n. 59.

[102] Gregory X, *Registres*, no. 497; Purcell, *Papal Crusading Policy*, p. 61.

[103] Gregory X, *Registres*, nos. 494–568.

[104] See below, ch. 2.

that 'had the pope lived all would have been well with the Holy Land'.[105] Salimbene of Adam, a disciple of Joachim of Fiore, was more sceptical. He declared that 'it is not the Divine Will that the Holy Sepulchre should be recovered; when Pope Gregory X wished the recovery [his] death became an obstacle'. And he added: 'It seems that the recovery of the Sepulchre is not Divine Will since the great number [of men] attempting it all laboured in vain.'[106]

Conclusion

The achievements of Pope Gregory X's crusading policy prove that under the right guidance much of the contemporary criticism of the crusade and its strategies could have been transformed into positive and constructive advice on crusade-planning and on the means to save the crusader kingdom. The bulk of the memoirs submitted to Pope Gregory in response to his request for advice on the eve of the Second Council of Lyons contained not only criticism but also sober and even expert advice based on a careful evaluation of the various aspects of a crusade in the Levant. On the whole the crusade-planning during Gregory X's pontificate reflects a new strategy of which the main features were the creation of a permanent garrison in the Latin Kingdom and the launching of small manageable expeditions manned by professional soldiers which would periodically succeed each other in the East. This, however, found only little response in papal crusading policy. First of all, Pope Gregory ignored all expert advice which planned the traditional *passagium generale*. Secondly, in view of the *de recuperatione Terrae Sanctae* treatises, of which so many were to appear from 1291 onwards, the council and the pope made very little of European sea-power as a factor which would be in their favour in any war against Islam. Though Pope Gregory was informed by experts like the master general of the the Temple, by Humbert of Romans, and it seems also by Fidenzio of Padua[107] of the weakness of the Mamluks at sea, he envisaged for this

---

[105] 'Estoire', p. 473.
[106] Salimbene, 'Chronica', pp. 494, 495.
[107] For Fidenzio of Padua see below, ch. 3.

crusade a fleet of twenty ships to be used solely for the war proper. It was therefore only after the loss of the Holy Land that the idea of an economic blockade of the sultan's ports was coupled with that of a special maritime police and of a *passagium particulare*. Moreover, it was only then that serious attempts were made to translate the recommendations into terms of action and to enforce the maritime blockade of Egypt. Gregory X, though well aware of the tremendous importance of the naval superiority of the Christians which would enable them to enforce it, did very little to profit from such an advantage. This is most surprising, as the naval superiority of the West and its importance in the struggle against the Mamluks was by 1262 stressed even by the Mongol il-khans of Persia. Gregory's measures to stop illicit trade with the Saracens amounted to no more than an almost verbatim repetition of the regulations made by the Fourth Lateran Council (1215) and the First Council of Lyons (1245).[108]

Conspicuously absent from crusade-planning after 1270 was also the thesis of the conquest or utter weakening of Egypt as the key to the reconquest of the Holy Land. Absent from both the memoirs presented to Gregory X, as well as from the deliberations of the Council, this thesis appears only on the eve of the fall of Acre, in 'La Devise des chemins de Babiloine', a military memoir addressed to the West by the Order of St John between the years 1289 and 1291. This was, possibly, the reaction to the disastrous crusade of St Louis to Tunis, the last made in the thirteenth century to conquer Egypt. Nevertheless, the strategy advocated at the Second Council of Lyons proves that the beginning of a new period in crusade-planning has to be placed in 1274 and not, as it is often stated, at the time following the loss of the Holy Land.[109]

---

[108] See above, n. 71, and the *Ad liberandum* (1215) and the *Afflicti corde* (1245) in Purcell, *Papal Crusading Policy*, app. A, pp. 187–95; Meyvaert, 'Unknown Letter', pp. 249, 259.

[109] 'Devise des chemins de Babiloine', in *Itinéraires à Jérusalem et descriptions de la Terre Sainte*, ed. H. Michelant and G. Raynaud (Paris, 1882; repr. 1966), 239–52; see also C. Schefer, 'Étude sur la *Devise de Babiloine*', *AOL* 2 (1884), 89–101 and P. V. Laurent, 'Croisade', pp. 136–7. For a different opinion see Atiya, *Crusade*, p. 46.

2
1276–1291
BETWEEN *CRUX CISMARINA* AND *CRUX TRANSMARINA*

Pope Gregory X's plan to launch a general crusade was brought to an abrupt end by his death on 10 January 1276. His short-lived successors had neither his ability nor his all-consuming desire to recover the Holy Land. Innocent V (1276), Hadrian V (1276), and John XXI (1276–7), died too soon to accomplish anything, whereas Nicholas III (1277–80), Martin IV (1281–5), and Honorius IV (1285–7) had none of Gregory X's zeal for the cause of the Holy Land. Paradoxically of all these popes the only one who gave some aid to the Holy Land was Hadrian V, whose pontificate lasted for only about fifty days; he granted the patriarch of Jerusalem Thomas Agni (1272–7) the sum of 20,000 *livres tournois* 'pro galeis construendis vel aliis magis bonorum virorum iudicis opportunis'. The money, however, did not reach the crusader kingdom until 1286.[1] During the years 1276–1290 no other aid was offered by the papacy to the struggling crusader outposts in Outremer.

Yet, though neither Louis IX nor Pope Gregory X had true successors, the realization of the crusade-plan of the Second Council of Lyons was discussed again and again. The need to solve conflicts and maintain the peace in Europe as a *sine qua non* of a crusade seemed to be the constant concern of those who in quick succession occupied the see of St Peter. Nor was interest lacking among the monarchs and nobility. The Cross was taken and retaken, and preparations were made more than once to depart on a crusade as soon as one was proclaimed. The Lyons' tithe collection went on and it seems that the papacy laboured

---

[1] Nicholas IV, *Registres de Nicholas IV*, ed. E. Langlois (Paris, 1886–91), no. 2259; 'Estoire d'Eracles empereur et la conqueste de la Terre d'Outremer', *RHC Hist. occ.* ii. 477; Marino Sanudo Torsello, 'Liber Secretorum Fidelium Crucis', in *Gesta Dei per Francos*, ed. J. Bongars (Hannau, 1611; Jerusalem, 1973), ii. 227; R. Röhricht, 'Syria Sacra', *ZDPV* 10 (1887), 10 n. 3; B. Hamilton, *The Latin Church in the Crusader States: The Secular Church* (London, 1984), 278, 290.

ceaselessly to fill its coffers; its proceeds were frequently granted to monarchs who took the Cross on condition that they go on crusade. None of the monarchs, just as none of the money, had reached the Holy Land during the three last decades of the existence of the crusader kingdom; Edward I of England who left Acre in 1272, was the last. Crusades were being proclaimed but their goal was the recovery not of Jerusalem, but of Aragon, Byzantium, and Sicily. King Philip III of France, considered an heir to his father as the *spiritus movens* of the crusading movement, died like his grandfather and father on a crusade. This, however, was a crusade against neither heretics nor infidels, but against a Catholic kingdom, Aragon.

On the whole, in the decade following Gregory X's pontificate the desire to aid the Holy Land somehow lost its urgency. Without a vicar of St Peter able or willing to continue the crusading policy of Pope Gregory, the tragedy of Louix IX became increasingly felt. The crusader kingdom in Outremer continued its dreary existence through truces with its enemies and it was hoped that it would go on until a *passagium generale* would come to its relief. As time went by the heads of Europe continued to deplore the sad plight of the Holy Land, and recognized the need to aid it in the form of a traditional general crusade, but their political exigencies nearer home came first; even the goal of crusading was to be dictated by political considerations and there were now no resources even for aid in the form of *passagium particulare*. It was only the elevation to the papal throne of a pope who shared Gregory X's sentiments towards the Holy Land and who was ready to pursue the latter's policy that won back for the crusade for the recovery of the Holy Land its place at the centre of European politics. The pontificate of Nicholas IV (1288–92) could have been a turning-point in both the history of the crusading movement and that of the crusader kingdom of Outremer, had it not come too late.

Popes and Monarchs

Gregory X's successor was Innocent V (1276), a French Dominican from Burgundy, whose natural affiliations were with the Franco-Angevin party at the Curia. Indeed, he made major concessions to Charles I of Anjou, but they did not include the authorization of an Angevin 'crusade' against Byzantium.[2] Immediately following his coronation (25 February 1276) he called upon the Christian princes to set off for the recovery of the Holy Places in Palestine from the infidels.[3] Thus resuming the crusading policy of his predecessor Gregory X, he proceeded to the organization of the *passagium generale*. On 23 May 1276 he wrote to the Byzantine emperor that the organization of the expedition for the aid of the Holy Land was the most urgent and important of his tasks as a pope. He informed the emperor that Rudolph of Habsburg, Philip III of France, Alfonso of Portugal (who took the Cross in 1268), and Prince Charles of Salerno, as well as many nobles and magnates, had already taken the Cross. The pope continued that the date of the expedition, the success of which depended on the aid of the Byzantine emperor, had not yet been determined.[4] As a matter of fact in 1276 none of the monarchs mentioned above, all of them *crucesignati*, were ready to depart on a crusade for the recovery of Jerusalem. Rudolph of Habsburg was prepared to go on a crusade but only after being crowned as emperor by the pope in Rome.[5] Philip III of France was engaged in a quarrel with Alfonso X of Castile, who ignored the French claims to succession in both Navarre and Castile. When he took the Cross in 1275,[6] it is probable that he did so reluctantly under the massive pressure of Pope Gregory. At that time it was believed that as son of Louis IX and as king of France Philip would continue the policy of his father. Writing in 1275 the troubadour Raymond Gaucelin appealed to the king in favour of a new crusade.[7] Philip III must indeed have felt some obligation towards both the crusading movement and the Holy

---

[2] On 2 Mar. 1276 Charles was nominated the vicar of Tuscany and senator of Rome: Potthast, no. 21103.

[3] M. H. Laurent, *Le Bienheureux Innocent V (Pierre de Tarentaisse) et son temps* (Vatican, 1947), no. 69 and pp. 228–30.

[4] For this letter see *Acta Romanorum Pontificum ab Innocentio V ad Benedictum X*[3], ed. A. Tantu, vol. v (Vatican, 1954), pt. 2, no. 2, pp. 2–4.

[5] For Rudolph's intentions at that time see Thomas Tuscus, 'Gesta Imperatorum et Pontificum', *MGH SS* xxii. 525.

[6] See above, ch. 1 n. 87; 'Chronicon Lemovicense', *RHGF* xxi. 786.

[7] M. Raynouard, *Choix de poësies originales de troubadours*, iv (Paris, 1819), 137–8.

Land. This is reflected by the aid he rendered the kingdom in form of *passagia particularia* in the years 1272, 1273, and 1275. Although it is most probable that, on the whole, King Philip favoured this form of crusading strategy rather than that of *passagium generale*, he nevertheless committed himself to the papal crusade-project and during the years 1275–84 it was expected that he would be the leader of the planned expedition to the East. In March 1276 he notified the pope that he would be ready to depart for the East in two years on 24 June 1278.[8] By that time, however, Philip was too involved in the war against Castile. In the following years Philip found himself still involved with Castile and then with the war against Aragon. Unlike his father, Philip put the welfare of France before that of the Holy Land.

Pope Innocent attempted as unsuccessfully as his predecessor to move King Edward I of England to engage himself in the *negotium Terrae Sanctae*. Edward remained as elusive as he had been with Pope Gregory X. As mentioned above, Edward was granted by Pope Gregory the Lyons' tenth of England, Wales, and Ireland on the condition that he took the Cross; no money was to be delivered to him until he was ready to depart.[9] Writing in February 1276 to Pope Innocent V he wished the pope success in the task of the liberation of the Holy Land, but claimed that at present he was unable himself to go to its aid.[10] In a further letter to the pope on 12 December 1276, Edward still hesitated about pledging himself and suggested to the pope that he was ready to promise only that he would go or send his brother Edmund in his place. A year later in 1277 Edward demanded from Pope John XXI that his expedition to Tunis should be considered as a crusade. The pope declined on the basis of a decision taken on this subject by Gregory X.[11]

During the pontificate of Innocent V the crusade was still preached, at least in certain parts of Europe, and the Lyons'

---

[8] M. H. Laurent, *Innocent V*, no. 80, p. 236.

[9] See above, ch. 1 n. 90.

[10] *Foedera, conventiones, litterae et acta cuiuscunque generis acta publica inter reges Angliae et alios quosvis imperatores, reges, pontifices, principes vel communitates 1101–1654*, ed. T. Rymer *et al.* (Hague, 1745), I. ii. 152.

[11] Ibid. I. ii. 155, 159; Gregory X, *Registres de Grégoire X*, ed. J. Guiraud *et al.* (Paris, 1892–1906), no. 539; Nicholas III, *Registres de Nicholas III 1277–1280*, ed. J. Gay (Paris, 1898–1938), nos. 10–11.

tithe collected,[12] but the rapid succession of popes was hardly favourable to the success of a crusade-policy. Innocent V died after five months. Hadrian V was pope for little more than seven weeks. His successor John XXI, a physician, logician, and theologian made an attempt to maintain Pope Gregory X's crusade-policy. He called upon the new king of Aragon Peter to help the Holy Land; and, as in his opinion the realization of the projected crusade depended largely upon King Philip III, he tried to negotiate in the quarrel between the king and Alfonso X of Castile. Following an unsuccessful expedition of Philip against Castile, the pope wrote (3 March 1277) to his French legate that King Philip must be warned that another attempt against Castile would bring upon him the spiritual punishment ordered at the Second Council of Lyons for those who broke the peace necessary for the recovery of the Holy Land.[13]

The successor of John XXI, Pope Nicholas III (1277–1280), resumed the policy of his predecessors as far as the Byzantine ambitions of Charles I of Anjou and the union of the Churches were concerned. His main interest, however, lay in the realization of his own aims in Italy, namely upon weakening the Franco-Angevin influence there and upon promoting the interests of his own family the Orsini. Alarmed by the increasing strength of Charles I of Anjou, this pope took from him (16 September 1278) his offices of vicar of Tuscany and senator of Rome.[14] While on the whole sacrificing the interest of the crusade to his Italian policy, this pope did not favour the cause of the Capetians and the Angevins as his successors were to do and this explains his attitude towards the various demands of King Philip III. He even made Philip III the scapegoat for the failure of the papal crusade-project, blaming the quarrel between Philip of France and Alfonso X of Castile for preventing the general crusade. In June 1279 he informed both the kings that he held them accountable for the misfortunes of the Holy Land of whose chaotic condition and suffering his legate (probably the patriarch of Jerusalem Elias) was an eye-witness. He also

---

[12] See e.g. 'Annales Basilenses', *MGH SS* xvii. 200; M. H. Laurent, *Innocent V*, pp. 228–55.

[13] *AE* ad an. 1277, no. 3; Potthast, no. 21229.

[14] Nicholas III, *Registres*, nos. 303–4; K. M. Setton, *The Papacy and the Levant 1204–1571*, i (Phil., 1976), 128–34.

pointed out that the quarrel was all the more dreadful because circumstances were singularly favourable for the renewal of the Holy War. Never had there been a better opportunity for Christians to recover the Holy Sepulchre. The greatly feared sultan of Egypt Baibars was dead and the Holy Land could easily be won back now that the Moslems were fighting among themselves.[15] And indeed, as other sources confirm, Nicholas III was quite correct in his opinion that the conditions in the East were at that time exceedingly favourable for the reconquest of the Holy Land.[16] The monarchs of France and Castile, however, remained deaf to the papal appeals to put an end to their quarrel for the sake of the Holy Land. Nor was Pope Nicholas III himself prepared to sacrifice his own interest as he so often admonished others to do. It was the calculations of his Italian policy that, for example, prevented Nicholas III from close co-operation with Rudolph of Habsburg in the interest of the crusade. He refused to carry out Pope Gregory X's promise to crown the emperor in Rome fearing that this act would increase the emperor's prestige in Italy. He must have been at the same time well aware that Rudolph, a *crucesignatus* since 1275, would not consider the crusade until he was properly crowned.[17]

Nicholas III was most active in administering the Lyons' tithe; the six years for which it had been granted expired during his pontificate in 1280.[18] Its administration suffered from many irregularities, due both to the opposition of the taxpayers and the quick succession of popes. Because of this, such a large amount remained unpaid at the termination of the sexennial period (24 June 1280) that collectors were continued in office. In England, for example, £110,890 had been collected by 1282, but £18,000 or more was still missing, even though large numbers of clergy were put under sentences of excommunication, interdict, or suspension for non-payment.[19] The pope also

---

[15] Nicholas III, *Registres*, nos. 758, 761–2, see also nos. 222, 261–4, 273, 275, 385–7, 677–8, 680.   [16] See below, n. 44.
[17] P. A. Throop, *Criticism of the Crusade: A Study of Public Opinion and Crusade Propaganda* (Amsterdam, 1940), 278. For the taking of the Cross by Rudolph of Habsburg see above, ch. 1 n. 86. See also Martin Oppaviensis, 'Chronicon', *MGH SS* xxii. 442; Thomas Tuscus, 'Gesta', pp. 524–5.
[18] Nicholas III, *Registres*, nos. 3, 8, 14, 42, 43, 62, 80–3, 100, 112, 125, 131, 193–6, 199–200, 447, 449, 480, 481, 491, 537, 539–40, 542–4, 546, 548, 763, 1026, 1027, 1033.
[19] Ibid., nos. 545, 549; W. E. Lunt, 'Papal Taxation in England in the Reign of Edward I', *EHR* 30 (1915), 406–7.

appealed to monarchs to aid him and his collectors in their work, and ordered stern measures against those among them who, like Alfonso III of Aragon, appropriated funds from the tithe.[20] The efforts made by Nicholas III reflect on the whole a growing reluctance on the part of the clergy to pay the tithe and, on the part of the monarchs, to see the funds leave their states. The German clergy, rebuked by Pope Nicholas for their negligence in paying the tenth, claimed that they objected to the collection on the ground that the funds were diverted to other causes than that of the Holy Land. In England the opposition was strong enough for a chronicler to claim that, in the public view, the three popes Innocent V, Hadrian V, and John XXI deserved their early death 'quod noluerunt dictam decimam relaxare'.[21] Similar opposition to the Lyons' tithe was voiced also by the French clergy. There were complaints from the bishops about restriction of crusaders' privileges in royal courts. Due to frequent abuses of those privileges—according to Philip III some of the crusaders in his kingdom abused their liberty by committing crimes like theft, murder, and rape— Pope Nicholas III forbade the French prelates (19 September 1278) to interfere with the king and his officials when they took action against lay crusaders who were guilty of crimes. John XXI ordered that lay crusaders who delayed their departure for the Holy Land should be subject to their lay lords.[22]

It was probably during the pontificate of Nicholas III that the Lyons' tithe began to be diverted to uses other than the *negotium Terrae Sanctae*. According to the Bolognese Dominican Francesco Pipino as well as to German chroniclers, it was from this fund that Pope Nicholas rebuilt the Vatican palace.[23] Even

---

[20] Nicholas III, *Registres*, nos. 3, 199, 539, 542–9, 553, 544, 1026–7; M. Purcell, *Papal Crusading Policy: The Chief Instruments of Papal Policy and Crusade to the Holy Land from the Final Loss of Jerusalem to the Fall of Acre 1244–1291* (Studies in the History of Christian Thought, 11; Leiden, 1975), 156–7.

[21] Martin IV, *Registres de Martin IV*, ed. F. Olivier-Martin *et al*. (Paris, 1901–35), no. 244; 'Annales prioratus de Dunstaplia', in *Annales monastici*, ed. H. R. Luard (RS, London, 1864–9), iii. 267; Lunt, 'Papal Taxation', pp. 398–408; Purcell, *Papal Crusading Policy*, pp. 154–7.

[22] Nicholas III, *Registres*, no. 171; Potthast, no. 21201; G. J. Campbell, 'Clerical Immunities in France during the Reign of Philip III', *Speculum*, 39 (1964), 415–17.

[23] Francisco Pipino, 'Chronicon', *RIS* ix. 724; 'Chronicon Imperatorum et Pontificum Bavaricum', *MGH SS* xxiv. 225; see also Martin Oppaviensis, 'Chronicon', p. 443.

if this accusation is correct this pope, unlike his immediate successors, actually refused firmly to divert the Lyons' tithe to a crusade against Christian lay powers. In 1278 Philip III requested Pope Nicholas to grant a plenary or partial indulgences to those—'non cruce-signati aut voto crucis astricti' —who made a contribution for the French expedition against Castile. Writing on 3 December 1278 Nicholas III tempered his refusal with a statement to give his utmost help should a general crusade become imminent.[24]

If during the pontificates of Gregory X's immediate successors, the preparations for a crusade to the Holy Land lost their momentum, the accession of Martin IV (1281–5) meant the diversion of papal crusading policy from the Holy Land to other places—Byzantium, Aragon, and Sicily. Simon of Brie, the new pope, had been an old friend of the Capetian family and had served Louix IX as chancellor and treasurer of St Martin of Tours. A cardinal-priest of St Cecilia and a papal legate in France since 1261, he was responsible for the promotion of the crusade in France as well as the administration of its instruments; he was thus actively involved in both the preaching of the second crusade of Louis IX (until 1268) and the crusade of Gregory X.[25] It was from his hands that King Philip III, his brothers, and brothers-in-law took the Cross in 1275. Indebted to King Charles I of Anjoy for his election to the papal throne, he favoured Franco-Angevin interests. Under pressure from King Charles, he reversed papal policy towards Byzantium and lent the Church's support to the king's plan for an expedition against Constantinople.

Charles I of Anjou, the younger brother of King Louis IX of France and from 1266 the secular ruler of the papal fief of Sicily, had since then even made plans for the restoration of the Latin Empire in Byzantium. By the Treaty of Viterbo (1267) he gained most of the rights of the deposed Latin emperor Baldwin II, including suzerainty over the Frankish principality of Achaea. He also succeeded having himself elected (c.1271–2) as king of Albania. During the years 1267–82 he tried by diplomatic activity to unite all the potential opponents of the Palaeologi as well as to obtain the support of Hungary, Serbia,

[24] Nicholas III, *Registres*, no. 392; Purcell, *Papal Crusading Policy*, pp. 94–7.
[25] J. Richard, *Saint Louis* (Paris, 1983), 384–5, 527, 533–4.

and Bulgaria, and the assistance of the Venetian fleet. However, Charles was forced to postpone his plan for a number of reasons. The first was Conradin's invasion in 1268. Then came the crusade of Louis IX to Tunis (1270). The pontificate of Gregory X produced another and most serious obstacle and during the period 1272-82 the papacy considered Charles's plan for the conquest of Byzantium to be fatal to the chances of winning back Jerusalem.[26]

It was while Gregory X was pope that Charles, frustrated in his ambition to conquer Byzantium and thus become an emperor, began to negotiate the purchase of the crown of Jerusalem from Maria of Antioch. Pope Gregory X supported the claims of Maria for several reasons. First of all he had probably formed a negative impression of King Hugh III of Lusighan when he met him in the crusader kingdom in 1271 and concluded from his own experiences in Outremer that little could be hoped from the Lusignans' government. He also encouraged Maria to sell her claims to Charles I of Anjou as he wished him to take a more active interest in the fate of the crusader state, not only for its own welfare but also to divert Charles from his Byzantine ambitions. The deal was completed on 18 March 1277 after Gregory X's death, and confirmed by Pope John XXI; Charles acquired from Maria her rights to the crown of Jerusalem for 11,000 Saracen besants and the promise of annuity of 4,000 *livres tournois*. Charles at once assumed the title of 'Rex Jerusalem' and sent out Roger of San Severino, count of Marsico with an armed force to be his bailli (*vicarius*) at Acre.[27]

[26] Setton, *Papacy*, i. 106-39; J. R. Strayer, 'Political Crusades of the Thirteenth Century', in *Setton's History*, ii. 367-70; D. Geanakoplos, *Emperor Michael Palaeolgus and the West 1258-1282* (Cambridge, Mass., 1959), 258-334.

[27] Röhricht, *Regesta*, no. 1411; 'Estoire d'Eracles', pp. 464, 475-6; 'Templier de Tyr', in *Gestes de Chiprois*, ed. G. Raynaud (Geneva, 1887), 206-7; Potthast, nos. 20532, 20632; M. L. de Mas-Latrie, *Histoire de l'île de Chypre sous la règne des princes de la maison de Lusignan* (Paris, 1852-61), i. 454-7; S. Runciman, *History of the Crusades* (Cambridge, 1955), iii. 342, 345. Maria was the daugter of Melisande (the daughter of Queen Isabel I of Jerusalem) and Bohemond of Antioch. When, following the death of Conradin (1268), Hugh of Lusignan (the grandson of Alice the daughter of Queen Isabel I) assumed the title king of Jerusalem, Maria declared that as she was a generation nearer than Hugh to their common ancestress Isabel I, she took precedence over him. The case was brought before the High Court, which decided in favour of Hugh. Maria then issued a formal protest and hurried off to plead her cause at Rome. See J. Riley-Smith, *Feudal Nobility and the Kingdom of Jerusalem 1174-1277* (London, 1973), 220-4; G. Hill, *History of Cyprus* (Cambridge, 1948-52), ii. 190-2.

Ironically it was during the period 1269–1280, most of which fell before he became the king of Jerusalem, that Charles took a vivid and active interest in the welfare of the crusader state. During this period he encouraged export of foodstuffs (mainly grains) and war-materials (including horses and mules) from Sicily to Outremer.[28] The Sicilian monarch also provided assistance to the crusader kingdom by providing transport to the Levant for important personages,[29] as well as by allowing the military orders to supply their establishments in the Holy Land from those in the kingdom.[30] Additionally, from 1277, large amounts of foodstuffs and war-materials[31] and above all troops were sent directly to the king's bailli in Acre: in 1277 about 2,000, and in the following year about 4,000, *stipendiarii* were dispatched from Sicily to Acre.[32] At this period Charles also tried to fortify this position among the nobility of Outremer by marriage alliances. In 1278 Margaret of Beaumont (a granddaughter of John of Brienne and a relative of Charles) was sent by the king to the East and there married off to Bohemond VII, the titular prince of Antioch. The latter's sister Lucia was married to Charles's admiral Narjon of Toucy.[33]

This policy of Charles changed in 1281, following the accession to the papal throne of Martin IV. This pope, unlike his predecessors, was the king's ally, who under pressure from Charles lent the Church's support to the recovery of the Byzantine Empire. Charles's planned expedition against Constantinople was sanctioned now by Pope Martin as a 'crusade against schismatics and usurpers'.[34] Following the Treaty of Orvieto of 31 July 1281 between Charles I, Philip of Courtenay the titular Latin emperor of Constantinople, and the

---

[28] *i Registri della cancelleria angioina*, ed. R. Filangieri di Candida *et al.* (Naples, 1950–80), xiii. 30–1; xiv. 53; xxvi, 101–2; xxvii. 138, 221, 255.

[29] Ibid. i. 290; vi. 192; vii. 199; xi. 122; xix. 23, 66–7, 69, 70, 241, 245; xxiii. 36.

[30] Ibid. ii. 24; iii. 189, 192; vii. 17–18, 45, 62; viii. 202; xiv. 50–1; xv. 36–7, 41; xix. 39, 50–1, 170–5, 181–2, 194; xxiv. 45.

[31] Ibid. xv. 55–6; xviii. 31–2, 106–8, 181–2, 194–8, 373–4, 399–400; xix. 17, 23–6, 51, 65, 76, 171–5, 181–2, 194–8; xx. 156, 218; xxi. 41; xxii. 114; xxiii. 38, 62, 65, 126, 129–31, 155, 256, 310; xxiv. 110, 126, 127–8, 159; xxvi. 101, 107.

[32] Ibid. xv. 55–6; xviii. 208; xix. 24–5, 253.

[33] Ibid. xviii. 414, 415; xix. 50, 151, 176–8; xxi. 73–4; xxii. 320; Röhricht, *Regesta*, no. 1422.

[34] Martin IV, *Registres*, nos. 269, 278; Potthast, no. 21815; *AE* ad an. 1281, no. 25; Geanakoplos, *Emperor Michael*, pp. 340–1; Setton, *Papacy*, i. 138 and n. 65.

republic of Venice,[35] the pope excommunicated (18 October 1281) the Byzantine emperor, thus disrupting the union of Lyons.[36] For the crusade against Constantinople, Charles was granted by the pope (18 March 1282) the crusading tithe for six years in the island of Sardinia and in the kingdom of Hungary. This was in accord with the decree of the Second Council of Lyons as Charles was a *crucesignatus*.[37] With the preparations for the expedition to Byzantium in progress, Charles's interest in his kingdom in Outremer diminished. Because of the necessity to provide for this expedition there had been a sharp decline since 1281 in the export licences granted to private merchants as well as to the military orders. Also the aid for the royal forces in the crusader kingdom declined considerably after 1280,[38] and following the Sicilian Vespers (30 March 1282) the bailli Roger of San Severino found himself without backing at all. Summoned to return to Sicily, he left Acre towards the end of 1282, confiding his position as bailli to his seneschal Ido Poilechien, who married the widow of John of Ibelin Lord of Arsuf, Lucia of Gouvain.[39]

It is most probable that Pope Martin IV, who claimed as late as 12 January 1284 that 'the business of the Holy Land is especially close to our heart', hoped that Charles, after realizing his lifelong plan to re-establish the Latin Empire in Constantinople, would set out on a crusade for the rescue of Outremer.[40] His grants of tithes to Charles for the recovery of Byzantium reflect a certain uneasiness and an attempt to conceal their real destination; Charles is described in the bull which granted him for six years the tenth of Sardinia and Hungary as one who,

---

[35] Geanakoplos, *Emperor Michael*, pp. 355–40. For the rest of this treaty see *Urkunden zur Ältern Handels- und Staatsgeschichte der Republik Venedig*, ed. G. L. Tafel and G. M. Thomas (Vienna, 1856–7; repr. Amsterdam, 1964), iii. 287–95.

[36] *AE* ad an. 1281, no. 25. (The text of this bull does not appear in Martin IV, *Registres*.) The sentence of excommunication was renewed on 7 May and 18 Nov. 1282: Martin IV, *Registres*, nos. 269, 278.

[37] Martin IV, *Registres*, no. 116. For the taking of the Cross by Charles, see above, ch. 1 n. 88.

[38] This is shown by the evidence of the Angevin registers. Whereas in 1277 there were three shipments, in 1278 seven, in 1279 one, in 1280 five, there were three in 1281 and one in 1282 and in 1283. On the whole the largest shipments were in 1277–8 and 1280. See above, n. 31, and *i Registri della cancelleria angioina*, xxiv. 126–8, 159; xxv. 5; xxvi. 101, 107.

[39] 'Templier de Tyre', p. 214; Runciman, *History*, pp. 392–3.

[40] Martin IV, *Registres*, no. 433; Setton, *Papacy*, p. 142.

inspired by zeal of faith, took the Cross for the promotion of *subsidium Terre Sancte*; the tithes were to be used against the infidels.[41] The same view must have been shared by other supporters of the Angevin project. This is reflected by the official historian of the French court, William of Nangis, who declared that King Charles was going to fight the Saracens and reconquer the kingdom of Jerusalem.[42] The truth was that the object of Charles's crusade was Byzantium and the reference to the king's intention to go to the rescue of the Holy Land was intended to link the two and so to legitimize the proclamation of a crusade against Byzantium. The recovery of Byzantium became thus a *sine qua non* for that of the Holy Land and thus deserved the status of a crusade.

In practice Martin IV's policy of supporting the project of King Charles meant at least temporary diversion of the instruments of the crusade from the shores of Palestine to those of the Bosphorus, and moreover the loss, by disrupting the union of the Churches, of a potential ally for the plan of the recovery of the Holy Land, the Byzantine emperor. It is true that the outbreak of the Sicilian Vespers and the conquest of Sicily by Peter III of Aragon (1282) put an end to the planned crusade against Byzantium. But still, Martin IV's policy meant that from 1281 the papacy concentrated all its efforts on the crusading plans of Charles I of Anjou. It meant also that a pope like Martin IV would be more easily prepared to divert the funds, collected since 1274 for the *negotium Terrae Sanctae*, to the recovery of the papal fief Sicily. In September 1283 Pope Martin IV warned that unless the rebellion in Sicily was quickly suppressed, 'all the work done . . . by . . . Pope Gregory X . . . for the preparation of the aid for the Holy Land will be denied any fulfilment'.[43] And indeed so it was.

The Crusade against Sicily and the Holy Land

Precisely at the time of the outbreak of the so-called Sicilian Vespers on 31 March 1282, a Hospitaller, Joseph of Cancy,

[41] Martin IV, *Registres*, nos. 116, 117.
[42] William of Nangis, *Chronique latine de Guillaume de Nangis de 1113 à 1300 avec les continuations de cette chronique de 1300 à 1368*, ed. H. Géraud, i (Paris, 1843), 516.
[43] Martin IV, *Registres*, nos. 457.

could write to King Edward I of England that 'Sachez sire, la Terre-Sainte ne [fu onques] legère à conquèrir si bonnes gens y venissent et viandes, comme elle est hui au jor, tout soit a que onques mais ne veimes meins de gens d'arme en la terre, ni meins de bon conseil.'[44] However, though the conditions for a crusade at that time were indeed excellent, the papacy had to focus its attention, efforts, and resources upon the recovery of its own fief of Sicily, and the project for a crusade to the Holy Land was more remote than ever.

On 13 January 1283 Pope Marin IV declared a crusade against the Sicilians. At first the preaching of this crusade was restricted to the kingdom of Sicily but on 5 April 1284—two years after the outbreak of the Sicilian revolt—it was extended to northern Italy and some time later—on 4 June 1284—to France.[45] During the short pontificate of Honorius IV (1285–7) the war against the Sicilians and their Aragonese allies not only continued, but indirectly led to a French crusade against Aragon (1285). The failure of this crusade, in the course of which the French king Philip III died, had a disastrous and long-standing impact on the attitude of his successor Philip IV the Fair to the crusade.[46] The latter refused up to 1305 to get involved with it. In 1285 the king promised Pope Honorius IV to take upon himself, together with his brother Charles of Valois, the nominal king of Aragon, the *negotium Aragonie*. However, unlike his father, Philip IV showed little interest in foreign affairs and after Charles of Valois renounced in 1290 his claim to Aragon, his interest in the crusade against Aragon diminished, though as late as the autumn of 1291 he insisted that this crusade should precede the one for the recovery of the

---

[44] *Cartulaire général de l'ordre des Hospitaliers de St Jean de Jérusalem 1100–1310*, ed. J. Delaville le Roulx (Paris, 1894–1906), no. 3782. For the conditions in the East, see J. Prawer, *Histoire du royaume latin de Jérusalem* (Paris, 1969), ii. 518–24.

[45] Martin IV, *Registres*, nos. 301, 570, 591. For the Sicilian Vespers see S. Runciman, *Sicilian Vespers: A History of the Mediterranean World in the Later Thirteenth Century* (Cambridge, 1961), *passim*.

[46] N. J. Housley, *Italian Crusades: The Papal–Angevin Alliance and the Crusade against Christian Lay Powers 1254–1343* (Oxford, 1982), 87–8, 102–3, 131. For the crusade against Aragon see J. N. Hillgarth, *Spanish Kingdoms 1250–1516*, 1 (Oxford, 1976), 255–9; J. R. Strayer, 'Political Crusades of the Thirteenth Century', in *Setton's History*, ii. 371–2.

Holy Land.[47] Moreover, at that time King Philip tried to be relieved from the custody of the crusader kingdom which he had inherited from his father. On 9 December 1289 he demanded that the *conservatio reliquiarum Terre Sancte* would be financed by tithes collected in other kingdoms than France. In 1289 he complained to Pope Nicholas IV that the money granted to his father for this aim was insufficient and in 1290 he asked to be relieved from this task altogether.[48]

The crusade against Sicily deeply affected the fate of that to the Holy Land. It caused an unavoidable collision of interests so far as papal policy, resources, and manpower were concerned. During the two crucial decades from 1282 to 1302 papal efforts on behalf of the crusade were divided between the attempt to maintain suzerainty over the kingdom of Sicily and the attempt to hold or, after 1291, to recover the Holy Land. The popes were well aware of the problem. They were criticized for devoting to the needs of the conflicts in Europe resources intended for the recovery of the Holy Land. Their reply was that a crusade could not be organized against the Moslems in the East until the Church was secure in the West. They justified the *crux cismarina* by insisting that it was an essential prelude to the *crux transmarina*. The papacy claimed that it was not the Holy See but its enemies who were impeding the crusade to the East by creating the need for crusades in Italy. The facts of geography made the threat to the Church and the Faith from rebel Christians in Europe greater than that posed by the Moslems in Asia. The nearer problem must be dealt with first.[49]

To realize this aim, funds which had been originally raised for Outremer were transferred to finance the Sicilian war. By May 1284, for example, this had happened to money given for the Holy Land in the form of oblations and legacies by the inhabitants of the kingdom of Sicily. Both Martin IV and

---

[47] G. Digard, *Philippe le Bel et le Saint-Siège* (Paris, 1936), vol. ii, no. 2, pp. 218–23; J. R. Strayer, *Reign*, pp. 386–9; S. Schein, 'Philip the Fair and the Crusade: A Reconsideration', in P. W. Edbury (ed.), *Crusade and Settlement* (Cardiff, 1985), 121–3.

[48] Nicholas IV, *Registres*, nos. 1005, 4409; Digard, *Philippe le Bel*, vol. ii, no. 2, p. 222. See also ibid., vol. i, pp. 71, 123; below, ch. 3.

[49] Martin IV, *Registres*, nos. 452, 457, 482, 570, 583, 587, 590; Nicholas IV, *Registres*, no. 6859; *Foedera*, I. iii. 25; Purcell, *Papal Crusading Policy*, pp. 94–8; Housley, *Italian Crusades*, pp. 35–70.

Honorius IV granted Philip III of France legacies which traditionally were to be employed for the crusade in the East.[50] To finance the Church's war, use was also made of the tenth which Gregory X had levied at the Second Council of Lyons (1274) for the use of his planned *passagium generale*. On 13 December 1282 the pope asked Philip III to devote 100,000 *livres tournois* of the French tenth, deposited at the Paris Temple, to the needs of his war against Romagna.[51] In the period April–June 1283 over 15,000 gold ounces were paid to Charles of Salerno, the son and successor of Charles I of Anjou, from the tenth which had been collected in Hungary, Sicily, Sardinia, Corsica, Provence, Aragon, and the imperial dioceses. In November 1283 16,000 gold ounces were given to Charles from the same source. In the following year crusader tithes collected in Scotland, Denmark, Sweden, Hungary, Poland, and other parts of eastern Europe, were used for the war for the recovery of Sicily.[52] Moreover, new tithes were levied and crusader indulgences used to promote the Franco-Angevin crusade against the rebellious Sicilians and Aragonese. On 2 September 1283 Philip III of France was granted by Pope Martin IV the triennial tithe of France 'in subsidium belli contra Aragonum' on behalf of Charles of Valois (who was granted by the pope the kingdoms of Aragon and Valencia on 5 May 1284). This was converted in May 1284 into a four-year tenth.[53] The French soldiers assisting Philip against Peter of Aragon were granted full crusading indulgences; such indulgences were also granted on 5 April 1284 to all those who aided the king of Sicily against the Sicilians and Aragonese.[54] At the same time, it should be mentioned, Martin IV refused the request of Peter to grant his expedition against Tunis the status of a crusade, as he suspected that the king of Aragon's real aim was not northern Africa, but the invasion of Sicily.[55]

Indulgences were granted to clergy as well. In June 1284 Martin IV offered them to all Italian clergy who paid in one

[50] Honorius IV, *Registres d'Honorius IV*, ed. M. Prou (Paris, 1886–8), no. 484; Housley, *Italian Crusades*, p. 102.
[51] Martin IV, *Registres*, nos. 272–3.
[52] Housley, *Italian Crusades*, p. 103; Setton, *Papacy*, i. 142–3.
[53] Martin IV, *Registres*, nos. 457–8, 580–1, 583–6; Housley, *Italian Crusades*, p. 176.
[54] Martin IV, *Registres*, nos. 457, 570; Housley, *Italian Crusades*, p. 149 and n. 18.
[55] Potthast, no. 21877; Purcell, *Papal Crusading Policy*, pp. 87–8.

year the whole of the triennial tenth levied at this time, as well as to the French clergy who paid the tithe in 1283. King James of Majorca was granted on 2 January 1285 a triennial tithe of his kingdom for his assistance to Philip III in the war against Aragon.[56] In June 1284 a triennial tithe of Sicily and Provence was granted to Charles I and in March 1285 a triennial tithe of northern and central Italy.[57] Honorius IV granted Charles (30 April 1286) the triennial tithe of all churches in Italy. Nicholas IV granted (September 1288) a triennial tithe to Philip IV of France who had undertaken the continuation of the Aragonese crusade. In August that year he asked the clergy and the orders of Sicily to give up a year's tenth to the defence of the realm and declared that if they did not do so voluntarily, they would be compelled to do this. In the following year he levied, on behalf of Charles II of Anjou,[58] a similar tenth in Italy and in the imperial dioceses not ruled by Philip IV. Even royal marriages were arranged for both the *crux cismarina* and *crux transmarina*. So in 1289 Charles II of Sicily married Margaret of France for 'succursus Terre Sancte ... et recuperatio terre nostre Sicilie'.[59]

It was Pope Nicholas IV (1288–92), during whose pontificate Acre fell, who made it his main ambition to revive Gregory X's crusade-policy. He aimed accordingly at the general pacification of Europe, the union with the Greek Church, and the recruitment of western Christians for a massive and co-ordinated action in the East.[60] Pope Nicholas was well aware of the desperate position of the Latin Kingdom and of the need of a general crusade to save Outremer; but even he felt it necessary to settle the Sicilian question before embarking upon any large-scale crusade to the Holy Land. Thus on 10 August 1290 and later on 23 March 1291 he promised crusade-indulgences to the people of Gaeta.[61] On 7 May 1291, the

---

[56] Martin IV, *Registres*, nos. 549, 591; Honorius IV, *Registres*, no. 12.

[57] Martin IV, *Registres*, nos. 587–90; Honorius IV, *Registres*, no. 12; Housley, *Italian Crusades*, p. 176.

[58] Honorius IV, *Registres*, nos. 395, 399; Nicholas IV, *Registres*, nos 613, 615, 617–18, 999–1004; 1142–52, 2136–7; Digard, *Philippe le Bel*, vol. i, pp. 61–2.

[59] Digard, *Philippe le Bel*, vol. ii, no. 10, p. 275.

[60] F. Heidelberger, *Kreuzzugsversuche um die Wende des 13. Jahrhunderts* (Leipzig and Berlin, 1911), 1–9; below, ch. 3.

[61] Nicholas IV, *Registres*, nos. 3017, 4711.

month of the fall of Acre, he renewed the bull of Martin IV of 1283 proclaiming crusade-preaching throughout the kingdom of Sicily against the Sicilians and their allies.[62]

At the same time Pope Nicholas IV worked for the aid of Outremer. Alarmed by the fall of Tripoli (26 April 1289), and informed by the embassy headed by John of Grailly about the desperate state of the kingdom, he appealed to the West in September 1289 urging immediate aid for the crusader East.[63] The pope also transferred (9 September 1289) to the patriarch of Jerusalem Nicholas de Hanapes a sum of 4,000 *livres tournois* as a loan to be spent on repairing the fortifications of Acre, building war-machines, and redeeming prisoners of war.[64] Obviously at that time, as at others when news was received about losses in the East, there was pressure on the pope from both East and West to concentrate upon the struggle in the Holy Land rather than that against his enemies in Europe. The Sicilian Ghibelline Bartholomew of Neocastro, a 'familiaris' of James II of Aragon, showed the prevalence of such opinions among the enemies of the Holy See. He recounts a story of a Templar messenger from the Holy Land (possibly a member of John of Grailly's embassy) who arrived at the Curia following the loss of Tripoli and allegedly addressed Pope Nicholas as follows: 'You could have relieved the Holy Land with the power of kings and the strength of the other faithful of Christ . . . but you preferred to attack a Christian king and the Christian Sicilians, arming kings against a king to recover the island of Sicily.' Moreover, the Templar is said to have added that unless the pope made peace with the Sicilians and sent help at once, Acre too would fall.[65]

---

[62] Ibid., no. 6702. For Martin IV see above, n. 45.
[63] Nicholas IV, *Registres*, nos. 613, 615, 617–18, 991–1004, 1142–52, 2136–7, 2251; Marino Sanudo Torsello, 'Liber', p. 230.
[64] Nicholas IV, *Registres*, nos. 1357, 1495; Röhricht, *Regesta*, nos. 1495, 1500; S. Schein, 'Patriarchs of Jerusalem in the Late Thirteenth Century; *Seigners espirituales et temporeles?*', in B. Z. Kedar et al., *Outremer: Studies in the History of the Crusading Kingdom of Jerusalem presented to Joshua Prawer* (Jerusalem, 1982), 303.
[65] Bartholomew of Neocastro, 'Historia Sicula', *RIS* ns xiii. 108–9; Housley, *Italian Crusades*, p. 77.

The Reaction to the Fall of Tripoli and the Crusade of 1290

However, although Tripoli had been a Christian possession of major importance, the reaction to its fall was rather mild. By the end of the 1280s the Europeans had doubtless become accustomed to defeats and territorial losses in the crusader kingdom, and Tripoli was not a name to attract their attention like those of holy places or of a thriving emporium like Acre; it was just 'une cite d'oultre mer'. Only a few European chroniclers even mention the disaster and even fewer record the event in detail. Most of them bitterly deplore the loss of Christian lives, estimated by them at forty thousand, as well as the desecration of the city's churches and relics.[66] Among the mainly brief accounts that of the so-called *Chronicon de Lanercost*, composed by an anonymous Franciscan from Durham, is outstanding. It records two miracles that occurred in the city on the eve of the fall of Tripoli, one to an English Franciscan and the other to Luceta, the abbess of the Convent of the Poor Clares. The chronicle attributes these stories, as well as many others, to Hugh the Franciscan bishop of Byblos, who after the fall of Tripoli (1289) went to England and spent two years in Durham.[67]

Moreover, in around 1290, immediately after its fall, the city became the scene of one of the most popular visions of the Late Middle Ages, the so-called 'Vision of Tripoli': according to most of its versions it was revealed to a monk in the last days of crusader rule in the city. Placed in such a dramatic setting, the vision became highly popular after the loss of the Holy Land in 1291. It predicted the fall of Tripoli, as well as other disasters to be suffered by Christendom; but it also promised the destruction of the Moslem enemy. The vision's opening lines

[66] William of Nangis, *Chronique*, p. 572; Girard of Frachet, 'Chronicon cum anonyma eiusdem operis continuatione', *RHGF* xxi. 9; *Grandes Chroniques de France*, ed. J. Viard, viii (Paris, 1834), 136–7; 'Annales Colmarienses Maiores', *MGH SS* xvii. 217; Bohemond of Trier, 'Gesta', *MGH SS* xxiv. 474; Nicholas Trevet, *Annales sex regum Angliae*, ed. T. Hog (London, 1845), 314–15; 'Annales Hibernie', in *Chartularies of St Mary's Abbey, Dublin*, ed. J. T. Gilbert, ii (*RS*, London, 1884), 320; 'Annales monasteri de Waverleia', in *Annales Monastici*, ii. 408; William Rishanger, *Chronica et Annales*, ed. H. T. Riley (RS, London, 1865), 116; Bartholomew Cotton, *Historia Anglicana necnon ejusdem Liber de archiepiscopis et episcopis Angliae*, ed. H. R. Luard (RS, London, 1859), 172.

[67] *Chronicon de Lanercost 1201–1346*, ed. J. Stevenson (Edinburgh, 1839), 128–31; see also Golubovich, i. 326–7; John Elemosina, Chronicon, in Golubovich, ii. 108, 127.

commemorated crusader Tripoli: 'An extraordinary vision occurred in the year 1287 in the monastery of the Cistercians in the city of Tripoli. A monk saying mass . . . saw a disembodied hand writing in gold letters . . . "The High Cedar of Lebanon will fall [Ps. 29: 5; Zach. 11: 1], and Tripoli will soon be captured and destroyed . . ."'.[68]

Though Pope Nicholas regarded a general crusade as necessary to save Outremer, circumstances dictated different measures which were easier to organize and quicker to accomplish. Hampered by the Sicilian question he decided, following the fall of Tripoli, upon a *passagium particulare*. At that time Europeans should have been well aware of the desperate position of the remaining crusader outposts in Outremer now that castle garrisons (Tripoli, Botron, and Nephim) from the north of the country were no longer available to them. Their response to the papal appeal for aid following the fall of the county of Tripoli was, however, disappointing, since it was limited to Venice and northern Italy. The Venetians, concerned about their commercial hegemony in Acre, agreed to provide, at the pope's expense, twenty galleys. This fleet was to be commanded by the doge's son Nicholas Tiepolo, assisted at the pope's request by John of Grailly and Roux of Sully. Each was entrusted with a thousand pieces of gold from the papal treasury. The fleet departed from Venice in the summer of 1290 in order to see to the defence and custody of the city of Acre, as well as of other Christian localities in the Holy Land. It carried a force of about 1,600 north Italians, mainly from Lombardy and Tuscany. The fleet that reached Acre in August 1290 numbered only thirteen galleys, of which five had been provided by James II of Sicily.[69]

---

[68] See R. E. Lerner, *Powers of Prophecy: The Cedar of Lebanon Vision from the Mongol Onslaught to the Dawn of the Enlightenment* (Berkeley, Calif., 1983), 37–61, 203–12, and below, ch. 4.

[69] Nicholas IV, *Registres*, nos. 2252–7, 2269–70, 4385–90; Marino Sanudo Torsello, 'Liber', p. 230; *AA*, vol. i, no. 1, pp. 1–2. It is known that 500 of the crusaders came from Parma and 600 from Bologna: see Digard, *Philippe le Bel*, vol. i, pp. 125–6. The fleet was partly financed with the sum of 20,000 *livres tournois* bequeathed by Pope Hadrian V (1276) for the aid of the Holy Land. See above, n. 1; Nicholas IV, *Registres*, no. 2259; see also R. Röhricht, 'Untergang des Königreiches Jerusalem', *MIÖG* 15 (1894), 14–17; D. Jacoby, 'Expansion occidentale dans le Levant: Les Vénitiens à Acre dans la seconde moitié du treizième siècle', *Journal of Medieval History*, 3 (1977), 225–65, esp. pp. 230–1; Schein, 'Patriarchs', p. 303.

The response of the north Italians to the papal appeal to take the Cross for the aid of the Holy Land, at a time when a crusade for the recovery of Sicily was being organized in other parts of Europe,[70] is important, as it clearly shows that the *crux transmarina* still held for some more attraction than the *crux cismarina*; that even at a time when the same crusade-privileges could be obtained within the borders of Italy, some considered the *crux transmarina* as the more worthy form of crusading; or perhaps that the 'defence of the faith' by fighting against Moslems was considered more just than defending it against fellow-Christians!

European chroniclers describing the participants of the crusade of 1290, mainly in the framework of their own reaction to the loss of the Holy Land following the fall of Acre, tend to refer to them as a rabble of peasants and unemployed town folk.[71] Such descriptions were, it seems, influenced by the consequences of the ill-fated attack by Italian crusaders on Moslem merchants in Acre; the incident provided the sultan with a plausible excuse to break his truce with the Franks.[72] In fact there is clear evidence that not all the participants were of low social status and that some of them were honourable north Italian citizens. It is known that some were from Bologna, Este, and Parma, and that some of them managed to escape the massacre in Acre and to return to their home-towns.[73] Yet, the north Italian crusade proved fateful. It arrived in Acre in August 1290, and when its participants realized that the threat of the sultan's attack upon Acre was over, some of them went back to Italy. The others were left unemployed in Acre, and it was they who attacked Moslem merchants in the city's market-

[70] See above, n. 61.

[71] Francisco Pipino, 'Chronicon', p. 733 refers to the Italian crusaders as 'pseudo-Christian'. Ptolemy of Lucca, 'Annalen', ed. B. Schmeidler, *MGH SRG* viii (Berlin, 1955), 220 and Bernard Gui, 'Flores', p. 201 refer to them as 'fools'. Modern historians followed suit. See e.g. Grousset, *Histoire des croisades*, iii. 409–10; Stickel, *Fall von Akkon*, pp. 25–8.

[72] See above, n. 71. See also 'De Excidio Urbis Acconis II', in *Veterum Scriptorum et Monumentorum Amplissima Collectio*, ed. E. Martène and U. Durand (Paris, 1724–33), v. 761–84; Bartholomew Cotton, *Historia*, pp. 431–2; 'Chronicon Sampetrinum', ed. B. Stübel, in *Geschichtsquellen der Provinz Sachsen*, i (Halle, 1870), 126; Hermann Corner, 'Chronica novella', in *Scriptores rerum Suecicarum medii aevi*, ed. E. M. Fant (Uppsala, 1818–71), iii. 126.

[73] 'Chronicon Estense', *RIS* xv. 50; 'Chronicon Parmense', *MGH SS* xviii. 708–9; 'Corpus Chronicorum Bononiensum', *RIS* NS xviii. 232–4.

places, and so provided the sultan with a *casus belli* and won them the undeserved reputation of causing the fall of Acre.[74]

The Revival of the Plan for a General Crusade

Though Pope Nicholas IV revived the policy of Gregory X for a general crusade, he gave preference to his plans for Sicily. He did not start to organize a general crusade until after the signing of the Treaty of Brignoles (12 February 1291), which settled the Sicilian question. On 29 March 1291 a general crusade was proclaimed, after Edward I, who was to lead it, agreed that the date of departure should be the Feast of John the Baptist, 24 June 1293.[75] The king decided to undertake the crusade following the successful conclusion of the Welsh war. In 1284 he began negotiations with Pope Martin IV over the tenth, and because of Edward's excessive financial demands, they dragged on through the pontificate of Honorius IV and into that of Nicholas IV. During the period 3 April 1287 to 22 February 1288 he took the Cross. According to the agreement between him and Nicholas IV he was granted for his crusade (16 March 1291) the proceeds of the tithe of the British Isles for a period of six years.[76] Edward's project was based on a large-scale military action with the co-operation of France, Sicily, and Aragon. Although he favoured the strategy of a general crusade, the fall of Tripoli and the immediate threat to Acre caused Edward to give his support to the current papal policy of small expeditions. When news arrived of the sultan's preparations to march against Acre, Edward, though then engaged in Scotland, dispatched in July 1290 a force commanded by Otho of Grandison to prepare the way for his own coming. Otho's company, however, included more priestly pilgrims than men-at-arms.[77]

[74] See above, nn. 71–2; S. Runciman, 'Crusader States 1243–1291', in *Setton's History*, ii. 593–5; Prawer, *Histoire*, ii. 541.

[75] Nicholas IV, *Registres*, nos. 6683–701; *Foedera*, I. iii. 76–83.

[76] Nicholas IV, *Registres*, nos. 6664–5, 6666–701. See also W. E. Lunt, 'Collectors' Accounts for the Clerical Tenth levied in England by Order of Nicholas IV', *EHR* 31 (1916), 102; id., 'Papal Taxation in the Reign of Edward I', *EHR* 30 (1915), 411–12.

[77] Mas-Latrie, *Histoire*, ii. 81–3; B. Beebe, 'Edward I and the Crusades', Ph.D. thesis (St Andrew's, 1970), 257; see also A. J. Forey, 'Military Order of St Thomas', *EHR* 92 (1977), 481–503, esp. pp. 488–9; C. L. Kingsford, 'Sir Otho de Grandison', *TRHS* 3 (1909), 125–95, esp. pp. 137–8.

When, with the Sicilian war over, the general crusade was proclaimed, Nicholas IV could count on the participation not only of the English king, but also on that of his own protégé Charles II of Anjou, as well as on that of Alfonso III of Aragon and of his brother James of Sicily. Alfonso, when signing the Treaty of Brignoles, offered to depart on a crusade as a penance for his sins. He promised to remain in the Holy Land with a good following, but died four months later. His brother James of Sicily stated as early as 1288 that he would go in person to the East with thirty galleys, 300 knights, and 1,000 foot-soldiers, if he were granted the tithe of Sicily for a period of three years, a truce in Italy, and an assurance that he was to hold his conquests in the East as the king of Jerusalem. Late in 1290 or early in 1291, on the eve of the fall of Acre, James repeated his offer. This time he talked about twenty galleys and 1,000 *almogavers*. In return he now demanded 40,000 gold ounces per year and absolution from the sentence of excommunication which he had incurred. Additionally he requested the crown of Jerusalem and the papal consent to a marriage between his brother Alfonso of Aragon and Edward I's daughter.[78]

Qalawun's plan to take Acre was already known in the West by April 1290.[79] The Western powers promised at that time to depart on a crusade but also sought to safeguard their Levantine interests by concluding treaties with the sultan of Egypt. On 25 April 1290 Alfonso of Aragon and James of Sicily concluded a treaty which safeguarded their kingdoms against attack by the sultan, while they for their part promised to safeguard the sultan's present and future conquests in Syria. They also undertook to denounce all military enterprises prepared against the sultan by either the pope, the kings of England and France, the emperor of Byzantium, or the Mongols. They even promised to attack the assailants and refrain from any aid to the Franks if the latter ever broke their truce with the sultan. When the Aragonese–Sicilian envoys arrived in Cairo they found there embassies of the German and Greek emperors, as well as of the Genoese republic. Whereas the envoys of Rudolph of Habsburg aimed to inspire the sultan with

---

[78] *Foedera*, I. iii. 78; *AA*, vol. i, no. 2, pp. 2–7; Bartholomew of Neocastro, 'Historia Sicula', *RIS* NS xiii. 114–15.

[79] *AA*, vol. iii, no. 8, p. 12; J. Richard, *Royaume latin de Jérusalem* (Paris, 1953), 336.

a peaceful disposition towards the Franks, the others came to safeguard their own, now threatened, interests. The Genoese, who had suffered heavy losses at the fall of Tripoli, took reprisals by capturing an Egyptian merchant-vessel off the southern coast of Anatolia, and by raiding the undefended region of at-Tina in the Nile delta. However, when Qalawun barred them from Alexandria, they hastened to make their peace (13 May 1290). No wonder that after the fall of Acre, Genoa was accused of having helped the sultan in his conquest. A later Venetian historian even ascribed the Venetian–Genoese war of 1293–9 to the crusading spirit of an aroused Venice, which sought revenge for the fall of Acre.[80]

Acre fell on 28 May 1291. The remaining Frankish possessions, Tyre, Sidon, Beirut, Haifa, Tortosa, and Château Pélerin, were soon abandoned. The siege (6 April–18 May), and the battle of Acre which followed (18–28 May), put an end to the weary existence of the kingdom in the last quarter of the thirteenth century. However, the heroic defence of Acre provided a splendid epilogue to a history which had begun almost two hundred years earlier, with the conquest of Jerusalem by the First Crusade.[81] It is this last chapter of resistance that partly explains the positive character of Europe's reaction. The fall caused a deep stir; yet the amount of acerbity, recrimination, and criticism was surprisingly small when compared, for example, with that which followed the Second Crusade, or the crusades of the thirteenth century. The loss of the Holy Land was hardly regarded by its contemporaries as final. As a temporary episode it did not immediately or fundamentally transform the concept of the crusade.

[80] Marino Sanudo the Younger, 'Vitae Ducum Venetorum', *RIS* xxii. 578; W. Heyd, *Histoire du commerce du Levant au Moyen Âge*, trans. F. Raynaud (Leipzig, 1936), i. 407–8, 415–24; G. Caro, *Genua und die Mächte am Mittelmeer 1237–1311: Ein Beitrag zur Geschichte des 13. Jahrhunderts* (Halle, 1895–9), ii. 133; Digard, *Philippe le Bel*, vol. i, p. 126; G. I. Bratianu, 'Autour du project de croisade de Nicholas IV: La Guerre ou le commerce avec l'infidèle', *RHSE* 22 (1945), 250–5.

[81] See Runciman, *History*, iii. 420–1; Prawer, *Histoire*, ii. 539–57.

3
1291–1292
THE LOSS OF THE HOLY LAND AND THE FIRST ATTEMPTS AT ITS RECOVERY

The Policy of Nicholas IV

The news of the loss of the Holy Land reached Pope Nicholas IV after 1 and before 13 August 1291. The encyclical *Dirum amaritudinis calicem* dated 13 August 1291 was the first papal acknowledgement of the news from the East. As a journey from Syria to Europe took between six to eight weeks, the pope could have learned about the fall of Acre at the end of July, but this was not registered until 13 August 1291. The encyclical is perhaps surprising in one aspect; it offered no ideological explanations of the disaster, except for the military one. From this point of view the *Dirum* is a far cry from the *Audita tremendi* (29 October 1187) issued by Gregory VIII following the fall of Jerusalem. Significantly enough even the classic phrase *nostris peccatis exigentibus* is absent. The *Dirum* announced the 'expugnatio civitatis Aconensis . . . per Babilonicum persecutorem'. Acre, the principal refuge of Christendom in the East, it was said, was taken by siege with terrible loss of life and the ruin of the city; it was a cause of most profound grief to the Church. Its appeal was apologetic and exhortatory: apologetic since it stressed the courage and devotion of Acre's defenders as well as the efforts of the Church to prevent its fall; exhortatory, since it called the believers to revenge the injury to Christ and his host, and to join the crusade which was at that time preached in Europe: 'ad recuperationem celerem dictae terrae'.[1]

The widely diffused encyclical the *Dirum*, as well as another encyclical the *Dura nimis* of 18 August 1291, ordered the convocation of provincial councils to deliberate and to formu-

---

[1] Nicholas IV, *Registres de Nicholas IV*, ed. E. Langlois (Paris, 1886–91), no. 7625; J. Prawer, *Histoire du royaume latin de Jérusalem* (Paris, 1969), ii. 11–13.

late remedies for the recovery of the Holy Land. The *Dura nimis* went further than the *Dirum* in putting on the councils' agenda the union of the orders of the Hospitallers and the Templars. This was explained by the concern of the Church for the conservation of the Holy Land after its recovery, as well as by the current public opinion (*vox communis*), that the two orders should be merged. The deliberations of the councils, composed of regular and secular clergy, were to be sent to Rome by the Feast of the Purification (2 February 1292), by two of the participants of each council. The *Dura nimis* stressed that although the Holy Land was the common concern of all the faithful, the prelates of the Church of God should take the lead. The archbishops were thus asked to intervene in bringing back peace and concord among Christians and to unite in avenging the injury inflicted upon the Faith. They should show such zeal in their efforts that Christ whose land was at stake, might eternally reward them. Nicholas evidently tried to reach the prelates first. He also sought advice from laymen, and particularly from the secular rulers of Europe, 'kings, princes, and magnates'. One result of Nicholas IV's appeal for advice was, it seems, that Ramon Lull presented him with his *de recuperatione* treatise. Moreover, it is possible that Fidenzio of Padua presented him on this occasion with the treatise which he had already completed before the loss of the Holy Land.[2]

One may ask why Pope Nicholas did not convene a general council especially as public opinion at that time was in favour of such an assembly. It is plausible to assume that in view of the major crusade he planned for 1293, Nicholas looked for more immediate remedies. To organize a general council would necessarily demand time; it was more convenient therefore to debate the means of aiding the Holy Land at provincial councils. This method of procedure, it is worth mentioning, had been employed by Pope Alexander IV in 1261 regarding remedies against the Mongol menace in the Holy Land, Hungary, and Poland. On the eve of the Second Council of Lyons, this sort of procedure was recommended by Humbert of Romans as a preliminary to the meeting of the general council. In fact there was little new in the appeal of Nicholas IV for

[2] Nicholas IV, *Registres*, nos. 6793–9, see also nos. 6778, 6782–3, 6849, 7381.

general advice. Previous losses in the Holy Land had resulted in similar appeals. Innocent III and Gregory X did the same in their respective summons for the Fourth Lateran Council (1215) and Second Council of Lyons (1274). Pope Nicholas followed the *Salvator noster in* of Gregory X, both in style—in his description of the conditions of the Holy Land and the grief of the Church—as well as in his instructions.[3]

The military orders had been subject to criticism almost since the early years after their foundation. The measures urged by reformers came to include the union of all the military orders, or at least the merging of the two main orders of the Hospital and the Temple. The latter demand, raised during the reign of St Louis, was actually discussed at the Second Council of Lyons. The proposal to unite all military orders was dropped when it became known that the Spanish kings would not consent because of the threat to their national orders. If Nicholas came back to the idea, it must have been because after the loss of the Holy Land the reformers became more demanding and even outright hostile. On the other hand, by putting the issue of the reform of the military orders on the agenda of the Church synods, Nicholas made their fate a topic of an almost public debate. Now as never before they became open to criticism. This perhaps partly explains the fact that the Hospital and Temple came to be widely blamed for the final disaster in the Holy Land.[4]

Absent from the treatise of Fidenzio of Padua (written on the eve of the fall of Acre), the issue of the orders was raised by both Ramon Lull and Charles II of Anjou as well as many of the chroniclers and synods, who declared themselves outspokenly for the unification of the Hospital and the Temple. However, as Nicholas IV died on 14 April 1292 before their replies reached Rome, the problem of the orders was left unsolved, to be raised again by Boniface VIII, and finally to be solved in the most unexpected and drastic way by Clement V (1312).[5]

[3] P. A. Throop, *Criticism of the Crusade: A Study of Public Opinion and Crusade Propaganda* (Amsterdam, 1940), 7–18, 200–1, and below, ch. 4.

[4] A. J. Forey, 'Military Orders in the Crusading Proposals of the Late Thirteenth and Early Fourteenth Centuries', *Traditio*, 36 (1980), 317–18; S. Schein, 'Image of the Crusader Kingdom of Jerusalem in the Thirteenth Century' *Revue belge de philologie et d'histoire*, 64 (1986), 705–7.

[5] James of Molay, 'Concilium super negotio Terre Sancte', in *Vitae Paparum Avenionen-*

From August 1291, when he learned of the fall of Acre, the policy of Nicholas IV regarding the crusade focused on the following: immediate aid for the Christian strongholds in the Levant, now Cyprus and Lesser Armenia; the enforcement of a maritime blockade of Egypt; the organization of a general crusade due to depart on 24 June 1293.

With the loss of the Holy Land the *subsidium Terrae Sanctae* turned to Cyprus and Lesser Armenia. They had to be defended, especially as they could become bases for the recovery of the Holy Land. Moreover, the loss of Acre accentuated their commercial importance. The Armenian Ayas (Lajazzo) was already becoming a major trading centre after the fall of Antioch (1268). After 1291 Cyprus replaced Acre; the colonies of maritime cities like Genoa, Venice, Pisa, and Barcelona crowded now into Famagusta and Nicosia.[6] Both Cyprus and Armenia were directly exposed to the sultan's attacks. It was feared that God, who had forsaken Acre because of the sins of the people, might abandon the island of Cyprus and Lesser Armenia, which would be inevitably devoured by the fury of the Egyptians under their raging sultan. Immediate action had to be taken to ensure their defence until the general crusade. A flurry of activity followed. Nicholas ordered ten galleys to be fitted out in Ancona and as many in Genoa. This papal fleet under Roger of Thodino was detached for Cyprus and Armenia. The financing was done by appropriating funds of the Temple and of the Hospital. The warriors who manned the galleys were granted full crusader indulgences. After the death of Pope Nicholas the plan failed. The cardinals, according to James Doria, appointed Manuel Zaccaria and Tedisio Doria as the admirals of the fleet fitted out in Genoa. Together they commanded twenty galleys (Manuel Zaccaria provided twelve and Tedisio Doria eight galleys). In this they violated a Genoese law, which forbade her citizens to fight for or against a

---

sium, ed. G. Mollat (Paris, 1914–27), vol. iii, p. 150; A. Luttrell, 'Emmanuelle Piloti and Criticism of the Knights Hospitallers of Rhodes 1306–1444', in id., *Collected Studies* (London, 1978), 7–8; J. Riley-Smith, *Knights of St John in Jerusalem and Cyprus c.1050–1310* (London, 1967), 201–3.

[6] D. Jacoby, 'Rise of a New Emporium in the Eastern Mediterranean: Famagusta in the Late Thirteenth Century', in *Meletai kai Hypomnemata*, i (Nicosia, 1984), 145–79, and in Jacoby, *Studies on the Crusader States and the Venetian Expansion* (Northampton, 1989), no. 8.

foreign ruler without specific permission. The two were fined by the republic but after the cardinals' intervention, they were exempted from the fine because they had been acting in the service of the Church for the relief of the Holy Land. The fleet departed in 1292. It was joined in Cyprus by fifteen galleys of Henry II of Cyprus, and the combined force sailed for the Karamanian coast and attacked the fortified port of Scandelore (Alaya). It took the tower which was on the sea-front, but failed at another. Unable to make any further progress the fleet abandoned its plan and sailed for Alexandria and then back to Cyprus, having accomplished nothing but the irritation of the sultan. Al-Ashraf Khalil (1290–3) summoned a meeting of his emirs and ordered a hundred ships to be built for the conquest of Cyprus. The scheme came to nothing because the emirs, unwilling to enter this vast enterprise, assassinated the sultan on 13 December 1293. He was followed by a sequence of claimants, who were murdered one after another, while Egypt was visited by a severe famine and plague. These disasters were ascribed by the Europeans to divine vengeance.[7]

In 1293 the Templars equipped in Venice six galleys for the protection of Cyprus. In July of that year, four Venetian ships equipped for the same purpose and two ships of the Templars fell in on their way to Cyprus with seven Genoese merchant vessels. A battle ensued, the Venetian ships were captured, and more than 300 of their men killed. This episode was an unpromising beginning to the efforts to defend Christendom in the East. The situation of Lesser Armenia was particularly serious. Since the fall of Antioch it had come under almost constant Mamluk attack. Its king Hetoum II (1289–93, 1295–6, 1299–1301; regent 1301–7) tried to appease the sultan by offering a large ransom. The sultan accepted but merely to postpone his inva-

---

[7] Nicholas IV, *Registres*, nos. 6432, 6778, 6854–6; James Doria, 'Annales Ianuenses', in *Annali genovesi de Caffaro e dei suoi continuatori*, ed. C. Imperiale de Sant'Angelo, v (Rome, 1927), 143–4; 'Templier de Tyr', in *Gestes de Chiprois*, ed. G. Raynaud (Geneva, 1887), 261–72; Bartholomew of Neocastro, 'Histoira Sicula', *RIS* NS xiii. 133; Marino Sanudo Torsello, 'Liber Secretorum Fidelium Crucis', in *Gesta Dei per Francos*, ed. J. Bongars (Hannau, 1611; repr. Jerusalem, 1973), ii. 232–3; Francesco Amadi, *Chronique, Chroniques d'Amadi et de Strambaldi*, ed. M. L. de Mas-Latrie (Paris, 1891), 228–9; P. M. Holt, 'Sultanate of al-Manṣūr Lāchīn 696–8/1296–9', *Bulletin of the School of Oriental and African Studies, University of London*, 35 (1973), 521–32; W. Heyd, *Histoire du commerce du Levant au Moyen Âge*, trans F. Raynaud (Leipzig, 1936), i. 365–72; ii. 1–23, 73–92.

sion till after the conquest of the Frankish territories in Syria. Hetoum II appealed then for help to Edward I and to the pope. After receiving his envoys headed by the Franciscan Thomas of Tolentino in January 1292, Nicholas IV sent them to the kings of France and England. On 23 January 1292 he ordered the preachers of the general crusade to exhort potential crusaders to proceed at once to help Armenia, 'placed in the very midst of perverse nations like a lamb among wolves'. He offered them the same indulgences as those granted for the general crusade. On the same day the pope wrote to Thibaud Gaudin the master general of the Temple, and to John of Villiers the master general of the Hospital, ordering them to depart with the papal fleet for the aid of Armenia. Roger of Thodino the captain of the papal fleet was ordered to depart immediately with his galleys for Lesser Armenia. As it had been predicted, Lesser Armenia was invaded by al-Ashraf (spring 1292). The castle of Abeldjes surrendered to him and in 1293 Hetoum II agreed, as the price of peace, to surrender the fortress of Bethesni. It is unknown whether the papal fleet ever reached Lesser Armenia, but the masters-general of the orders as well as Otho of Grandison are known to have stayed there during the years 1292–3.[8]

The papal fleet was intended for yet another purpose. According to James Doria it had to harass the Saracens as well as those who visited the territories of the sultan.[9] It seems that after the loss of the Holy Land, the blockade of Egypt as a preliminary and necessary step for the general crusade, became

[8] Nicholas IV, *Registres*, nos. 6850–6; S. Der Nersessian, 'Kingdom of Cilician Armenia', in *Setton's History*, ii. 655–6; M. L. Bulst-Thiele *Sacrae Domus Militiae Templi Hierosolymitani Magistri* (Göttingen, 1974), 293–4; J. Riley-Smith, 'The Templars and the Teutonic Knights in Cilician Armenia' in T. S. R. Boase (ed.), *Cilician Kingdom of Armenia* (Edinburgh and London, 1978), 117; A. Luttrell 'Hospitallers' Intervention in Cilician Armenia 1291–1375', in Boase (ed.), *Cilician Kingdom*, p. 121. For the letter of Hetoum II to Edward I see *Chronicon de Lanercost 1201–1346*, ed. J. Stevenson (Edinburgh, 1839), 487–9; Bartholomew Cotton, *Historia Anglicana: necnon ejusdem Liber de archiepiscopis et episcopis Angliae*, ed. H. R. Luard (RS, London, 1859), 219–23. The critical situation in Lesser Armenia was emphasized in 1291 by letters, obviously fabricated in Lesser Armenia and allegedly written by the sultan of Egypt to Hetoum II following the fall of Acre. For the letters, see John of Pontissara, *Register*, ed. C. Deeds (Canterbury and York Society, 19, 30; 1915–24), vol. ii, pp. lxxxiii–lxxxiv, 481–2; Bartholomew Cotton, *Historia*, pp. 215–19; E. Sivan, *Islam et la croisade: Idéologie et propagande dans les réactions musulmanes aux croisades* (Paris, 1968), 184.

[9] James Doria, 'Annales Ianuenses', p. 143.

an integral part of the policy of Nicholas IV. The idea itself was not new. As far back as the Third Lateran Council (1179), the popes had been trying to stop western merchants supplying war-materials to the Saracens, which could be used by them against the crusaders. The Fourth Lateran Council (1215), and the First and Second Councils of Lyons (1245 and 1274) pronounced excommunication on those supplying the Saracens with slaves, arms, iron, or timber for the construction of ships. The same punishments were pronounced against those who served the Saracens as pilots or who gave them counsel or aid of any sort, especially in the Holy Land. The goods of the traitors were to be confiscated and used for the crusade. Those caught would be reduced to slavery. This had to be made public in all maritime cities on every Sunday and all feast days. Those who confessed were not to enter a church unless they gave up their gains. But not only direct traffic in arms was affected. Since the Fourth Lateran Council the papacy had tried to stop all commerce with Saracen lands and especially with Egypt. The Fourth Lateran Council and the First Council of Lyons prohibited, under pain of anathema, the dispatch of ships to Moslem lands for a period of four years. This prohibition was repeated in the Second Council of Lyons for a period of six years. The intention was to increase the number of ships at the disposal of the crusaders and to deprive the Saracens of considerable revenues. Such measures, if successful, would have tremendously increased the importance of the Christian ports in the Holy Land. Tense as the relations between Christendom and Egypt were in 1215–16, there were 3,000 western merchants in Alexandria. Later, when Damietta was conquered (5 November 1219), the ports of Tyre and Acre were almost entirely abandoned by merchants and even by pilgrims, who flocked to the newly conquered city. It was clear that if Egyptian ports were put out of bounds, Christian domination of the Syrian littoral would become an indisputable necessity for the Italian, Provençal, and Spanish maritime cities.

This conciliar legislation did not succeed. Trade with the Saracens was so profitable that it was worth risking excommunication, especially as merchants could escape the penalties by paying a fine, and were thus contributing towards the defence of the Holy Land! Such practices went back at least to the time of Clement IV (1265–8), and they became a regular procedure under John XXII (1316–34). The repetition of prohibitions proved that the popes were fighting a losing battle. As they depended on the maritime cities for fleets to carry a crusade, they had to give in and relax the prohibitions. Even Innocent III had to withdraw his prohibition in the face of Venetian pressure, and limit it to exports of war-materials. In the last resort the enforcement of the prohibitions was entirely dependent on the maritime republics. Their profitable trading with the Saracens often made the crusade appear as a nuisance, if not an outright obstacle, to their calculations. And yet the maritime republics were not opposed to crusading schemes nor reluctant to take part in them. They were Christians and they were merchants. The question was how to balance the obvious clash of interests. One way of doing it was to maintain friendly relations with Egypt and at the same time to strive to destroy the Moslem monopolies, which they hoped to inherit. As their main interests were in Egypt, they were more likely to participate in a plan which aimed at the conquest of Egypt than in plans aimed at other Moslem countries. As to the strategy of economic blockade, they would agree to it and even support it when their position and subsequently their prosperity were at stake. This happened, for example, after the Fifth Crusade (1218–21), when the Italian merchants could not recover their former position in Alexandria; when the Emperor Frederick II planned the conquest of Egypt, Venice in order to support him went as far as prohibiting all commercial intercourse with that country (1224–7). But Frederick's crusade became one of diplomacy, and the policy of commercial alliances with the sultans was resumed. Here was also another lesson to be learned. The sultan depended on the Italians no less than the Italians on the sultan. Therefore their participation in economic restrictions or even in a military attack, did not compromise their standing in Egypt for long. Sooner or later they would recover their position, though this might require large-scale bribery and they might have to be content with reduced privileges.[10]

---

[10] M. Purcell, *Papal Crusading Policy: The Chief Instruments of Papal Policy and Crusade to the Holy Land from the Final Loss of Jerusalem to the Fall of Acre 1244–1291* (Studies in the History of Christian Thought, 11; Leiden, 1975), 187–99; W. Heyd, *Histoire du commerce*, i. 385–405.

Pope Gregory X clearly recognized the role which the economic blockade of Egypt could play in the reconquest of the Holy Land. He also recognized the importance of the naval superiority of the Christians. This was heavily stressed during the Second Council of Lyons by William of Beaujeu master-general of the Temple, by Humbert of Romans, and perhaps by Fidenzio of Padua. They may have been strengthened in their convictions by the complete failure of Baibars' naval attack on Cyprus in 1271. Gregory X, however, wanted the fleet of seventeen to twenty vessels to be used for a war proper.[11] It is not clear whether the suggestion to create a special maritime police to enforce the economic blockade was made during the Second Council of Lyons. If the idea was discussed at all, it was discarded by Gregory X. It seems more likely that the idea was an innovation of Nicholas IV. His decision could have been influenced by Fidenzio of Padua who advocated it in his 'Liber recuperationis Terrae Sanctae' written on the eve of the fall of Acre. Both were Franciscans; both Nicholas IV, then master-general of the order (elected on 2 May 1274) and Fidenzio, the provincial vicar of the Holy Land, took part in the Second Council of Lyons. It is thus plausible to assume that Nicholas IV was familiar with Fidenzio's opinions before 1291. Both envisaged the crusade as an all-European *passagium generale* under a single leader, supported by the forces of the Mongol il-khan of Persia, the Armenians, and the Georgians. A sea-blockade of Egypt carried out by a special fleet was to precede the attack. Nicholas was aware that to make the blockade effective, the police-fleet should be provided from sources other than the maritime republics against whom it would be primarily directed. Accordingly he commanded the Temple and the Hospital to build a war-fleet. Charles II of Anjou, who wrote *c*.1292–4, estimated that the Hospital, the Temple, and the king of Cyprus then held in the island ten galleys each. According to James Doria the Templars had by 1293 only two galleys. In 1300 the Hospital, the Temple, and the king of Cyprus were able to raise only sixteen galleys between them.[12]

---

[11] See above, ch. 2, and Throop, *Criticism of the Crusade*, pp. 246–8.
[12] Fidenzio of Padua, 'Liber recuperationis Terrae Sanctae', in Golubovich, i. 46–50; Charles II of Anjou, 'Conseil du Roi Charles', ed. G. I. Bratianu, *RHSE* 19 (1942), 355; James Doria, 'Annales Ianuenses', p. 167; *Cartulaire général de l'ordre des*

Nicholas's growing awareness of the importance of commercial war is also reflected by his decrees against commerce with Egypt. Immediately after the fall of Acre, Nicholas IV renewed the ban ordered at the Second Council of Lyons and already repeated by himself in 1289. His bull *Olim tam in generali* (23 August 1291) forbade the export of arms, horses, timber, and food, as well as some other merchandise, to the territories of the sultan and moreover declared a general and absolute embargo by prohibiting the dispatch of any vessels to the lands occupied by Saracens. The sanction was now as before, excommunication and additionally, perpetual infamy and loss of standing.[13] The prohibitions were primarily directed against Genoa and Venice. Both were strongly criticized after the loss of the Holy Land not only because of their conflicts in Acre, but also because of their commerce with Egypt. Such criticism was mainly voiced by the 'experts' familiar with the situation in the Levant, like Thadeo of Naples, Ricoldo of Monte Croce, Fidenzio of Padua, and Charles II of Anjou. Both Genoa and Venice were warned already in *Dirum amaritudinis calicem* against the transgression of the ban. Genoa, however, who in 1291 had followed the admonishments of Nicholas IV, renewed its relations with Egypt in 1304. Venice had already done so in 1302. Since 1305 Pisa had kept a consulate in Alexandria. James II of Aragon renewed in 1292 his treaty of alliance with the sultan of Egypt. After regaining the favour of the Church in 1295 he published an order which banned generally all commerce with the lands of the sultan. This did not, however, keep him from regularly renewing his alliance with the sultans of Egypt. Thus in a period ranging from one to thirteen years the grandiose plan of undermining the Egyptian economy and thereby weakening the arch-enemy of Christendom was falling to pieces.[14]

Hospitaliers de St Jean de Jérusalem 1100–1310, ed. J. Delaville le Roulx (Paris, 1894–1906), nos. 4177, 4183; A. Luttrell, 'Hospitallers in Cyprus after 1291', *Acts of the First International Congress of Cypriot Studies* (Nicosia, 1972), 163 n. 3, *contra* Riley-Smith, *Knights of St John*, pp. 200–1, 330.

[13] Nicholas IV, *Registres*, nos. 6784–9; G. I. Bratianu, 'Autour du projet de croisade de Nicolas IV: La Guerre ou le commerce avec l'infidèle', *RHSE* 22 (1945), 250–5.

[14] Nicholas IV, *Registres*, nos. 6782–3; Heyd, *Histoire du commerce*, ii. 6–30; A. E. Laiou, *Constantinople and the Latins: The Foreign Policy of Andronicus II 1282–1328* (Cambridge, Mass., 1972), 101–2.

Nicholas did not fare better in his attempts to pacify Europe or to create a united and common front against Islam, a coalition which would supply the human and financial reservoir of the future general crusade. A month after the fall of Acre (18 June 1291), Alfonso III of Aragon died without a direct heir. The Aragonese called in James of Sicily. Thus the union of Aragon and Sicily was re-established, and the Treaty of Brignoles signed half a year earlier (19 February 1291), repudiated. In the autumn of 1291, therefore, Nicholas IV found himself once again busy with the affairs of Sicily. In August 1291 he excommunicated James II of Aragon as well as the Sicilians and started to work for the restoration of Sicily to Charles II of Anjou. Such a restoration would facilitate the recovery of the Holy Land. Thus, at least, he argued when requesting Philip IV on 1 October 1291 to aid Charles II of Anjou. Yet, whereas before the fall of Acre Pope Nicholas was convinced that the question of Sicily should be solved before the launching of a general crusade, he now adopted a different position. Writing to Philip IV on 13 December 1291, Nicholas argued that the loss of the Holy Land changed the whole situation and that now the Holy Land must become the first care of every Christian. All else must wait till the decisions of the Church synods reached Rome. A crusade against Aragon would handicap the recovery of the Holy Land and cause a scandal in Christendom; it might even ruin completely the crusade of Edward I. Accordingly the pope refused the demand of Philip IV that he should be granted a six years' tithe for a crusade against Aragon and that such a crusade should be preached in France.[15]

From August 1291 Nicholas showed even greater energy in his efforts to gain support for the general crusade of 1293; those efforts were concentrated on England, France, Genoa, and Venice. Accordingly, on 10 September 1291 Bernard, bishop of Tripoli was dispatched to England. The Dominican Gerard Picalotti, bishop of Spoleto, was dispatched on 23 August 1291 to France to announce the sad tidings and to exhort Philip IV to supply forces for the crusade. If Philip refused, the bishop was

---

[15] Nicholas IV, *Registres*, no. 6849; Potthast, no. 23842; G. Digard, *Philippe le Bel et le Saint-Siège* (Paris, 1936), vol. i, pp. 129–36; E. G. Léonard, *Angevins de Naples* (Paris, 1954), 179–80.

to demand that the tenth granted to his father Philip III the Bold should be handed over to the French crusaders. The papal envoy accomplished nothing. Philip IV remained as deaf to this summons of the pontiff to aid the Holy Land as he had to all the others made since 1288. At that time, as pointed out by G. Digard, King Philip was uninterested in a crusade to the Holy Land and moreover believed that 'de terra ipsa sinistrum aliquod eveniret'. On 9 December 1290 he even demanded that the pope should release him from responsibility for the custody of the Holy Land which he had inherited from his father. In the autumn of 1291 he not only refused to become involved in the papal plans for the recovery of the Holy Land but, as said above, demanded that the pope should launch a crusade against Aragon.[16]

At the same time the archbishop of Reggio was sent to Genoa and to Charles II of Anjou. The Venetians and Genoese were asked on 13 August 1291 to make peace with each other, to prepare ships, and to send to the pope experienced men to advise him in the meanwhile on the best measures to be taken. It was probably for the arrangement of these matters that the archbishop of Reggio was sent to Genoa and the bishop of Urbino to Venice. As mentioned above only the Genoese provided galleys for the papal fleet. The mission to Charles II of Anjou, a *crucesignatus* since 1275, was more successful; Nicholas apparently obtained an informal promise from Charles to take part in the general crusade. Charles, however, like the pope himself, saw the recovery of Sicily as the prerequisite for recovery of the Holy Land. In his negotiations with the Genoese for naval aid against Sicily Charles argued (1292) that the aid for the crusade against Sicily should be considered as aid for the Holy Land, 'for without the recovery of the said island . . . the Holy Land cannot be recovered'.[17]

From among other leading figures in Europe, Nicholas could also count on the participation of Edward I of England, who was to depart as the leader of the planned crusade, and of Guy of Flanders.[18] Following the fall of Acre Edward abandoned his

---

[16] Nicholas IV, *Registres*, nos. 4409–14, 6779–81, 7396; Digard, *Philippe le Bel*, vol. i, pp. 123–4; above, ch. 2.   [17] James Doria, 'Annales Ianuenses', p. 157.
[18] Nicholas IV, *Registres*, nos. 5739–40, 6836–7; Bratianu, 'Autour', p. 250 n. 3.

plan for a co-ordinated crusade. He feared that the endless disputes among the European powers would only delay his crusader projects still further. The renewal of the war over Sicily after the death of Alfonso III of Aragon was certainly another factor in the abandonment of the plan of a co-ordinated crusade. Therefore, in 1291 Edward had already shifted from seeking European alliances to the recruitment of his own baronage. He also looked for allies outside Western Europe, in Norway, Hungary, and Achaea, and even among the Mongols of Persia. The death of Pope Nicholas in April 1292 and the resulting vacancy (1292-4) somewhat delayed Edward's preparations, though they continued until 1293. The preaching of the crusade in England went on until 1295. But whatever Edward's personal feelings about the loss of the Holy Land, domestic issues outweighed his desire for a crusade. In the years 1293–1307 the crusade was in conflict with English political requirements. The increasing French pressure on Gascony and clashes in the Channel made it impossible for Edward to launch his expedition. A decade of struggle in Wales, Scotland, France, and Flanders followed during which his agreement with the papacy could not be carried out. By the time these hostilities ceased (1303), Edward had abandoned all crusading activity. Yet his concern for the Holy Land continued to manifest itself through letters and contacts and his fame as the ideal of a crusader knight remained intact. In his will, it was alleged, he directed that 'his heart should be sent to the Holy Land and that a hundred mercenaries should be paid for one year to serve there the Cross of Christ'. Edward's devotion to the Holy Land and his crusader zeal found expression also in the 'Commendatio Lamentabilis in Transitu Magni Regis Edwardi', written probably by a minor canon of St Paul's, where the pope referred to Edward's efforts to pacify Christendom so that it could take the Cross, and the bishops mentioned in their lamentation his war against the Saracens.[19]

[19] B. Beebe, 'Edward I and the Crusades', Ph.D. thesis (St Andrew's, 1970), 14–172, 290–308; Nicholas Trevet, *Annales sex regum Angliae*, ed. T. Hog (London, 1845), 413–14; 'Commendatio Lamentabilis in Transitu Magni Regis Edwardi', in *Chronicles of the Reigns of Edward I and Edward II*, ed. W. Stubbs (RS, London, 1882–3), ii. 7, 11, 16; *Annales Angliae et Scotiae*, ed. H. T. Riley (RS, London, 1865), 378; *Anonymous Short English Metrical Chronicle*, ed. E. Tettle, Early English Text Society (London, 1935), 106.

The death of Rudolph of Habsburg (15 July 1291) probably interfered with Nicholas IV's plans, as Rudolph, who had already taken the Cross on 18 October 1275, had been ready to consider departing on a crusade if properly crowned as emperor. His death was followed by a vacancy till the election of Adolf of Nassau (5 May 1292) and internal anarchy for six years before the election of Albert I of Austria in 1298. Adolf of Nassau, it is worth mentioning, undertook upon his coronation the 'subsidium Terrae Sanctae'.[20]

At the same time as he was making almost frantic efforts to launch a crusade, Nicholas was also active in dealing with the infidels and schismatics. The first Franciscan to be elevated to the papacy, he devoted much effort throughout his pontificate to missionary work. His efforts to convert the Mongols as well as to reach a union with the Ethiopian, Georgian, Armenian, and Byzantine Churches were redoubled in view of the crusade of 1293. On 23 August 1291 he appealed to Andronicus II Palaeologus the Byzantine emperor (1282–1328) and to the kings of Lesser Armenia and Georgia 'ad baptisma suscipiendum . . . ut facilius Terrae Sanctae recuperationi incumbant'. As far as the Byzantine Empire was concerned, Nicholas who obviously followed Gregory X, sought a peaceful solution through a diplomatic marriage between Catherine of Courtenay, the titular empress of the Latin throne, and the Greek heir-apparent Michael Palaeologus. His successors, however, were to think in terms of restoring the Latin Empire by military action rather than of reaching an *entente* with the Greeks.[21]

As for the Mongols, Nicholas's correspondence with Arghun the il-khan of Persia (1284–91), pursued, till the loss of the Holy Land, the objectives of their conversion and permission for Franciscan missions to enter Persia. The reign of Arghun marked the apogee of friendly relations between the Mongols and the West. His four embassies to Europe (1285, 1287–8, 1289–90, 1290–1) aimed at a military alliance against Egypt. Arghun's great interest in the projected alliance is reflected in

---

[20] Throop, *Criticism of the Crusade*, pp. 255–6, 270–1; *MGH: Legum Sectio IV, Constitutiones*, ed. J. Schwalm (Hanover, 1904–6), vol. iii, no. 62, p. 53; no. 239, p. 231; no. 474, p. 460.
[21] Nicholas IV, *Registres*, nos. 6809–14, 6825–32; Laiou, *Constantinople*, pp. 49–50.

his letter addressed to Philip IV dated 1289. It was presented to the king of France by the third embassy of Arghun (1289–90), led by a Genoese, Buscarel de Ghizolfi. Arghun notified the king of France of his intention to begin a campaign against the Mamluks in the spring of 1291. He urged Philip IV to send his own army, as agreed with the embassy of Rabban (Mar) Sauma. Following the expected victory of the allies, Jerusalem would become a French possession. The il-khan wrote that 'if fulfilling thy sincere word, thou sendest thy troops at the time agreed upon, and if, blessed with good fortune by Heaven, he conquer these people [i.e. the Mamluks], we shall give you Jerusalem'. The technical details of the joint campaign were dealt with in a separate memorandum prepared by Buscarel and written in French. It proposed a meeting in the plain of Damascus for February 1291. Among other points, the Mongol il-khan promised to provide the French king with twenty to thirty thousand horses, either as a gift or at a reasonable price.[22]

Pope Nicholas sent two letters to Arghun which were dated 15 July 1289 (this was delivered by the embassy led by the Franciscan John of Monte Corvino) and 21 August 1291. In these letters the pope followed the traditional pattern of papal correspondence with the Mongols, appealing to Arghun to accept Christianity and the sacrament of baptism. But writing again to Arghun only two days later on 23 August, the pope used totally different language. It was not only mission or conversion which were being discussed. Far more weighty and urgent matters came to the fore. The pope informed the il-khan of the loss of Holy Land and of his appeal to the Catholic rulers of the West to unite their efforts for its recovery. He also informed the il-khan of the crusading plans of Edward I and of the preaching of a crusade throughout Christendom. This crusade (*negocium*), the pope wrote, could be successful only if supported by the 'powerful arm' of the il-khan. It is actually here that the papacy accepted for the first time the Mongol

[22] Nicholas IV, *Registres*, nos. 571, 577, 2240; J. B. Chabot, 'Notes sur les relations du roi Arghoun avec l'Occident', *ROL* 2 (1894), 566–629; A. Mostaert and F. W. Cleaves, *Lettres de 1829 et 1305 des il-khāns Argun et Oljeitu à Philippe le Bel* (Cambridge, Mass., 1962), 17–54; D. Sinor, 'Mongols and Western Europe', in *Setton's History*, iii. 513–34. For the letter of Arghun to Philip IV see Mostaert and Cleaves, *Lettres*, pp. 17–18; J. A. Boyle, 'Il-Khans of Persia and the Prince of Europe', *Central Asiatic Journal*, 20 (1976), 34–5.

First Attempts at Recovery

offers of military assistance for a crusade. One can agree with J. B. Chabot that the sudden change in the papal attitude to the Mongols, as reflected in the difference between two letters written within two days, was due to some new information obtained by the pope. It is however unlikely, as Chabot and recently J. A. Boyle assumed, that this information was 'l'émotion que cause parmi les princes croisés la nouvelle de la prise de Saint-Jean-d'Acre'. It is more likely that the change was caused precisely because Nicholas IV found less enthusiasm than he expected for his crusade in other quarters. The change in the papal attitude to the Mongol offer of a military alliance was possibly also influenced by other considerations. First of all Arghun's proposals for a joint campaign in Syria were most concrete. As an ally his forces could also be employed for the defence of Armenia whose *subsidium* had in 1291 to be ensured.[23] Moreover, Fidenzio of Padua, whose plan, it seems, Nicholas adopted, pronounced himself strongly in favour of Mongol participation in the crusade. Since the conversion of the members of the Mongol embassy at the Second Council of Lyons, that of the Mongols of Persia was eagerly expected. Philip IV of France and Edward I of England were both inclined to accept the Mongol offer. The growing credibility of the Mongol proposals was already reflected in the European accounts of Arghun's invasion of Syria in 1281. Generally, it seems that public opinion was in 1291 more inclined than before to accept the participation of the Mongols. This disposition of the Europeans can be deduced from their reaction to the rumours of a Mongol recovery of the Holy Land nine years later in 1300. Besides, in 1291 the protagonists of the crusade were desperate for manpower.[24]

The diplomatic relations with the Mongol il-khans of Persia broke down with the deaths of Arghun (10 May 1291) and Nicholas IV (4 April 1292) and were not to be resumed until

[23] Nicholas IV, *Registres*, nos. 6722, 6814; Chabot, 'Notes', pp. 619–23, followed by Boyle, 'Il-Khans', p. 36; Fidenzio of Padua, 'Liber', p. 57; J. Richard, 'The Mongols and the Francks', *Journal of Asian History*, 3 (1969), 51–7.
[24] Boyle, 'Il-Khans', pp. 31–2; P. Meyvaert, 'Unknown Letter of Hulagu, Il-Khan of Persia, to King Louis IX of France', *Viator*, 11 (1980), 245–59, esp. p. 257; S. Schein, '*Gesta Dei per Mongolos 1300*: The Genesis of a Non-Event', *EHR* 95 (1979), 808–9. For Armenia, see above, nn. 7–8. For the friendly relations between Arghun and Armenia, see Boase, 'History of the Kingdom', in id. (ed.), *Cilician Kingdom*, pp. 28–9.

1299. With the failure to organize a crusade in 1293, the great chance of the West for a joint expedition with the Mongols was missed. Gone was also the chance to convert them, if it ever existed. Already in 1291 Ramon Lull had stressed the fact that the Saracens were doing everything to convert the Mongols and the latter, victims of their simplicity, could be easily assimilated by the 'Moslem sect'. This, argued Lull, would place Christendom in grave danger. Lull had foreseen that if the Mongols were not converted to Christianity they would be converted to Islam. This happened in 1295.[25]

The Mongol policy of Nicholas IV influenced, as we shall see, not only the policy of his successors but had a major impact on the attitude of the Europeans at large. The papal crusader policy under Nicholas IV as a whole, seems to have been well received by Christendom. The criticism of the papacy after the fall of Acre was unexpectedly insignificant. The chroniclers described the papal reaction in terms of the pope's profound anguish and of his immediate action to help the Holy Land. Those descriptions are sometimes even exaggerated in favour of the papacy. For example, an anonymous Dominican annalist of Colmar who noted for 1290 (*sic*) not only the fall of Acre, but also that of Jerusalem, reported that Nicholas IV sent, after these disasters, 60,000 soldiers at his own expense to the Holy Land. He also allegedly granted the kings of England and France a tithe of their kingdoms for the crusade. The death of the pontiff was ascribed to his sorrow over the loss of the Holy Land; some chroniclers ever present it as the main reason for the failure of the action taken by the papacy and others for the recovery of the Holy Land.[26]

In evaluating the pontificate of Nicholas IV it is possible to agree with A. S. Atiya, who saw in it 'the birth of a new epoch in the history for the crusades', as well as with P. V. Laurent, who believed that a new beginning had been made at the Second

---

[25] See Ramon Lull, 'Tractatus de modo convertendi infideles seu Lo Passatge', ed. J. Rambaud-Buhot, in *Beati Magistri R. Lulli opera latina*, iii (Palma de Mallorca, 1954), 96, 105–6.

[26] 'Annales Colmarienses Maiores', *MGH SS* xvii. 217–18; John Elemosina, 'Liber Historiarum', in Golubovich, ii. 109; Weichard of Polhaim, 'Continuatio Weichardi de Polhaim', *MGH SS* xi. 813; Eberhard of Regensburg, 'Annales', *MGH SS* xvii. 594.

Council of Lyons.²⁷ If one introduces the distinction between planning and realization, or at least attempts at realization, the picture becomes clearer and certainly more balanced. As far as plans were concerned there is little in Nicholas IV's crusade-planning which did not exist in those of the Second Council of Lyons. Things look different, however, when we consider the attempts at realization. Those were actually pursued by Nicholas IV. The creation of a police-fleet to blockade Egypt as a necessary prologue to a general crusade, the pressure on military orders to reform or unite, the active pursuit of a Mongol alliance, all so many-faceted aspects of a crusader policy, concentrated and concerted by a vigorous pope, gave a new momentum to crusade-planning. This crusade-policy was to be closely followed for more than a century.

The Birth of a New Literary Genre: The *de recuperatione Terrae Sanctae*

One of the unintended results of Nicholas's vigorous crusader policy was the inauguration of a new epoch in crusade literature. It will not be an exaggeration to say that his request for advice in the absence of a general council stimulated the creation of a new branch of literature, the *de recuperatione Terrae Sanctae* memoranda, which since 1291 occupied in terms of bulk an important place in the literary output of the period. This new branch of literature has apparently much in common with its predecessors, namely the crusade-treatises from the thirteenth century like the 'Opus tripartitum' of Humbert of Romans, the 'Collectio de Scandalis Ecolesiae' of Gilbert of Tournai, and the 'Tractatus de Statu Saracenorum' of William of Tripoli. All three were inspired by the papal request for advice and both contain an exhortatory element, which links them to crusade *excitatoria*, common throughout Europe since the First Crusade.²⁸ Yet there is a dividing line between them. The early treatises can be described as working papers

---

[27] A. S. Atiya, *Crusade in the Late Middle Ages* (London, 1938), 45–6; P. V. Laurent, 'La Croisade et la question d'Orient sous le pontificat de Grégoire X', *RHSE* 22 (1945), 136–7; F. Pall, 'Croisades en Orient au Bas Moyen Âge: Observations critiques sur l'ouvrage de M. Atiya', *RHSE* 19 (1942), 536.
[28] See above, ch. 1.

submitted for conciliar discussions and therefore mainly concerned with such topics at the ideology, preaching, and organization of the crusade. They were concerned with what can be best described as the European approach to the crusade. In this they closely corresponded to the advice dispatched in 1292 by the provincial councils to Rome. Both were little, if at all, concerned with crusade-strategy. The memoranda composed after 1291, but including Fidenzio of Padua who wrote slightly earlier, are practical guidelines and, as such, largely concerned with general strategy as well as with detailed plans to be followed. Their contents recall 'La Devise des chemins de Babiloine', composed between the fall of Tripoli (26 April 1289) and that of Acre (18 May 1291) by the Order of St John. This was basically a politico-military report concerned with details of forces and detailed plans to be followed by the West in the envisaged conquest of Egypt.[29] 'La Devise' is unique, but it is not too hazardous to assume that other similar memoirs were prepared in the Christian East to serve the purpose of crusade *itineraria* in the future. The *de recuperatione* memoirs combine the characteristics both of a treatise like that of Humbert of Romans and of 'La Devise'. The loss of the Holy Land was obviously the catalyst of the combination, yet, though the fall of Acre was decisive, the great number of such memoranda is still puzzling. Moreover, they reflect a basically new attitude to the crusade. As their authors were usually familiar with the strategy of war, the memoirs tended to transform the intended crusade into a minutely planned expedition. Though intended as propaganda, they were also characterized by their very secular, highly professional, and practical strategic concepts of the crusade. As a rule we can distinguish two major types of treatise. On the one hand, the more theorizing treatises of a Pierre Dubois, William of Nogaret, or Galvano of Levanto, that is, of Europeans less acquainted with the Levant; on the other hand, the more empirical treatises composed by people familiar with the Levant, like Fidenzio of Padua, Marino Sanudo Torsello, Hayton, the masters of the military orders, and Charles II of Anjou.

The three first crusader memoirs which inaugurated the *de*

[29] P. V. Laurent, 'Croisade', pp. 121–2, 136.

recuperatione Terrae Sanctae treatises originated in Nicholas's appeal for advice. Their importance should not be underestimated. Not only do they bear testimony of the European mood and aspirations following 1291, but a memoir like that of Fidenzio of Padua seems to have had a direct impact on actual papal policy. A rather detailed analysis of the three earliest of the *de recuperatione* treatises is made here for several reasons. As already remarked, they bear testimony to European ways of thinking at the moment of the loss of the kingdom of Acre, and as such are of major importance. But their importance transcends in many aspects the temporal framework in which they were written. It will be argued that they became models which were to dominate the whole period dealt with in this study. There were to be in the next generation a plethora of similar treatises, some of them far more famous than the three indicated, but their fame was more the result of accident or of the standing of their authors, than of their originality. More often than not they were to repeat earlier ideas, and though some showed changes, basically they were to be variations of the same theme. This in itself is important not only because it shows a lack of originality, but because it also proves the continuity in the realm of *remedia*, as formulated around 1291. It also proves the continuity of the concept of the crusade, though this goes back to the previous generation, to the Second Council of Lyons. Last but not least, a detailed analysis is necessary, because existing historiography does not always do justice to the treatises in points of detail and especially in placing them in the right context of events.

Fidenzio of Padua

The earliest in point of time of the *de recuperatione Terrae Sanctae* memoirs was the treatise of Fidenzio of Padua, which probably had a direct impact on actual papal policy. There was too much correspondence between Nicholas's policy and the ideas of Fidenzio's plan to assume coincidence only. If the crusade of 1293 had taken place, it would probably have followed Fidenzio's ideas. Fidenzio's 'Liber recuperationis Terrae Sanctae', apparently inspired by Gregory X's request for advice on the eve of the Second Council of Lyons, was not drawn up until shortly before the fall of Acre. The most recent date mentioned was February 1290.[30] Written before the loss of the Holy Land, the 'Liber' presented views partly based on a situation which no longer existed after the fall of Acre. Thus it offered a glimpse into opinions which prevailed between the fall of Tripoli and that of Acre. Fidenzio's plan was not only the most influential but also the most detailed of those submitted to Nicholas IV, and its author was better qualified than any other to produce a plan for a crusade. His participation in the Second Council of Lyons where he met Nicholas IV and Charles II of Anjou; his prolonged residence in Syria as the vicar of the Franciscan province of the Holy Land in the 1260s; his acquaintance with the sultan and his court; his experience with the Mamluk army; his knowledge of Arabic as well as his travels in the East all added weight to the arguments of the man who had since 1266 been the Franciscan provincial vicar of the Holy Land.[31]

Fidenzio's treatise was divided into two parts. The first consisted of a history of the Holy Land from time immemorial; the second dealt with the means of the liberation and preservation of the Holy Land. There was little new in the historical introduction. Already Jacques de Vitry, prefacing his *Historia Orientalis*, had expressed the wish that his book should be of use in strengthening the faith of the crusaders. Since Gregory X, and possibly even earlier, popes had urged their clergy to read history so as to increase their devotion to the Holy Land. In the *de recuperatione Terrae Sanctae* treatises, however, the historical accounts served, as remarked by P. Riant, not only as propaganda, but as a kind of systematic scrutiny of past errors. The character of the first part of Fidenzio's treatise was, as noted by J. Delaville le Roulx, as much moral as historical.[32] His criticism was directed against the 'Franks' of the Holy Land and against the Christian West, including the papacy. Fidenzio's

---

[30] Fidenzio of Padua, 'Liber', pp. 9, 25; Atiya, *Crusade*, p. 38, points out that the mention of the Hijra year narrows down the completion to the first three days of Jan. 1291, but he mistakenly assumes that it was written before the fall of Tripoli.

[31] Golubovich, ii. 1–7; Atiya, *Crusade*, pp. 36–8; P. V. Laurent, 'Croisade', pp. 121–3.

[32] Fidenzio of Padua, 'Liber', p. 27; P. Riant, 'Description du *Liber Bellorum Domini*', *AOL* 1 (1881), 289–93; J. Delaville le Roulx, *France en Orient au XIVe siècle: Expeditions du maréchal Boucicaut* (Paris, 1885–6), i. 20.

view of the state of the crusader establishments in Syria almost predicted the end of Outremer in 1291. Except for the issue of the military orders, which he left out, his criticism of Europe's attitude to the Holy Land did not vary from the self-criticism current in Europe after 1291. Possibly, had he written after the loss of the Holy Land he would have recommended also the reform of the orders.

Fidenzio, who apparently borrowed from Jacques de Vitry, described the Franks as effeminate, imperceptive, indifferent fools, with little inclination for fighting. It was *inter alia* their unawareness and imprudence that caused the fall of Tripoli, and the attack of the sultan on Lesser Armenia (1290). Other faults he found with the Christians were internal strife and lack of leadership: 'not charity and true love but discord flourished among the Christian inhabitants of the Holy Land'. The kingdom was in a constant state of war of all against all: 'the Venetians, Genoese, and Pisans fought each other and they all quarrelled with the king. The confraternities of the *Suriani* fought each other. Neither was their charity between the military orders. The barons frequently disagreed. Whereas one city was at peace with the sultan, others were at war. When one of the military orders was fixing a truce with the infidels, the others started war on them'. Division and strife dominated. A sultan had allegedly said to the Christians that whereas he was a serpent with a single head whose tails followed, they were a serpent with many heads, and scarcely a tail to follow them all. Some Christians were *Latini*, others *Suriani*. The former were still as they had been at the time of the first conquest, a mixed bunch. Those who crossed the seas did more harm than good. Those who came with great fervour to the Holy Land 'returned with greater fervour to their own homes'. Those who promised great things did little or nothing. They did great harm to the Christians by going on pilgrimage to Jerusalem, in spite of the general sentence of excommunication, since they paid large amounts of money to the Saracens, which the latter used to fight the Christians. Still worse, Latins sold to Moslems iron, arms, timber, and other war-materials thus giving them the equipment to kill Christians. Finally, there were the 'many Christians among the Latins' who crossed to the Holy Land only to join the Saracens. For 'love of riches and carnal desire', they gave up their Christian faith and fought against their former co-religionists. The Christians were the best of the sultan's soldiers, and actually they supplied the major part of his army.³³

Christendom was blamed for its indifference to the causes of the Holy Land: 'the European Christians did not help Outremer as they ought to but rather abandoned it in most hard conditions; and if anybody rendered aid he later forsook it and returned to his country. How could so few Christians resist so many thousands of infidels? The Curia had caused long ago to help as it should; other affairs occupied its heart and thus the affair that should have occupied its heart above all others, was not taken care of or was little considered'. The Christian kings fought and mutually destroyed and annihilated each other. 'As they esteem temporal possessions,' asked Fidenzio, 'why did they neglect acceptance of the largest and richest territory?'³⁴

So much for criticism. The second part of Fidenzio's treatise has been described by a prominent historian as 'dealing with the practical aspects of a successful crusade for the recovery of the Holy Places as well as the ways by which it may be retained under Christian dominion'.³⁵ This description is better suited to the memoir of Charles II of Anjou than to that of Fidenzio. Fidenzio's second part, like the first, was not devoid of moralizing. Moral virtues should replace the presently prevalent vices. The basic virtue of goodness (*bonitas*), which consists of all the other virtues, namely charity, chastity, humility, piety, unity, sobriety, legality, perseverance, patience, should be one of the three most important qualities of the Christian host. The two other qualities were the size of the host and its leadership (*sufficentia exercitatis et presidencia capitis*). The sultan, Fidenzio estimated, had more than 40,000 knights. Consequently, the Christian host should consist of 20,000 to 30,000 knights and a considerable body of infantry. It should be strongly armed, well disciplined, and united. The Christians should explore the

---

³³ Fidenzio of Padua, 'Liber', pp. 13–16, 60; Jacques de Vitry, *Historia Orientalis: Iacobi de Vitriaco Libri Duo* (Douai, 1577), 73–5, 128–9, 133–7; Throop, *Criticism of the Crusade*, p. 98n. 137. Fidenzio's description of the vices of the Franks as fighters was, however, somewhat inconsistent with his description of the battles of which he was an eyewitness. While referring to the fall of Safed and Antioch, he eulogized Christian warfare: see Fidenzio of Padua, 'Liber', pp. 14–15, 24–5, 26–7.

³⁴ Fidenzio of Padua, 'Liber', pp. 15–16.   ³⁵ Atiya, *Crusade*, p. 39.

Saracen tactics and their use of bows, arrows, and light horse, and adopt some of them. Such effective exploration should be done by special agents (*exploratores*) such as the sultan used. The latter kept himself informed by his spies of all the actions of the Christians, not only in the parts adjacent to his own country but also in remote regions. The frequent references to the 'sultan' meant sometimes Qalawun, sometimes Baibars. As was justly observed by N. Daniel, the description of the sultan served Fidenzio for his image of the perfect crusader leader, whose qualities were partly compounded out of the virtues of the successful Moslem leaders. His description of the perfect leader was drawn less from an actual figure than from abstract ideas. Such an ideal leader, the *capitaneus*, is a composite figure of the practical war-leader and statesman, in whom were embodied some of the main chivalric virtues as well as the Aristotelian ideal of a leader.[36]

Fidenzio's opinion of the Saracens directly contradicted the optimistic hopes of William of Tripoli. The Saracens, he asserted, were most stubborn in their faith. Like Humbert of Romans earlier and Ricoldo of Monte Croce later, he admitted that the only conversions in the Holy Land were those of Christians to Islam. However, while Humbert of Romans emphasized the psychological appeal of Islam, both Fidenzio and Ricoldo were conscious of the strong material motives for the desertion from Christianity. Fidenzio, Ricoldo of Monte Croce, and probably also Ramon Lull envied the wealth and economic prosperity of the Moslems. Thus the recommendations of an economic war, though they stemmed from other reasons, were to some extent the result of the conviction of the economic superiority of Islam. The concern about the wealth of the Moslem world was not new, but around 1291 it became more acute.[37]

The expedition itself, Fidenzio said, should be waged simultaneously by a sea- and land-force. The former should consist of

---

[36] Fidenzio of Padua, 'Liber', 27–33; 35–46; N. Daniel, *Arabs and Medieval Europe* (Edinburgh, 1975), 188–90.

[37] N. Daniel, *Arabs*, pp. 214–18, and below, pp. 98, 103. For Fidenzio's disillusionment with the prospects of mission to the Saracens in the East see B. Z. Kedar, *Crusade and Mission: The Interplay between Two European Approaches towards the Muslims in Medieval Times* (Princeton, NJ, 1984), 156.

a fleet of forty or fifty, but not less than thirty galleys, all well-equipped with men, war-materials, and provisions. Its naval bases should be Cyprus, Acre, Ruad, eventually Rhodes, with preferably a bridgehead in Egypt. The fleet would enforce the sea-blockade against Egypt. Its first duty would be to intercept Christians who 'in search of worldly gains and in defiance of the penalty of excommunication imposed by the Church, continue to trade freely with the enemy'. Fidenzio mentioned eleven advantages of the sea-blockade. The aim of the blockade would be to make the Christians masters of the sea, reduce the danger of Saracen naval attacks on Syria and Cyprus, and prevent the possibility of a Saracen blockade, which would cut off the Christian host from Europe during the crusade. Moreover, the blockade would impoverish Egypt. From Alexandria alone they get a daily income of more than 1,000 gold florins, which the sultan might spend on the equipment of his horsemen. If the Christians ceased to frequent Egypt, the Saracens would lose all that income and thus sustain a heavy blow. They would be deprived of the means of selling their own export products. Their textile industry of 10,000 weavers would be ruined. Additionally they would also lose the income from the commerce with the Far East, as the Indian trade would be deflected from the Red Sea and Egypt to Persia and Lesser Armenia. The blockade would deprive the Saracens of many necessary products including war-materials; it would also prevent the import of boys from the shores of the Black Sea to reinforce the enemy's army of Mamluks; destroy the ports of the Saracens, and put an end to the piracy in the Mediterranean.[38]

Referring to the sea-blockade, Fidenzio repeated an opinion, already current at the Second Council of Lyons, that it would be an effective weapon owing to the Saracens' ignorance of the art of navigation. In his opinion, however, an effective blockade of Egypt by a special fleet would suffice by itself for the liberation of the Holy Land. This opinion was to be adopted by Ramon Lull, Charles II of Anjou, Hayton, Marino Sanudo Torsello, Henry II of Cyprus, and many others. The strategy of the double attack as introduced by Fidenzio, was less

---

[38] Fidenzio of Padua, 'Liber', pp. 46–9.

popular, but nevertheless was to be followed by James of Molay. The originality of the plan of Fidenzio should not be underestimated. He was the first to introduce the idea of blockading Egypt from the sea, not just as an auxiliary weapon but as one equal in importance to land warfare.[39]

Fidenzio discussed three alternative expedition routes. The first was an overland expedition across Europe to Constantinople and hence, through Anatolia and Lesser Armenia, to the Holy Land. The second was the sea-route from Venice and Genoa. The third was by land as far as Brindisi, and then by sea across the Adriatic to Durazzo, across the Balkans to Constantinople, whence it would follow the latter stages of the first route. Fidenzio discussed their advantages and disadvantages without specifying this preferences. The stopping-points which he recommended for the host before penetrating Syria were Egypt, Acre, Tripoli, St Simeon, Portus Palorum, all equally accessible by the three routes. His preferences were for Portus Palorum and St Simeon (Soldinum). The bay of St Simeon was suitable for small vessels, which might then sail up the Orontes as far as Antioch; Portus Palorum could accommodate the larger galleys. From there the host was to march on Antioch and establish itself there. At Antioch the host should await its Armenian, Georgian, and Mongol allies. Syria was not heavily garrisoned. The danger would always come from Egypt and therefore Egypt should be kept from aiding its garrisons in Syria. The fleet should cut the communications between Egypt and Syria as well as carry the economic war against Egypt to a successful conclusion.[40]

Fidenzio concluded his plan with a number of recommendations regarding the maintenance of the Holy Land once it had been conquered. A sufficient garrison, the custody of the sea by ten galleys, the establishment of fortified points throughout the country, a stable government under one leader as well as Christian virtues like humility, were the most important factors for the 'conservatio Terre Sancte'. This permanent garrison

---

[39] Ibid. 46, 50; below, ch. 6.
[40] Fidenzio of Padua, 'Liber', pp. 51–8. Portus Palorum, Palli (Pals), Porto di Plas, portus de Pallibus, once situated a few miles south-west of Ayas on the gulf of Alexandretta, was already abandoned in the 1300s. See Atiya, *Crusade*, p. 42 n. 3; Heyd, *Histoire du commerce*, ii. 73–4.

should be financed by European prelates and cities. Bishoprics, abbeys, and cities should keep one, two, three, or more knights in the Holy Land. The necessity of adopting this last-mentioned measure for defending the kingdom was explained by the decline in the number of those willing to take the Cross.[41]

Fidenzio's pessimistic estimate of the chances of the traditional method of recruitment for the future kingdom in the Holy Land fitted well with his reluctance or even rejection of the traditional general crusade. This view, shared by Ramon Lull and Charles II of Anjou, became, after the loss of the Holy Land, the chief feature of crusade-planning. Another common strategic approach shared by the three men had already been present in the last century's planning. This was the necessity of conquering territories adjacent to the Holy Land. They were to serve as bases from which a final attack on the Holy Land could be launched. The fact that some of the theoreticians of the crusade included in their memoirs justifications of such conquests, suggests that this posed for them not only strategic but also moral problems. In other words, whereas the authors of the *de recuperatione Terrae Sanctae* treatises fully recognized the legitimacy of the recovery of the Holy Land and therefore did not feel any need to justify it, the conquest of other territories possessed either by infidels or schismatics needed legitimization.

From the First Crusade onwards such justifications were available, especially in the writings of Pope Innocent IV and of his pupil Hostiensis. Both these distinguished authorities justified the conquest of such territories, if they were formerly held by the Roman Empire, as an actual recovery of territories that rightfully belonged to Christians. Both also recognized the legitimacy of conquering countries which had been at no time under either Roman or Christian rule. Though the infidels possessed and ruled countries, the pope could interfere in certain specific cases *ex potestate* in the internal legislation of non-Christian countries. Hostiensis took this argument even further. With the advent of Christ, he argued, every human being, every dominion, principality, and jurisdictional power had been taken away from the infidels and handed over to the

[41] Fidenzio of Padua, 'Liber', pp. 58–60.

faithful, and all this 'de jure et ex justa causa'. This translation of power was originally made to the person of Christ who then transferred it to the papacy. This was Hostiensis's premiss for asserting that all the infidels should be subjected to the faithful. On the whole it was Innocent IV's opinions that became more accepted than those of Hostiensis.[42]

Among the crusade-planners Fidenzio of Padua, whose crusade-plan included the conquest of the kingdoms of Turkey, Damascus, Egypt, Carthage, and Tunis (all territories once possessed by the Roman Empire), is found referring to such conquests as the recovery of what was rightfully Christian property.[43] Writing early in 1292 Ramon Lull used the same reasoning involving the 'Donation of Constantine' which bestowed upon the Roman Church the right to all the territories that once belonged to the Roman Empire, in the East as well as in the West. Moreover, as the crusade was above all a war aimed to secure free activity for missionaries, Lull argued that the conquest of territories of non-Christians and schismatics was justified, because this was a necessary prelude to their conversion.[44] For Marino Sanudo Torsello the conquest of territories outside the Holy Land and particularly of Egypt was a military necessity for the recovery of the Holy Land, and was justified as the recovery of a once-Christian land.[45]

Yet the conquest of territories other than the Holy Land did not pose serious moral problems for all contemporaries. For Marino Sanudo Torsello, Charles II of Anjou (c.1292–4), Pierre Dubois (1308), and Henry II of Lusignan (c.1311–12) such conquests were military necessities if the Holy Land was to be recovered.[46] This kind of justification had already been

---

[42] W. Ullman, *Medieval Papalism: The Political Theories of the Medieval Canonists* (London, 1949), 120–3; J. Riley-Smith, *What were the Crusades?* (London, 1977), 18–33.

[43] Fidenzio of Padua, 'Liber', pp. 16, 50–1. For an earlier e.g. of such reasoning see Cafaro of Caschifelone, 'De liberatione civitatum Orientis', *RHC Hist. occ.* v. 63.

[44] Ramon Lull, 'Tractatus', pp. 102, 105; H. Wieruszowski, 'Ramon Lull et l'idée de la Cité de Dieu', *Estudis Franciscans*, 47 (1935), 90–2; ed.'s introd. in Ramon Lull, 'Liber de acquisitione Terrae Sanctae', ed. E. Kamar, *Studia Orientalia Christiana Collectanea*, 6 (1961), 48–53.   [45] Marino Sanudo Torsello, 'Liber', pp. 31–2.

[46] Charles II of Anjou, 'Conseil', pp. 353–61; Pierre Dubois, 'Oppinio cujusdam suadentis regi Franciae ut regnum Jerosolimitanum et Cipri acquireret pro altero filiorum suorum ac de invasione Egipti', in id., *Recovery of the Holy Land by Pierre Dubois*, trans. W. I. Brandt (New York, 1956), 154–62; Henry II of Lusignan, 'Consilium', ed. M. S. de Mas Latrie, in Mas Latrie, *Histoire de l'île de Chypre*, 118–25.

current during the invasions of Egypt in 1218 and 1249, when the conquest of Egypt was seen as necessary for the safety of the Latin kingdom.[47] The loss of the Holy Land in 1291 necessitated only a minor readjustment of justification which already existed. Secular and moral justification were found side by side.[48]

Ramon Lull

Ramon Lull's writings suggest that until the beginning of 1292 he had no interest in promoting a crusade. In his letters to an unnamed prelate and to the University of Paris (1287–9), as well as in his 'Liber contra Antichristum' (1289), Lull argued again and again for exclusively peaceful means as the most efficient for the recovery of the Holy Land. In his sociophilosophical, satirical book *Blanquerna* (c.1283–5), he went as far as saying that 'the pope himself, and the Christian kings, in their efforts to conquer the Holy Land, adopt the method of Mohammed, who took by force of arms all he wished, instead of adopting the method of Christ and his apostles, who by preaching and martyrdom converted the world.'[49]

But early in 1292, the man acclaimed as the 'greatest missionary who worked among the Moslems in the Middle Ages' presented Pope Nicholas with the 'Epistola ... pro recuperatione Terrae Sanctae' and the 'Tractatus de modo convertendi infideles seu Lo Passatge'. It is in that treatise that Lull for the first time declared himself explicitly a partisan of an armed crusade. The shift from purely pacific means to an armed crusade, even if intended to secure a free field for missionaries, suggests a certain despair, which can be assigned to the loss of the Holy Land, yet another victory for Islam. The 'Tractatus', written in Rome shortly after the fall of Acre, was certainly inspired by that event. Lull mentioned it in the 'Tractatus', saying that 'omnes sunt in tristicia de amissione Terrae Sanctae'. From that time on he would invariably combine the

---

[47] Prawer, *Histoire*, ii. 148–50, 324–6; Riley-Smith, *What were the Crusades?*, pp. 23–4.
[48] Ullman, *Medieval Papalism*, p. 128.
[49] R. Sugranyes de Franch, *Raymond Lulle, docteur des missions* (Schöneck and Beckenreid, 1954), 81, 99; Kamar's introd. to Ramon Lull, 'Liber de acquisitione', pp. 24 n. 34, 46–53.

two aims, peaceful mission and armed action, to recover the Holy Land. These seemingly divergent activities find a common denominator in the great and final end: the universal conversion. Though he concentrated upon the conversion of the Saracens, his ultimate aim was the conversion of all unbelievers, including the Mongols, and at the same time the union with the schismatics. It should be kept in mind, that though an armed crusade became an integral part of his thinking from 1292, he always regarded such an expedition as subordinate to missionary efforts and mainly intended to enforce and thus secure their success.[50]

In the 'Tractatus' Lull suggested a combined naval and land expedition. The maritime war should be waged under three admirals chosen and controlled by the Church. Each was to supervise a given area of the Mediterranean: one, attached to Spain, was to control the area from Tripoli in Barbery to Safi in Morocco; the second, the coasts of Syria from Tripoli to Armenia; the third, the coasts of the Byzantine Empire. The blockade would be aimed at banning all merchandise destined for the Saracens as well as the destruction of Saracen coastal settlements and their maritime bases. The destruction of the Saracen merchant class would result in a growing number of Christian merchants being able to move freely and securely on the seas. Only after the destruction of the Saracen bases would the Christians start the *passagium*. Then their enemies would be in no position to resist them.[51]

The *passagium* or 'land-war' was to be directed by the pope with a king and the masters-general of the Templars, Hospitallers, and the Teutonic Knights, so that simultaneous use would

[50] Ramon Lull, 'Epistola summo Pontifici pro recuperatione Terrae Sanctae', ed. J. Rambaud-Buhot, in *Beati Magistri R. Lulli opera latina*, iii. 96–7; id., 'Tractatus', p. 106; id., 'Vita Coaetenea', ed. H. Harada, *CCCM* xxxiv (1980), 2840 Kedar, *Crusade and Mission*, pp. 195–6; B. Altaner, 'Glaubenszwang und Glaubensfreiheit in der Missiontheorie des Raymundus Lullus: Ein Beitrag zur Geschichte des Toleranzgedankens', *Historisches Jahrbuch*, 48 (1928), 586; cf. Throop, *Criticism of the Crusade*, p. 139. Historians once dated this shift to 1294 with Lull's 'Petitio' to Celestine V (see e.g. Delaville le Roulx, *France*, i. 28; Atiya, *Crusade*, pp. 74–7). The recently discovered ms of the 'Tractatus' establishes the date of the work as the beginning of 1292, before the death of Pope Nicholas IV on 14 Apr. See J. Rambaud-Buhot in her introd. to Ramon Lull, 'Tractatus', pp. 93–5. For similar ideas regarding mission, i.e. coercive conversion, see Kedar, *Crusade and Mission*, pp. 183–9.

[51] Ramon Lull, 'Tractatus', pp. 99–100.

be made of the spiritual and the temporal swords.[52] The 'two swords' symbolize the duty of the Church to convert the infidels by means of arms and the faculty of word, knowledge, and devotion. Lull's political theory would have been acceptable to any theocratic contemporary of Gregory VII or even that of Bernard of Clairvaux: Lull assigned to the papacy absolute supremacy over the entire world; the pope and his cardinals were the true vicars of God. H. Wieruszowski has rightly emphasized the anachronistic character of Lull's ideas.[53] The visible sign of the temporal power is the 'Donation of Constantine', which Lull interpreted even more radically than the curialists. He deduced from it the right of the Roman Church over the Orient and Constantinople. His aim was not to augment the power of the pope as such, but to facilitate the mission. The power of the pope was therefore made conditional upon his duty to conquer lands from the infidels—a stage in the greater act of conversion of souls. The secular rulers commanding the crusade should remain subordinated to papal authority. Lull accepted the practical necessity of a civil authority, but like Gregory VII he saw in it an 'inevitable evil'. Some years later $c.$ 1305 Lull came to recognize the growing importance of secular states. The crusade without their aid was impossible. Their help was also indispensable for the conversion of Islam. But his appeals to the secular rulers, and above all to Philip IV the Fair, had but one end in view: the promotion of the glory of God, and not, as Pierre Dubois and William of Nogaret saw it, the glory of the monarch. J. N. Hillgarth has shown that Lull's alleged lack of interest in politics or in contemporary problems is without foundation. Ramon Lull, though he always remained loyal to his principles, was very practical in his choice of secular patrons and in his search for means to promote his plans. From 1305 onwards he turned from the papacy to place his main hopes in secular rulers. Implicitly he recognized the disintegration of the Holy Roman Empire and none of his appeals was addressed to the kings of the Romans, not even to Henry VII. He sought aid from James II of Aragon and from Philip IV of France, and

---

[52] Ibid. 100–2.
[53] Wieruszowski, 'Ramon Lull', pp. 90–9.

again from the maritime republics of Venice, Genoa, and Pisa, and the kings of Cyprus and Sicily.[54]

Instead of regarding peace in Europe as a pre-condition of the crusade, Lull assumed that internal wars would stop following the call for a crusade. About finances, he was rather vague. The Church should levy a tithe on ecclesiastical property. The Holy See would also collect revenues from indulgences and penitential fines. The Church should also use its own property, but on this it should use its own judgement.[55] For Lull, the papal universal government of Christendom carried with it duties and obligations in the context of the crusade. To show a good example the pope should participate in the crusade in person. He should be accompanied by churchmen and laity who knew Greek, were learned in philosophy and theology, and were capable of healing the schism. The Byzantine Empire appeared to Lull to be an obstacle in the propagation of the faith. It was necessary that Rome should possess Constantinople and that the return of the Greeks to the Roman Church should be assured, if necessary by force. Because the Greeks were Christians, the conquest of Constantinople appeared to Lull to be a disciplinary act and not a forced conversion. In his opinion the acceptance by the Greeks of the primacy of the pope as well as the incorporation of the Byzantine Empire into the Roman Church would contribute to the conversion of the Saracens, the Mongols, and other non-Christians. Moreover, to Lull Constantinople appeared geographically to be the port of the Orient and was indispensable to the recovery of the Holy Land. He feared that it would be conquered either by the Mongols or the Saracens. Therefore, if the Greeks, after discussion, refused to join the Church voluntarily, their land should be conquered and their troops should be made to join the crusade, marching ahead of the Latins. On the other hand if the Greeks consented voluntarily to the union, the pope should authorize their emperor to keep his possessions and accompany the expedition. He and his princes would take an oath of fealty to the pope and the Roman Church; hostages would be given

---

[54] J. N. Hillgarth, *Ramon Lull and Lullism in Fourteenth-Century France* (Oxford, 1971), 28–30, 47–63, 108–34; cf. R. Sugranyes de Franch, 'Projectes de creuada en la doctrina missional de Ramon Llull', *Estudios Lulianos*, 4 (1960), 275–9.

[55] Ramon Lull, 'Tractatus', pp. 109–12.

and a Latin garrison installed in Constantinople until the Holy Land was conquered.[56]

The conquering host would proceed through Lesser Armenia to Jerusalem and then to Tripoli (in Barbery) and the coast of north Africa. After the conquest of the Holy Land, the master-general of the Temple would lead his troops to Barbery (north Africa), whereas the master-general of the Hospital would take his into Turkey. The master-general of the Teutonic Knights would fight in Laodicea (Lycaonia) in Asia Minor with the aid of a king from that region, the ruler of Armenia. The distance put between the orders would assure concord and charity. Lull's strange ways of achieving harmony obviously showed that he had in mind their conflicts in the crowded Acre. Moreover, it showed that at that time Lull, though he was aware that the orders posed a problem, was not sure what the solution should be. In *Blanquerna*, as well as in his letter to Pope Nicholas IV and the cardinals, which accompanies the 'Tractatus', Lull advocated the merging of the orders. This was to be repeated in all his later plans.[57]

This grandiose scheme for a crusade leads up to the main aim of Lull, the conversion of the infidels. The third part of the 'Tractatus' provides a plan for mission through religious disputations conducted by men trained in special schools to be founded in Rome, Paris, Spain, Genoa, Venice, Prussia, Hungary, Caffa, and Taurus, as well as other places suitable for learning different oriental languages. The belief in conversion as a result of argument and discussion, and in educating cadres by studying oriental languages, that is to say milieux and cultures, was characteristic of the Dominicans rather than the Franciscans. Yet it was Lull's activity as a Franciscan tertiary that created the most famous of such language schools, Holy Trinity, founded by Prince James of Majorca at Miramar in 1274. At the time Holy Trinity was founded, five Dominican schools, whose *spiritus movens* was Ramon de Penyafort, were already operating at Valencia, Jativa, Murcia, Barcelona, and Tunis. This schooling programme, repeated by Lull in all his

[56] Ibid. 98, 100–1; id., 'Epistola', p. 98; Sugranyes de Franch, *Raymond Lulle*, p. 82.

[57] Ramon Lull, 'Tractatus', pp. 96, 101; id., 'Epistola', p. 96; id., *Blanquerna*, trans. Sugranyes de Franch in *Raymond Lulle*, p. 102; A. J. Forey, 'Military Orders in Crusading Proposals', pp. 320–1.

treatises, would later be adopted by the Council of Vienne (1312).[58]

The basic ideas of the 'Tractatus' were to be later developed in Lull's numerous treatises and especially the 'Liber de fine' (1305) and the 'Liber de acquisitione Terrae Sanctae' (1309). It is the combination of the crusade with missionary activity that marks the originality of Lull.[59] His strategy for a crusade did not, however, differ from that of his contemporaries. As far as his crusade-planning was concerned Lull borrowed both from other theorists and from actual crusade policies. P. E. Kamar has pointed out that Lull was possibly inspired by the plan of Fidenzio of Padua, as he could have easily come across Fidenzio's 'Liber recuperationis Terrae Sanctae' during his stay in Rome, where he composed the 'Tractatus'.[60] Both plans had much in common, although in that of Lull, as in that of Nicholas IV, the war at sea should precede the war on land. Actual crusade-policy was reflected, for example, in Lull's linking of the general crusade with the Byzantine Empire and the alternatives of its union with Rome or its armed conquest. The latter was the policy of Charles I of Anjou as well as of the popes Urban IV (1261–4), Martin IV (1281–5), and Honorius IV (1285–7). The launching of an anti-Byzantine crusade, perceived as 'helpful to the affairs of the Holy Land', was actually prevented only by the Sicilian Vespers (1282).[61]

Charles II of Anjou

Another of the plans inspired by Nicholas IV was that of Charles II of Anjou, a *crucesignatus* since 1275 who carried since

---

[58] Ramon Lull, 'Tractatus', pp. 102–5. See also R. I. Burns, 'Christian–Islamic Confrontation in the West: The Thirteenth-Century Dream of Conversion', *AHR* 76 (1971), 1386–434; E. R. Daniel, *Franciscan Concept of Mission in the High Middle Ages* (Ky., 1975), 68–9, 72–3.

[59] Ramon Lull's other treatises concerned with the crusade are the following: 'Petitio Raymund: pro conversione infidelium' (Naples, 1294); 'Petitio Raymundi pro conversione infidelium' (Rome, 1295); 'Petitio Raymundi in concilio generali ad adquirendam Terra Sanctam' (1311). The 'Tractatus', the 'Liber de Fine', and the 'Liber de acquisitione' are crusade-projects; the rest are petitions.

[60] Kamar in his introd. to Ramon Lull, 'Liber de acquisitione', pp. 59–60.

[61] D. Geanakoplos, *Emperor Michael Palaeologus and the West 1258–1282* (Cambridge, Mass., 1959), 305–71; id., 'Byzantium and the Crusades 1261–1354', in *Setton's History*, iii. 27–42, and below, ch. 6.

1285, among his other titles that of king of Jerusalem.[62] It is likely that this plan was completed after the death of Nicholas IV (4 April 1292) during the papal interregnum before the election of Celestine V on 5 July 1294.[63] Charles's ideas were those of a secular ruler with practical experience in warfare and European politics but with no particular attachment to the crusade. Charles II of Anjou, the king of Naples and Jerusalem, had little concern for a crusade to the Holy Land. Since his return from Spanish captivity (1284–9), he was primarily concerned with the recovery of Sicily, and in this he had the support of his cousin Philip IV of France. The statement of his intentions to set out on a crusade (9 February 1291) as well as the crusader plan he wrote were therefore mainly intended to please the pope. Still, Charles II bore the crown of Jerusalem and he should have felt a moral obligation at least to make the task of the Christian reconquest his own. But Jerusalem had no priority in his plans; at best it had to be integrated into the larger framework of his policy. For Charles, as he stated in 1292 the pre-condition of the recovery of Jerusalem was the retaking of Sicily. He was ready in 1289 to transfer his rights to Jerusalem to the crown of Aragon if he could get back Sicily. On 1 July 1300 when it was believed that the Mongols had conquered the Holy Land and were ready to hand it over to Christendom he appointed Mellorus de Ravendel, a noble refugee from the crusader East, as his vicar in the Holy Land. It seems that Charles as other European rulers in 1300 was attempting to benefit from the Mongol conquest and tried, through the dispatch of his vicar to the East, to ensure himself the transfer of the Holy Land from the Mongols to his hands.

[62] See above, ch. 2 n. 27.

[63] It is difficult to agree with Bratianu either that the 'Conseil' of Charles II of Anjou was the first to be submitted to Pope Nicholas after the loss of the Holy Land, or that it was the most influential. As a matter of fact, the plans of Fidenzio of Padua and of Ramon Lull preceded that submitted by Charles. The latter's plan was, it seems, stimulated by the papal request for advice in Aug. 1291, but it was finished only after Nicholas's death, during the papal interregnum, which lasted for two years till the election of Celestine V on 5 July 1294, as it includes the following: 'Item conseille li dis ros ... que quant pape sera quil ordenast concile general' (Charles II of Anjou, 'Conseil', pp. 360–1). It follows thus that a part of the treatise at least, was already written during the papal interregnum and not, as Bratianu assumed, between Aug. 1291 and 4 Apr. 1292; cf. Bratianu in his introd. to Charles II of Anjou, 'Conseil', pp. 293–7.

However, when the Sicilian question was temporarily solved by the Treaty of Caltabellota (31 August 1302) he showed but remote interest in the recovery of the kingdom whose crown he wore.[64]

Charles II's 'Conseil' begins with the declaration, wise in itself, that a *passagium generale* against the overwhelmingly strong sultan of Egypt would be a folly. The sultan of Egypt disposed of all the forces of Islam, and his prestige, after the expulsion of the Christians, and his triple victory over the Mongols, was immense. The Christians had no bridgeheads on the coasts of Egypt or Syria and even if the disembarkation were to succeed, the crusaders would perish as victims of the climate and disease. Consequently, the way to fight Egypt was to destroy its economy. It was necessary to destroy the coast and especially the port of Alexandria. Fifty galleys, fifty transports, and 1,500 men would suffice to enforce the blockade. Ten galleys each should be provided by the Hospitallers, the Templars, and the king of Cyprus; the rest, as well as the fighting men, should be provided by the Holy See. Only when the blockade of Egypt had been successfully completed should the *passagium generale* be launched. As the place of disembarkation Charles suggested Cyprus 'que lisle de chipre est planteurouse de tous biens ce est las plus prochaine isle que la terre sainte aie apertenans'. Alternatively, Acre or Tripoli could be used, but on that, experts should be consulted.[65]

The general crusade was to be led by the military orders reorganized after their unification under a single chief, 'home de grant valor', preferably the son of a king or at least 'preudoume et de haut lignage', appointed probably by the pope, and the master-general of the new united order. This warrior-king (*bellator rex*) appeared also in the plans of Fidenzio of Padua and Ramon Lull. The latter, in his 'Epistola' to Nicholas IV and the cardinals in 1292, perceived exactly as Charles did the crusading army as a new military order with its

---

[64] Bartholomew of Neocastro, 'Historia', p. 136; Digard, *Philippe le Bel*, vol. ii, p. 275; Léonard, *Angevins de Naples*, pp. 194–5. Charles II of Anjou inherited the crown of Jerusalem from his father Charles I of Anjou. See Pierre Dubois, 'Oppinio cujusdam', pp. 155, 161; S. Runciman, *History of the Crusades* (Cambridge, 1955), iii. 329; N. J. Housley, 'Charles II of Naples and the Kingdom of Jerusalem', *Byzantium*, 54 (1984), 527–35.

[65] Charles II of Anjou, 'Conseil', pp. 353–6, 359–60.

grand master a member of a royal family. It was, however, Charles who thought that the warrior-king, once the crusade was crowned by success, should be king of Jerusalem and grand master of the united order.[66]

The last part of the 'Conseil' was devoted to the means of preservation of the Holy Land. The new order, composed of 2,000 knights and 200 sergeants, was envisaged as a permanent army to be kept in the Holy Land, financed by the whole of Christendom. It should be granted all the territories and resources taken over from the infidels, as well as all the rents and possessions which belonged to the crown of Jerusalem. Extensive franchises should be granted to the Genoese, Pisans, Venetians, and Catalans, as well as 'toutes maniers de gens que peupler se vorroient en la terre sainte'. Obviously Charles was well aware of the difficulties of colonizing the Holy Land. Charles preferred to exclude from his actual military and naval operations the maritime republics, unfavourable to the blockade, but he assigned to them the role of the future settlers.[67]

The recent discovery of Ramon Lull's 'Epistola' and the 'Tractatus' (1292) makes the 'Conseil' far less original than was assumed by scholars and especially by its editor G. I. Bratianu. Some of the ideas considered to be original, like the maritime blockade of Egypt, the fusion of the orders, the establishment of a permanent army with a uniform organization, the warrior-king, and finally the insistence upon a general council as a necessary prologue to the crusade, actually appeared in Lull's treatises. The similarities between their ideas seem so close as to suggest that Charles was directly inspired by Lull's 'Tractatus' and 'Epistola'; later, however (after 1294), it was Lull who was inspired by Charles's treatise. As we find much of what

---

[66] Charles II of Anjou, 'Conseil', pp. 356; Ramon Lull, 'Epistola', p. 97; Fidenzio of Padua, 'Liber', pp. 41–6. The orders to which Charles referred were the following: Hospital, Temple, Teutonic Knights, Santiago, Calatrava, Roncevaux, St Anthony, Trinity, Altopasso, Premonstratensians, and Grammont. After the Teutonic Knights, Charles referred to 'dou cles', which was, it seems, the Spanish Order of Santiago, known as the Order of Caceres or Ucles. See J. F. O'Callaghan, 'Foundation of the Order of Alcantara 1176–1218', *Catholic Historical Review*, 47 (1962), 483.

[67] Charles II of Anjou, 'Conseil', pp. 356–60; S. Schein, 'The Future *Regnum Hierusalem*: A Chapter in Medieval State Planning', *Journal of Medieval History*, 10 (1984), 95–105.

Charles II said also in the decisions of the Church councils of 1292 and even in chronicles written c.1292–4, it is clear that he expressed prevailing rather than original opinions. None the less, Charles's *Conseil* became far more influential than those of Fidenzio or Lull. Its impact was to be felt in later crusade-planning and especially that of the French.[68]

The three plans written during the pontificate of Nicholas IV foresaw that Jerusalem would be won on the battlefields of Egypt, and they thus return to the thirteenth-century tradition. Whether they favoured land- or sea-routes, none suggested a direct attack on the Holy Land, but the establishment of bases from which a final attack should be launched. The actual crusade became an epilogue, a glorious one no doubt, of a pedestrian, economic victory achieved through a blockage of Egypt. It was Christendom's naval power and its supremacy on the seas which would assure victory. The three plans expressed the main features of crusade-planning after 1291. Plans composed later were to be mere variations of those of 1291–4.

[68] Charles II of Anjou, 'Conseil', pp. 361–2.

4

1291–1292

THE LOSS OF THE HOLY LAND IN PUBLIC OPINION

The loss of the Holy Land had profoundly stirred European society and this is reflected in the sheer volume of references in chronicles, in treatises tinged with apologetics, and in sources dealing with other affairs. Acre appeared, for example, in Marco Polo's 'Il Millione' (*c.*1298) and reappeared constantly in the mutual accusations of the French and English during their conflicts, in the correspondence of Pope Boniface VIII and Philip IV of France, and obviously in the trial of the Templars.[1] James of Verona, an Augustinian canon who went on pilgrimage to the Holy Land in 1335, noted in his 'Liber Peregrinationis' that the ladies of Cyprus wore black cloaks out-of-doors as a sign of mourning for the fall of Acre. The same custom was still noted by Nicholas Marthoni, a notary from the Italian town of Carinola, who visited Cyprus almost a hundred years later (1394).[2] Acre also appeared in Dante's indictment of Boniface VIII in the *Inferno*: 'But he, the Prince of the modern Pharisees, | Waging war near the Lateran—and | Not against Jews, nor Moslem enemies, | For his enemies were Christian | And none had been to conquer Acre, nor were they | Merchants within the realm of the Soldan'.[3] In around 1300 an Italian friar named John wrote that when on the vigil of Ascension of 1292

---

[1] According to Marco Polo, 'when the Soldan of Babylon went against the city of Acre and took it, the Soldan of Aden (Yemen) sent to his assistance 30,000 horsemen and full 40,000 camels'. This is not confirmed by other sources; Ibn Furat mentions forces from Damascus, Hamah (Hamath), the rest of Syria, of Egypt, and of Arabia. See Marco Polo, *Book of Ser Marco Polo*, ed. H. Yule and H. Cordier (London, 1903), iii. 438–9 and n. 4. See also *Political Songs*, ed. T. Wright (Camden Society, London, 1839), 249; *Grandes Chroniques de France*, ed. J. Viard, viii (Paris, 1834), 148.

[2] *Excerpta Cypria: Materials for a History of Cyprus*, ed. C. D. Cobham (Cambridge, 1908), 16–24. The same story also appeared in Ludolph of Suchem, *De Itinere Terre Sancte*, ed. F. Deycks (Stuttgart, 1851), 46.

[3] Dante, *Inferno*, xxvii. 85–90. See also Benevenuto of Imola, 'Excerpta Historica Ex Commentaris in Coemediam Dantis', *RIS* i. 1111–13.

he was reciting the Psalm 'Oh God, the heathen have come into thine inheritance; they have defiled the Holy Temple' (Ps. 79: 1), he felt a horrible anguish in his heart at 'that lamentable death' that had transpired overseas. This caused him such tears that he was unable to continue his reciting. Thus instead he prayed to God to have mercy upon his people. As late as 1305 the Italian preacher Giordano of Rivalto assured his flock that the loss of Tripoli and Acre should not cause loss of faith as they were God's punishments.[4]

The countless references to the fall of Acre as marking the end of the Christian rule in the Holy Land point to an unusual emotional involvement.[5] Not since the crusades of St Louis (1248–70) can so much involvement be discerned nor such fragmentation and diversity in public opinion. The following example is symptomatic of the controversial nature of the issue. During a *Quodlibet* session before Henry of Ghent at the University of Paris (Christmas 1291), a question was raised whether a knight who threw himself on the enemy and thus lost his life, had thereby performed a 'magnanimous' deed, or should have escaped to save his life. As an example, the question referred to the siege of Acre.[6]

The treatises, chronicles, and the decisions of Church councils ordered by Nicholas IV, reveal a general, sustained, and vivid interest in the fate of the Holy Land. There was an intensification of support for a crusade on the one hand, and the rise of eschatological expectations, linked to a peaceful recovery of the Holy Land, on the other. This dual phenomenon now became the pattern of European reaction to events connected with the Holy Land or with the crusade. Immediately following the loss of the Holy Land this was reflected by chronicles as well as by apologetic treatises which described and analysed the events with the clear purpose of explanation and justification. Of the latter, three have survived, possibly a remnant of a larger

---

[4] R. E. Lermer, *Power of Prophecy: The Cedar of Lebanon Vision from the Mongol Onslaught to the Dawn of the Enlightenment* (Berkeley, Calif., 1983), 38–9.

[5] Characteristically enough, the only two accounts that can be described as impartial are those of two eyewitnesses: John of Villiers the master general of the Hospital and the so-called 'Templar of Tyre'. See John of Villiers in *Cartulaire général de l'ordre des Hospitaliers de St Jean de Jérusalem 1100–1310*, ed. J. Delaville le Roulx (Paris, 1894–1906), 4157; 'Templier de Tyre', in *Gestes de Chiprois*, ed. G. Raynaud (Geneva, 1887), 256–75.

[6] Henry of Ghent, *Aurea Quodlibeta* (Venice, 1616), fo. 394ᵛ.

amount which have disappeared. The treatises in question are: 'De Excidio Urbis Acconis', Thadeo of Naples's *Hystoria de desolacione civitatis Acconensis*, and the 'Epistolae' of Ricoldo of Monte Croce.

The Apologists: 'De Excidio Urbis Acconis', Thadeo of Naples, Ricoldo of Monte Croce

These treatises, with the exception of that of Ricoldo of Monte Croce, have been neglected by students of history. Yet they can be regarded as fully representative of the opinions prevalent among various classes and milieus of European society. The 'De Excidio' is the sole and lonely witness of opinions which were probably current among the popular and lower classes; the *Hystoria* of Thadeo of Naples was intended for the higher ranks of society; the 'Epistolae' of Ricoldo of Monte Croce emanated from a missionary milieu. It is usually contended that the three treatises were written as propaganda pamphlets.[7] However, an attentive re-examination and a comparison of these treatises with the *de recuperation Terrae Sanctae* group lead to different conclusions. Though the 'De Excidio' was explicitly written as an *exhortatio* and elements of propaganda were included in the two other treatises, their main concern was to explain the loss of the Holy Land to Christendom. The need for such explanations obviously became then more acute. Yet the problem did not differ from that of the beginning of the crusader movement: 'Ubi est Deus eorum?' (Ps. 115: 2.) To explain the catastrophe solely in terms of the phrase *nostris peccatis exigentibus* was not satisfactory and the authors of the treatises tried to see their way between this traditional justification and what might be called logical or material causes. Yet none of them really abandoned the biblical reasoning. The materialistic explanations were always accompanied by the basically religious exegesis. Crusader propaganda, though it came to the fore in the 'De Excidio' and the *Hystoria*, was secondary to the main aim which is apologetic and directed against the criticism of the Church and even of Providence.

[7] A. S. Atiya, *Crusade in the Late Middle Ages* (London, 1938), 30–4, 158–60.

'De Excidio Urbis Acconis'

The 'De Excidio' is known through a large number of Latin and French manuscripts of the last decade of the thirteenth and the beginning of the fourteenth centuries. As it was also copied by numerous chroniclers, by some even before 1300, it must have enjoyed a wide circulation especially in France and in England. Nothing is known about the author. His class-based and anti-aristocratic opinions suggest that he was a commoner, possibly a minor cleric. He stated that he was not an eye-witness of the fall of Acre, but that his information came from various people who were personally present. His sources included the account of John of Villiers and several papal bulls regarding the event.[8]

In the first part he dealt with the events prior to the siege of the city; in the second and last he described the siege and the fatal end. His explicit intention was to exhort believers to take revenge for the insults inflicted by the infidels upon Christ through the humiliation of the Holy Land. The whole treatise was pervaded by an Isaiah-like gloom; it had this in common with some other chroniclers and the *Hystoria* of Thadeo of Naples. The most interesting part of 'De Excidio' is, in our context, the epilogue, a pathetic exhortation to set out for the recovery of the Holy Land. It was actually an indictment of the heads of Christendom:

> Cry the daughter of Sion over this dear city . . . Cry over your chiefs, who abandoned you. Cry over your pope, cardinals, prelates, and the clergy of the Church. Cry over the kings, the princes, the barons, the Christian knights, who call themselves great fighters, but fell asleep not in the Valley of Tears, but in the Valley of Sin, who left this city full of Christians without defence and abandoned it, leaving it alone like a lamb among wolves.[9]

---

[8] 'De Excidio Urbis Acconis Libri II', in *Veterum Scriptorum et Monumentorum Amplissima Collectio*, ed. E. Martène and U. Durand (Paris, 1724–33), v. 758; S. J. Van den Gheyn, 'Note sur un manuscrit de *l'Excidium Acconis*, en 1291', *ROL* 6 (1898), 550–4; V. M. Le Clerq, 'Relation anonyme de la prise d'Acre en 1291', *HL* xx (1842), 79–89; J. F. Michaud, *Histoire des croisades* (Paris, 1824–9), v. 421–2.

[9] 'De Excidio', cols. 757, 759, 783.

Paraphrasing the apocryphal Book of Baruch, the author mercilessly exposed the heads of Christendom:

Some, mounted on their handsome horses, forgot the tribulations and the distress of Acre. Others, living in the midst of their own glories and mundane vanities, sat in chairs of iniquity. Instead of governing the church in the spirit of devotion and humility, they took money destined for pious aims, erected superb towers, built magnificent palaces, and decorated them with expensive paintings. To pay for such expenses, they took away by means legal and illegal the substance of the poor. Yet they were chosen from among the people as wise and able to govern the church and to be the rulers of the goods of Christ and his believers. But, alas, those who should have been so foreseeing lost their sight because of the importunity of their desires, the burden of their riches, the anxiety of avarice ... Others, in the bloom of their age degraded their reason by the softness of their souls and the baseness of their vices. They keep themselves busy all day by pursuing wild beasts with packs of hounds ... and that for capturing a poor wild boar or shabby fawn. Meanwhile they neglect their affairs, they let the reins of their government float free, thus exposing themselves to danger of an unglorious death ... Others of this kind who pretend to have the cause of God [*Dei negotium*] at heart, and to want to revenge his injuries, actually accumulate gold and silver, which they extort by violent demands from their subjects and their poor churches. All they do is to suppress by force neighbouring kingdoms and principalities and add them to their own kingdom. But meanwhile they neglect the principal cause, with which they pretend to be occupied. Alas what else can be said? Everywhere I see only ambition, avarice, and ill-fortune of the poor.

This deeply-felt, though long-winded, peroration terminates with a prayer for the recovery of the Holy Land.[10]

---

[10] Ibid. 783–4. The author refers by mistake to Habakkuk instead of Baruch. This entire paragraph is a paraphrase of Baruch 3: 16–18. G. Digard (*Philippe le Bel et le Saint-Siège* (Paris, 1936), vol. i, p. 128 n. 1) remarks that the epilogue of 'De Excidio' as well as other passages in the introduction are almost a literal transcription of the declarations against the carnal Church in commentaries on Jeremiah and Isaiah ascribed to Joachim of Fiore, and describes the threatise as 'Joachimiste'. This is only partly correct. It is true that 'De Excidio' includes pseudo-Joachimite doctrines. These doctrines became during the second half of the 13th cent. an integral part of the social radicalism of popular movements. As Digard himself points out (*Philippe le Bel*, vol. i, pp. 127–8), the Joachimites complained that the celestial Jerusalem was sacrificed for the sake of the earthly Jerusalem. The author of 'De Excidio', though he shared some Joachimite views, still sought the earthly Jerusalem. The treatise can thus hardly be described as 'Joachimite'.

Hypocrisy and indifference, the selfish search for personal glory, wealth, and luxury, as well as the avarice of the rich and of the upper classes in general, were thus presented as the cause of the loss of the Holy Land. The responsibility for the fatal event was put squarely on the shoulders of the leading classes in Christendom. The importance of the 'De Excidio' should, however, also be evaluated in a different context. It seems to be the only testimony which presented the events from the point of view of the 'poor'. The all-pervading tone was its outspoken hostility against the 'mighty'. Doubting the crusader zeal of the leading classes, it emphasized that of the poor. Implicitly it glorified the latter and in this sense it foreshadowed a mood which would launch the Crusades of the Poor in 1309 and in 1320. It is probably not just a coincidence that one of the French manuscripts of the 'De Excidio' originated in Picardy. In the Picardian version the attack on the rich is even more acrimonious than in the Latin manuscripts. Picardy, it should be remembered, played a leading part in the *Pastoraux* movement of 1251 and it was destined to play a part in the Crusades of the Poor of 1309 and 1320.[11]

In another manuscript, that of the Benedictine monastery of St Jacques of Liège, the spirit of social and religious dissatisfaction was emphasized by its two annexed Sibylline prophecies: a post-1291 version of the so-called 'Vision of Tripoli', translated into French north-eastern dialect, and the 'Prophecy of Merlin'. The 'Vision' is an example of the argument that much of the contents of medieval religious prophecy should be read as the expression of dissatisfaction with the present and hope for the future. It appeared first around 1239 in Hungary, at a time of great eschatological hopes and fears catalysed by the dramatic appearance of the Mongols. It purported to contain a divine message seen in a vision by a Cistercian monk during mass. Matthew Paris inserted it under 1239, the year of the defeat of Beit-Hânûn near Gaza. According to another version written between 1289 and spring 1291 it had been revealed to a monk (Franciscan, Premonstratensian, or Cistercian) in Tripoli on the eve of the fall of the city (1289). It is in yet another

[11] N. Cohn, *Pursuit of the Millenium: Revolutionary Millenarians and Mystical Anarchists of the Middle Ages* (London, 1970), 89–107. The French MSS of the 'De Excidio' are still unedited; fragments were published by Le Clerq, 'Relation', pp. 79–98.

form that it became highly popular and throughout the period c.1291–1487 appeared frequently in French and English sources. In the period 1291–1312 it appeared in almost identical versions in the anonymous continuation of Menko, the Premonstratensian abbot of Bloemhof (Floridus Hortus) near Verum in Frisland; in the chronicles of Eberhard, archdean of Regensburg; Gilles le Muisit (Aegidius Li Muisis), abbot of the Benedictine St Martin of Tournai and Weichard of Polhaim, later archbishop of Salzburg.[12]

The 'Vision' was, as already said, that of a monk who saw during mass a hand writing in letters of gold on the Corporal:

'The high Cedar of Lebanon will fall' [Ps. 29: 5; Zach. 11: 1], the city of Tripoli will soon be captured and destroyed. Acre will perish, Mars will dominate Saturn. Saturn will lie in ambush against Jupiter. The bat will suppress the lord of bees.[13] For fifteen years there will be one God and one faith. The other God will vanish and the Sons of Jerusalem will be liberated from captivity. A leaderless race will appear, a misfortune of the clergy and Christianity. The ship of St Peter will be tossed about by the waves but it will escape and rule in the Last Days of the World. There will be many wars, disasters, cruel famines, great mortality, and political upheavals. Then the countries of the Barbarians [or Saracens] will be converted. The Mendicant Orders and all the other sects [sectae] will disappear. The Beast of the Occident and the Lion of the Orient will conquer the entire world. And thus for fifteen years peace will rule the Universe. Then there will be a great crusade [commune passagium; communs passages] to the Holy Land. Victorious, Jerusalem will be honoured and the Holy Sepulchre glorified. At that peak of glory and peace Antichrist will appear.[14]

---

[12] Van den Gheyn, 'Note', pp. 550–6; Matthew Paris, *Chronica Majora*, ed. H. R. Luard (RS, London, 1872–83), iii. 538; Menko, 'Chronicon et Continuatio', *MGH SS* xxiii. 567–8; Weichard of Polhaim, 'Continuatio Weichardi de Polhaim', *MGH SS* xi. 811; Eberhard of Regensburg, 'Annales', *MGH SS* xvii. 605; Aegidius Li Muisis (Gilles le Muisit), 'Chronica', *Corpus Chronicorum Flandriae*, ed. C. de Smedt, ii (Brussels, 1841), 141; R. E. Lerner, 'Medieval Prophecy and Religious Dissent', *Past and Present*, 72 (1976), 3–24. The 'Vision of Tripoli' appears also in a Latin MS of the 14th cent., probably from the Cistercian abbey of Himmerod in the diocese of Treves. See J. Leclercq, 'Textes et manuscrits cisterciens dans les bibliothèques des États-Unis', *Traditio*, 17 (1961), 168–9; and see now Lerner, *Powers of Prophecy*, pp. 9–83, 213–19.

[13] According to Lerner (*Powers of Prophecy*, pp. 45–6), the victory of the bat over the bees is the victory of the Saracens over Christians.

[14] Van den Gheyn, 'Note', pp. 550–3; Lerner, *Powers of Prophecy*, pp. 72–83.

There was thus a mixed message in the 'Vision'. It predicted dreadful as well as wonderful future happenings. The fall of Tripoli and Acre would be followed by days of suffering: great battles, massacres, famines, plagues, and the fall of kingdoms; clergy and Christianity would be greatly threatened, the ship of St Peter would be rocked by the waves but would escape destruction; the mendicant orders would, however, be annihilated. But in the end the Holy See would triumph. There would be a successful crusade, the Holy Sepulchre would be venerated, and during this time of tranquillity news would be heard of Antichrist. This mixture of pessimism and optimism reflected much of the character of the prevailing views after the loss of the Holy Land: a certain anxiety due to the approach of the end of a century, and an idyllic vision of a peaceful return of Christianity to the Holy Land.

The 'Prophecy of Merlin' ('Prophetia Merlini'), which had circulated in the West since the twelfth century, belongs to the genre of political prophecies. Inserted in the 'De Excidio', it was a Latin adaptation of the prophecy concerning the Last Emperor as it appeared in the 'Romance of Merlin' (c.1120 –35). The latter was actually a version of the 'De Ortus et Tempore Antichristi'(c.950) of Adso of Montier-en-Der or a version of this version, since Adso's treatise was widely popular as well as extensively adapted and revised during the Middle Ages.[15] The prophecy told the story of a French king who was to become emperor. He would rule in peace and would go to Jerusalem and receive the *sceptrum et corona* on the Mount of Olives. The nations shut up behind the Black Sea by Alexander the Great would break their bonds and invade the Empire, but would be defeated by the Emperor. Victorious, his eyes glued to the words: 'Rex Romanorum vindicet sibi omne regnum terrarum', he would invade the land of the Saracens, lay it waste, and destroy their temples. The Saracens would be converted and the Cross of Christ would be raised on the temples of idols. The Jews would be converted and henceforth saved, and Judah and Israel would live in peace. After twelve

---

[15] See P. J. Alexander, 'Byzantium and the Migration of Literary Works and Motifs: The Legend of the Last Roman Emperor', *Medievalia et Humanistica*, 2 (1961), 47–82; B. McGinn, *Visions of the End: Apocalyptic Traditions in the Middles Ages* (New York, 1979), 66–87, 180–5.

years the emperor would return to Jerusalem, deposit his crown, and surrender his 'regnum et imperium' to God the Father and Jesus Christ, and die. His glorious sepulchre would be in Jerusalem; there would be no emperor and Antichrist would be born. In the original version, it should be mentioned, the emperor dwelt in Jerusalem until the appearance of Antichrist. He then climbed the hill of Golgotha, hung his crown upon the Cross, and died. Antichrist ruled for a time, but was slain by God before the Day of Judgement.[16] Thus the 'Prophecy of Merlin' stood for French claims to the imperial crown and the leadership of the crusading movement. Both were to become an integral part of the policy of Philip IV. While the claims to the empire were raised during the imperial vacancies of 1308 and 1313, the claims to lead the crusade had appeared by 1305.[17]

Both prophecies predicted the conversion of the Saracens. In the 'Vision of Tripoli' the Saracens were converted peacefully, apparently of their own free will. The prophecy concerning the Last Emperor made the Saracen conversion the result of conquest. In both prophecies the heathen were converted before the coming of Antichrist, leaving the world with the horrible anxiety of events which would occur after the Christian victory. Of the two theories of conversion the more popular in the last decade of the thirteenth century was that of the Last Emperor. As a matter of fact it was to be Ramon Lull's idea of the aims of the crusade.[18]

The presence of the two well-known and popular prophecies in a treatise like the 'De Excidio' shows that the loss of the Holy Land, like other catastrophies, created an atmosphere of unusually high eschatological pressure. In 1291 as in 1300, when it was believed that the Holy Land had been recovered by the Mongols for Christendom, the eschatological expectations were coloured by the close of the century: this was not a propitious time for launching a new crusade. A fierce protagonist of the crusade, Thadeo of Naples, would follow Humbert of

---

[16] Van den Gheyn, 'Note', pp. 555–6; Alexander, 'Byzantium', pp. 47–8.
[17] Below, ch. 6.
[18] R. E. Lerner, 'Refreshment of the Saints: The Time after Antichrist as a Station for Earthly Progress in Medieval Thought', *Traditio*, 32 (1976), 97–144, esp. p. 110.

Romans in his condemnation of those who believed in prophecies.[19]

Thadeo of Naples

Highly rhetorical and didactic, the *Hystoria de desolacione civitatis Acconensis* was intended for a different public and offered a more elaborate interpretation of the events. Though it was contended by A. S. Atiya that Thadeo of Naples wrote the earliest propaganda tract after the fall of Acre, a rather doubtful claim in itself, it seems more likely that, whatever the cause, his main interest was elsewhere. As the title indicates, the author attempted to explain God's fatal decision. The exhortation for the recovery of the Holy Land covered only three of its sixty-six pages, which suffices to invalidate the argument of Atiya that it is mainly a homily and an exhortation, with the events of 1291 as a mere excuse.[20]

The author refers to himself as 'magister civis Neapolitanus', who for several years resided in Syria. He wrote the *Hystoria* in Messina in December 1291. P. Riant assumed that he was possibly a mendicant friar, but the identity of Thadeo of Naples is still an enigma. His treatment of Joachim of Fiore makes it doubtful if he belonged to any monastic order and he was certainly not a mendicant.[21]

The *Hystoria* was presented as a letter addressed to Christendom. Its first chapter seems to be an inaccurate transcription of the encyclical *Dirum amaritudinis calicem*. Intending probably to convey a complete account of the siege, the author was soon distracted by his own rhetoric; the treatise therefore lacked any sort of structure. It certainly did not provide a chronological account of the events. On the contrary, Thadeo eulogized, deprecated, and described selected and non-related episodes of the last days of Acre. His sources of information were mainly

[19] For Humbert of Romans see P. A. Throop, *Criticism of the Crusade: A Study of Public Opinion and Crusade Propaganda* (Amsterdam, 1940), 174–5.
[20] Atiya, *Crusade*, pp. 32–3; but see P. Riant in his pref. to Thadeo of Naples, *Hystoria de desolacione civitatis Acconensis*, (Geneva, 1873), p. xix, and F. Pall, 'Crisades en Orient au Pas Moyen Âge: Observations critiques sur l'ouvrage de M. Atiya', *RHSE* 19 (1942), 544–5.
[21] Thadeo of Naples, *Hystoria*, p. 39, and Riant's pref., ibid., pp. ix–xi; cf. Atiya, *Crusade*, p. 31.

Italian merchants. He was certainly not original and the only incident not met with elsewhere was that of two hundred priests and members of religious orders, who without other arms than the Cross, hurled themselves on the enemy, only to be cut to pieces. Thadeo emphasized their bliss in ascending 'celestial Jerusalem'. This episode was paralleled by his detailed descriptions of the details of Nicholas of Hanapes the patriarch of Jerusalem, William of Beaujeu the master-general of the Temple, Matthew of Clermont the marshal of the Hospital, the defenders of the Templar Tower, and the Teutonic Knights. The victims were martyrs. Death for Christ was glorious: it meant eternal life with Christ. All those who deserted the city were severely condemned and most of all John of Grailly and the members of Italian communes. The former was accused of deserting Tripoli (1289) and then Acre. The communes were blamed not for quarrels, but for their selfishness, avarice, and lack of crusader zeal. It was the moral weakness of the Christians, as compared to the determination and zeal of the Saracens, that decided the fate of Acre. Thadeo shared with Ricoldo of Monte Croce and Fidenzio of Padua this awareness of the superior qualities displayed by the Moslems in their conduct of the siege.[22]

The responsibility for the fall of Acre and the loss of the Holy Land was squarely and equally placed on the shoulders of the citizens of Acre and of Christendom. Elaborately adorned with scriptural citations, especially from Jeremiah and Isaiah, the fall of Acre, which, at least, was not an innocent lamb among wolves, was presented as the Day of Judgement. Sinning by quarrels, delusions, lust, jealousy, ambition, and dissent, it was Christendom which deserted Acre. Thadeo's explanation is therefore in terms of the traditional phrase *nostris peccatis exigentibus*. His message was that the recovery as well as the loss of the Holy Land were predestined by Providence. Christendom, he argued, was warned by the prophecies of Jeremiah, Daniel, and Isaiah. Their relevance was explained through the Joachimite theory of 'double sevens', a set of parallels between 'seven periods' of the Old Testament and 'seven ages' of the Church under the New Testament. Thadeo apparently knew two trea-

---

[22] Thadeo of Naples, *Hystoria*, pp. 3–5, 15–16, 18, 25–8, and see above, pp. 96–7.

tises then attributed to Joachim of Fiore, the genuine Joachimite 'Expositio magni Abbatis Joachim in Apocalipsim' and the pseudo-Joachimite 'Super Hieremiam Prophetam' (c.1248). The 'Expositio', using the imagery of Revelation 9: 20, developed a theory of two Antichrists: one the satanic 'beast', defeated before the reign of the Saints; the other, Gog, defeated after the reign of the Saints but before the Last Judgement. This age would end with a period of trials under Antichrist; battles between the armies of the beast and the heavenly host, which would be followed by the defeat and capture of Satan. Thadeo placed the loss of the Holy Land in the setting of this last period, before the seventh age of the Church. The time of the trials began, in Thadeo's opinion, in the year 1290. This was according to Daniel 12: 11: 'And from the time that the continual burnt offering is taken away, and the abomination that makes desolate is set up, there shall be a thousand two hundred and ninety days', since on the basis of Ezekiel 4: 6, Daniel's 1,290 days should be reckoned as 1,290 years. The rule of the beast whom Thadeo identified with the sultan of Egypt, would last until the end of the century. Then it would be followed by the wordly sabbath, a period of perfect peace, the conversion of the infidels, the perfection of the ecclesiastical regime, as well as a new state of spiritual knowledge. In this perspective, integrated into the Joachimite seventh age of the Church, the fall of Acre and the loss of the Holy Land were inevitable, but, they were temporary losses only.[23]

Though influenced by the Joachimite concept of history, Thadeo did not accept its conclusions regarding the crusade. He was fully aware of its dangerous consequences and impact on the crusading spirit. Like other opponents of Joachimism he was not impervious to its influence, but did not accept its determinism. His views came close to those of the Joachimite Arnold of Villanova. The latter, a lay theologian and physician to Boniface VIII and then to Benedict XI, argued in his 'De Adventu Antichristi' (1297) that the Christian triumph over Antichrist would include the ultimate capture of the earthly Jerusalem. Like Arnold, Thadeo seems to have regarded the

---

[23] Thadeo of Naples, *Hystoria*, pp. 34, 39–48, 55, 59–60; Lerner, 'Refreshment', p. 119.

reform as an act of purification. They differed, however, in their plan of reform. Arnold of Villanova appealed to Philip the Fair (1297 and 1301) to become the divine instrument of the *renovatio mundi* and the recovery of the Holy Land. Thadeo believed that those functions lay with the pope. Accordingly he ended his *Hystoria* with a number of exhortations addressed to God, the pope, all the kings and princes, and finally Christendom at large. He prayed to God that the Christians might overthrow his cruel enemies; he exhorted the pope as the vicar of Christ to eradicate paganism and redress the injury to the Saviour so that his name might be honoured throughout the world and the Christians might return to the Holy Land; he then solicited kings and princes to cease their dissensions and act as one body in the bosom of the Church militant. The faithful were exhorted to avenge the bloodshed of Christians and to save by force of arms the Holy Land.[24]

The vehemence and passion with which the authors of the 'De Excidio' and the *Hystoria* exposed their views, were not only the result of individual temperaments, not even, or at least not only, an expression of deeply felt grief. They were also prompted by the urgent need to combat defeatist sentiments caused by the loss of the Holy Land. Such sentiments came to the fore with special force in the writings of Ricoldo of Monte Croce.

Ricoldo of Monte Croce

A Dominican of Santa Maria Novella in Florence, Ricoldo of Monte Croce (1243-1320) was one of the numerous mendicant friars who throughout the thirteenth and fourteenth centuries set out with enthusiastic faith, endless zeal, and sincere hope to convert the pagan or Moslem Orient. Author of numerous treatises, including his great polemical treatise 'Improbatio Alchorani' ('Refutation of the Koran') (*c.* 1310), and later one of the leading figures in the administrative hierarchy of his pro-

---

[24] Thadeo of Naples, *Hystoria*, pp. 47-8, 63-5; M. Reeves, *Influence of Prophecy in the Later Middle Ages: A Study in Joachimism* (Oxford, 1969), 314-15; Lerner, 'Refreshment', p. 134.

vince, Ricoldo was above all a missionary.[25] He set out on his mission in 1288, starting with a pilgrimage to the Holy Land. The news of the fall of Acre found him in Baghdad. His 'Epistolae commentatoriae de perditione Acconis' or 'Epistolae ad Ecclesiam Triumphantem' were written in 1291 during his wanderings in the East, which included Baghdad and Nineveh (Mossul).[26]

In Baghdad, where he was preaching, Ricoldo saw long convoys of Christians taken captive in Acre, making their way to remote parts of the Levant. He was told that more than 30,000 Christians were killed in one day in Acre. He learned from a nun 'worthy of belief', who had been captured in Acre, about the death of all the thirty Dominicans in the city. In Mossul he found on sale church utensils pillaged in Acre. Here he bought a missal and Gregory's 'Morals on Job'. All around him Islam was prospering the flourishing. Now, after the fall of Antioch, Tripoli, and Acre, the Saracens claimed in public the impotence of the Christian God. Jesus, they proclaimed, was only a man and could not aid the Christians against Mohammed. The fortunes of Mohammed surpassed the fortunes of Christ. Not only Saracens, but Jews and Mongols, mocked the Christians: 'evil is he who can help and yet does not want to help his own'. Many Christians drew extreme

[25] On his life and writings see R. Röhricht's introd. in his edn. of Ricoldo of Monte Croce, 'Epistolae V commentatoriae de perditione Acconis 1291', *AOL* 2 (1884), 258–63; P. Mandonnet, 'Fra Ricoldo de Monte-Croce, pelerin en Terre Sainte et missionaire en Orient', *Revue biblique*, 2 (1893), 46–66, 182–202, 585–607; A. Dondaine, 'Ricoldo de Monte Croce', *Archivum Fratrum Praedicatorum*, 37 (1967), 119–79, esp. pp. 133–4. The 'Refutation' appears under several titles: 'Tractatus contra legem Sarracenorum', 'Tractatus contra legem Mahometi'; 'Confutatio Alchorani'; 'Pro pugnaculum fidei'. See also J. Henninger, 'Sur la contribution des missionaires à la connaissance de l'Islam', *Neue Zeitschrift für Missionswissenschaft*, 9 (1953), 176–8; N. Daniel, *Islam and the West: The Making of an Image* (Edinburgh, 1960), 126–7, 196–8, 215–45, 404–5; id., *Arabs and Medieval Europe* (Edinburgh, 1975), 198–9, 215–17; R. Southern, *Western Views of Islam in the Middle Ages* (Cambridge, Mass., 1962), 68–70.

[26] The 'Epistolae' carry therefore the inscription 'data in Oriente'. According to Röhricht the letters were written in 1298 (see his pref. to the 'Epistolae', p. 260 n. 16). M. Monneret de Villard ('La Vita, le opere e i viaggi di frate Richoldo da Monte Croce OP', *Orientalia christiana periodica*, 10/11 (1944/5), 266–7) dated the letters as 1301. According to Mandonnet ('Fra Ricoldo', pp. 588–90), the prologue and the first letter were written in Baghdad and the rest in Nineveh. The 'Epistolae' consist of a short preface and five letters which are such in form only. The treatise is rather a kind of collection or record of reflections and prayers. In a sense it is also a personal diary presented in epistolary style.

conclusions and converted to Islam. For several years Ricoldo was alone in the remote parts of the Orient with no news from his order. He, 'miser et peccator', sent to preach the faith of Christ to the Saracens, saw around him not only Mongols and others accepting Islam but even Christians converting to that victorious religion. Nuns were made slaves or concubines; the sea near Tripoli and Acre turned red from Christian blood and in the place where God had walked as man, only the voice of Mohammed was heard. His outlook as to the prospects of Christianity was hardly optimistic: 'if the Saracens continue to do as they did in two years to Tripoli and Acre, in several years there will be no Christians left in the whole world'. Is it possible, he asked, that 'God wants the whole world to become Saracen?'[27]

Ricoldo experienced a crisis of faith. Desperate, he felt as if the entire world, including God, had turned against him. A devoted Christian, bent on converting the infidels to the religion of the only Almighty and True God, he came to doubt whether his mission and vocation were in accordance with the will of a god of whom he despaired, and who seemed to have abandoned Christendom in general and himself in particular. Like Job, he doubted Providence and divine justice. It is with this crisis of faith that his 'Epistolae' were concerned. They were addressed in 'amaritudine animae' to God and to the Celestial Court and sought an answer which he could not provide himself. 'Why has God permitted that such havoc, massacre, and eviction should befall Christendom and why does he permit the Saracens to enjoy such temporal prosperity?' The Saracens ruled in peace and without contradiction from India to the western outskirts of the world. They possessed the most fertile parts of the universe. 'Why should these beasts, who destroyed churches, murdered Christians and Christian missionaries, enjoy such power against Christendom for almost seven hundred years?' Ricoldo was unable not to voice his misgivings: 'I wonder, seeing that once you wanted to spare the

---

[27] Ricoldo of Monte Croce, 'Epistolae', pp. 264–6, 269, 270, 272, 277, 280–1, 284, 291, 293. There were about 30 Dominicans in Acre on the eve of its fall. The circumstances of the death of one of them, a convert Lupus de Cuscia, as recorded for 1291 in the *necrologia* of Santa Maria Novella of Florence, proves as true the story of Thadeo of Naples about the 'army of monks'. See Mandonnet, 'Fra Ricoldo', p. 53.

entire city of Sodom, if therein were found ten just men, could not ten just men be found in Tripoli or Acre?' This was a voice of protest and resentment. In Ricoldo's opinion Acre should have been saved for the sake of the Dominican brethren, Patriarch Nicholas of Hanapes, or William Beaujeu the brave master-general of the Temple alone.[28]

Ricoldo's monologue, his passionate and yet didactic and scholarly treatment of the Koran and the Scriptures, expressed his inability to accept the interpretation of *nostris peccatis exigentibus*. He found some consolation in Gregory I's 'Morals of Job' and he sought the truth in the Scriptures, where he learnt that salvation is not always granted. He begged God, the Celestial Court, and the Holy Virgin to put an end to the perfidy of Mohammed and the power of the Saracens and make possible their conversion. Bent on converting the infidels, Ricoldo was one of the best exponents of what many others must have felt after the fall of Acre—total bewilderment. His painfully probing questions touched a central problem of the crusader movement: 'Why did God allow holy men or preachers like St Dominic, St Francis, St Louis of France and others, pious kings and barons, to waste so many lives and efforts in vain?' None of them had succeeded in having victory over the beast Mohammed who was continuing to devour Christians and to force them to abandon their faith. This was as powerfully expressed by Ottokar von Steiermark, a German *knappe* who wrote after the fall of Acre in his 'Osterreichische Reimchronik': 'Sag herre got, sag an warumt hast du daz getan?' The questions were not new. They had already been asked after the Second Crusade. Ricoldo, another Job, was fully aware of his inability to provide answers.[29]

The three treatises written almost immediately after the fall of Acre and the loss of the Holy Land were an attempt to explain the recent events to Europeans. Each was at the same time the

---

[28] Ricoldo of Monte Croce, 'Epistolae', pp. 264–71, 276, 285, 287, 294.
[29] Ibid. 268–70, 288–92, 296; Ottokar von Steiermark, 'Oesterreichische Reimchronik', ed. J. Seemüller, *MGH SS Deutsche Chroniken*, v (Hanover, 1890–3), p. 698, ll. 52359–60; G. Constable, 'Second Crusade as seen by Contemporaries', *Traditio*, 9 (1953), 266–76; B. Z. Kedar, *Crusade and Mission: The Interplay between Two European Approaches toward the Muslims in Medieval Times* (Princeton, NJ, 1984), 199–200.

expression of a personal experience of the calamity. They did not actually explain, rather justified the course of events. Their explanations, except for very rare exceptions, did not add to the moralizing tone of preaching. Almost everything was said in the same tone, using the same arguments as after the Second Crusade. In the best of cases they expanded the list of sins and sinners.

Despite the fact that the fall of Acre had been foreseen for the last few decades, it is obvious that the event came as a shock. It was difficult to absorb it and, as there was little to add to the current criticism, two of the authors of our treatises, the author of the 'De Excidio' and Thadeo of Naples, not at ease with the common accusations, added to the orthodox justification of the events and his eschatological vision of the events supplied an element of metaphysical chronology in which the calamity became part and parcel of the history of all humankind, a tragic but necessary phase on the road to salvation. The loss of the Holy Land was thus justified and explained by the predestined history of the world. This latter, however, included Christian domination of the Holy Land as well as the conversion of the infidels. Although predestined, both had to be accomplished by human endeavours. The loss was temporary, its recovery would soon follow. This left the idea of the crusade, and even the idea of the mission, intact. Both were instruments to redress an ephemeral reverse. It was thus only Ricoldo of Monte Croce, who, though not without reservations, clung to orthodoxy. But even for him the loss was a temporary setback. This he deduced from his concept of divine justice.

Chroniclers

Of the three major outbreaks of 'pro-crusade' during the period 1274–1314 it was that which immediately followed the fall of Acre which was the most powerful. This is irrefutably proved not only by almost all extant chronicles which include references to the loss of the Holy Land but also by the quality of the accounts and interpretations which are elaborate and detailed, repeating time and again the texts of the papal bulls the *Dirum amaritudinis calicem* and the *Dura nimis*.[30] The event was described as a tragic disaster and the mood of the chroniclers was that of the prophets of doom of Israel. Some saw themselves as Jeremiah lamenting the ruin of Jerusalem.[31] Acre, they said, was like Jerusalem of old, a place of refuge. Jerusalem had been a refuge for proselytes and Jews; Acre was for all the Catholic nations.[32] This image, already current before the fall, was picked up by the papal propaganda in 1291 and thus, perhaps, transmitted to the chroniclers. In the *Dirum* Acre appears as the asylum of Christendom; the chronicles describe it, perhaps more appropriately, as the place of refuge for Christendom overseas.[33]

This positive image of Acre was not shared by all. Acre was frequently presented as a den of vice, a contemporary Sodom. Its inhabitants, seen as consumed by lust, were regarded as criminals, drunkards, jugglers, and conjurers collected from all over Christendom. They were accused of arrogance (*superbia*) and greed (*avaritia*) as well as gross extravagance (*luxuria*) and frivolous ways of life (*levitas*).[34] This unattractive image of Acre

[30] Bartholomew Cotton, *Historia Anglicana: necnon ejusdem Liber de archiepiscopis et episcopis Angliae*, ed. H. R. Luard (RS, London, 1859), 200–1, 203–4; Walter of Guisborough, *Chronicle*, ed. H. Rothwell (London, 1957), 231–2; William le Maire, 'Gesta', in *Spicilegium sive collectio veterum aliquot scriptorum*, ed. L. d'Achery (Paris, 1723), ii. 179–80.

[31] John of Winterthur, 'Chronica', ed. F. Beathgen, *MGH SRG* iii. (Berlin, 1924), 39–42; 'Annales prioratus de Dunstaplia', in *Annales monastici*, ed. H. R. Luard (RS, London, 1864–9), iii. 366–7; 'Annales monasterii de Osneia', in *Annales monastici*, iv. 331; 'Annales Hibernie', in *Chartularies of St Mary's Abbey, Dublin*, ed. J. T. Gilbert, ii (RS, London, 1884), 320; 'Annales monasterii de Waverleia', in *Annales monastici*, ii. 110; *Flores Historiarum*, ed. H. R. Luard (RS, London, 1890), iii. 74; Walter of Guisborough, *Chronicle*, p. 228; *Annales Angliae et Scotiae*, ed. H. T. Riley (RS, London, 1865), 400–1.

[32] 'Annales de Waverleia', p. 410; see also *Flores*, iii. 74; *Chronicon de Lanercost 1201–1346*, ed. J. Stevenson (Edinburgh, 1839), 138.

[33] William of Tripoli, 'Tractatus de Statu Saracenorum', ed. H. Prutz, in Prutz, *Kulturgeschichte der Kreuzzuge* (Berlin, 1883), 589; Nicholas IV, *Registres de Nicholas IV*, ed. E. Langlois (Paris, 1886–91), no. 7625; Balduin of Ninove, 'Chronicon', *MGH SS* xxv. 546; William of Nangis, *Chronique latine de Guillaume de Nangis de 1113 à 1300 avec les continuations de cette chronique de 1300 à 1368*, ed. H. Géraud, i (Paris, 1843), 278. See also Ludolph of Suchem, *De Itinere*, p. 44; Mandeville, *Mandeville's Travels*, ed. M. Seymour (Oxford, 1967), 21; D. Hay, *Europe: The Emergence of an Idea* (New York, 1966), 56–61. John of Victring 'Chronicon Carinthiae', in *Fontes Rerum Germanicarum*, ed. J. F. Bohmer (Stuttgart, 1854–68), i. 329) described Acre as 'civitas clarissima, Christianorum domicilium'.

[34] Bohemond of Trier, 'Gesta', *MGH SS* xxiv. 474–5; 'Annales de Dunstaplia', p. 366; Walter of Guisborough, *Chronicle*, pp. 228–9.

had also been current in the thirteenth century; it appears, for example, in John of Joinville's 'Life of Saint Louis'. Accusations of arrogance, greed, and extravagance were already to be found in the diatribes of Gerhoh of Reichsburg's 'Libellus de Investigatione Antichristi' after the Second Crusade (c.1161–2), and were a recurrent motive in official ecclesiastical propaganda for the reform of the Church on the one hand, and in the apologetics of the crusades on the other.[35]

The chronicles, just like the treatises which dealt with the loss of the Holy Land were dominated by the urgent need to explain and even to justify the events. And yet, despite the numerous references to the fall of Acre, only a few chroniclers made any serious attempt to confront the challenge. Their explanations and justifications were on the whole rather shallow if compared with the already analysed treatises of the apologists. They were more in the realm of emotional generalities, and were unsatisfactory as explanations of cause and effect. It is not easy to be conclusive about the qualitative difference between the apologists and the chroniclers, although there was a marked difference in specialized knowledge. The chroniclers, with a few exceptions discussed below, explained the tragedy in terms of traditional or self-evident causes. The apologists, more ambitious, attempted to analyse it in a more profound way and to draw far-reaching historical conclusions.

In addition to the traditional explanations in terms of *nostris peccatis exigentibus* and *Deus le vult*, the chronicles adduced the state of Outremer, the behaviour of the Italian crusades in Acre, the interests of the papacy and of other groups involved, and finally the events of the Last Days of the World as the causes of the fall. The condition of Outremer was most frequently described in terms of corruption and of the never-ending jealousies and conflicts between the military-orders, especially the Hospital and the Temple.[36] Additionally, the

[35] John of Joinville, 'Life of Saint Louis', in *Chronicles of the Crusades*, trans. M. R. B. Shaw (London, 1963), 316–17; J. Prawer, *Histoire du royaume latin de Jérusalem* (Paris, 1969), 391–2; Coh, *Pursuit of the Millenium*, pp. 101–2; Constable, 'Second Crusade', pp. 273–4.

[36] Weichard of Polhaim, 'Continuatio', p. 813; Eberhar of Regensburg, 'Annales', p. 594; 'Annales de Dunstaplia', p. 366.

weakness or the lack of central power and its corollary, the multitude of competing lordships, were also stressed, as were national differences, by men like the Dominican Ptolemy of Lucca, as the main causes of the calamity. He wrote that the seven lords of the city, namely the Templars, Hospitallers, the Teutonic Knights, the consul of the Pisans, Henry II King of Cyprus and Jerusalem, King Charles I of Sicily, and Patriarch Nicholas of Hanapes disagreed not only about the system of government, but even about the defence of the kingdom.[37] Similarly, for Bartholomew of Neocastro the city on the eve of its fall suffered from the quarrel of the Pisans with the Venetians and of the Hospitallers with the Templars, whereas the newly arrived Italian crusaders 'Baccho vacabant'.[38] Giovanni Villani, who wrote in the 1330s or 1340s, sums up in a sense the Italian views (although one can detect here an affinity with Jacques de Vitry), by claiming that Acre, on the eve of its fall, had seventeen different jurisdictions.[39]

Another aspect of the state of Outremer as projected by the chronicles is of special interest and relates to the attitude of the Franks to the Moslems. They were blamed for a policy of tolerance or appeasement towards the infidels; even more, they were accused of co-operation and connivance with and betrayal of the city to the sultan of Egypt. They concluded truces with the infidels, traded with them, and above all trusted their promises. Such accusations were not entirely new. Since Freidank's harsh verses at the time of the crusade of Frederick II, such accusations had commonly been made by European visitors to the Orient.[40] A very different chord was struck by the anonymous author of the 'Chronicon Sampetrinum'. Although he referred to the state of Outremer as the main cause of disaster, his was a rather more comprehensive view, because he pointed out the basic lack of balance of power between the Franks and their Moslem antagonists,

---

[37] Ptolemy of Lucca, 'Annalen', ed. B. Schmeidler, *MGH SRG* viii (Berlin, 1955); John Elemosina, 'Liber Historiarum', in Golubovich, ii. 109.
[38] Bartholomew of Neocastro, 'Historia', p. 131.
[39] Giovanni Villani, 'Historia Universalis', *RIS* xiii. 337–8.
[40] 'Chronicon Estense', *RIS* xv. 50; John of Victring, 'Chronicon', p. 328; Freidank, *Freidanks Bescheidenheit*, ed. Panniez (Leipzig, 1878), 126.

besides the determination of the sultan of Egypt to take Acre.[41]

Pondering the causes of the fall of the kingdom, another group of chroniclers assigned them to even more general (one may say, less obvious) but probably more deep-seated and long-term trends. They may be summarized as the estrangement of the European community from its progeny in Outremer. Accusations were hurled against the papacy, the cardinals, and the secular rulers of Europe. They were accused of being so preoccupied with their own ambitions, interests and conflicts that they forgot and neglected the Holy Land.[42] The Italian chroniclers, who particularly mourned the loss of Italian lives and the annihilation of Italian colonies, ports, and markets in Outremer, turned their grief into an indictment of the papacy for sending the Italians to Acre and thus provoking the wrath of the sultan.[43] The famous incident of the attack on Moslems in Acre's market-place was dealt with also by Bernard Gui, one of the most prominent Dominican historians at the turn of the century, who referred to the Italian crusaders as the 'fools' sent by Nicholas IV. Another Dominican, Francesco Pipino, described them as 'pseudo-Christians'.[44] William of Nangis, the keeper of the records (*custos cartarum*) of the French crown (1285–1300), who wrote the official version of the history of the French monarchy, stressed the fact that the Italians were sent to Acre by the pope 'against the advice of the military orders, the Temple, and the Hospital'.[45]

Accusations against the papacy become more general in the work of Bartholomew of Neocastro, who wrote a detailed account of the fall of Acre, based on the story of one Arsenius, a Greek monk of the Order of St Basil, who during a pilgrimage himself saw the calamity and presented the pope with the sad news of the event. On that occasion, we are told by

[41] 'Chronicon Sampetrinum', in *Geschichtsquellen der Provinz Sachsen*, ed. B. Stubel, i (Halle, 1870), 126–8. For the determination of the sultan to conquer Acre, see also John of Winterthur, 'Chronica', p. 41.

[42] Menko, 'Chronicon', p. 567; John of Victring, 'Chronicon', p. 329.

[43] James Doria, 'Annales Ianuenses', in *Annali genovesi de Caffaro e dei suoi continuatori*, ed. C. Imperiale de Sant'Angelo, v (Rome, 1927), 130, 140; Francesco Pipino, 'Chronicon', *RIS* ix. 733.

[44] Bernard Gui, 'Flores Chronicorum', *RHGF* xxi. 701; Francesco Pipino, 'Chronicon', pp. 733–4.

[45] William of Nangis, *Chronique*, i. 274.

Bartholomew, Arsenius addressed the pope thus: 'Holy Father, if thou hast not heard our sorrow, out of the bitterness of my heart will I reveal it. Would to God that thou hadst not been so intent on the recovery of Sicily'. The fall was due not only to the 'peccata populi' but also, to the 'Romanae sedis inconstantiam'. 'Thus it was a real miracle', said Arsenius to the pope, 'that God did not permit the island of Cyprus to be taken by the infidels'.[46] Bartholomew, it seems, was not alone in connecting the fate of the Holy Land with that of Sicily, a connection to be stressed later by Boniface VIII.[47] Ottokar von Steiermark went back to Constantine the Great, blaming the emperor for the fall of Acre on account of the 'Donation of Constantine', which had ensnared the papacy with Sicily.[48]

Yet, despite these particular accusations, the tone of criticism on the whole was unexpectedly mild, especially when compared with such criticism as followed the Second Crusade or with that prevalent in the thirteenth century.[49] This mildness is surprising in view of the evident emotional stir caused by the loss of the Holy Land. Perhaps the tragedy of the moment checked more severe criticism. This left, as always, the traditional justification in terms of *nostris peccatis exigentibus*. Either by itself or together with additional explanations, the phrase appeared in most of the chronicles and thus became the most current and popular explanation of the disaster.[50] The popularity of this old theme—the most frequently recuring in the

[46] Bartholomew of Neocastro, 'Historia', pp. 131–3.
[47] Boniface VIII, *Registres de Boniface VIII*, ed. G. Digard et al. (Paris, 1884–1939), nos. 3917, 4395.   [48] Ottokar von Steiermark, *Reimchronik*, p. 686, ll. 51589–92.
[49] For the criticism of the crusade, see esp. the works of Throop and Siberry cited in the Bibliography. For the criticism of the Second Crusade, see Constable, 'Second Crusade', pp. 213–81; J. E. Siberry, *Criticism of Crusading 1095–1274* (Oxford, 1985), 77–81.
[50] Peter of Dusberg, 'Cronica Terre Prusie', in *Scriptores rerum Prussicarum*, i. 207; John Elemosina, 'Liber', p. 104; Bartholomew of Neocastro, 'Historia', p. 133; Bohemond of Trier, 'Gesta', p. 475; John of Winterthur, 'Chronica', p. 42; Eberhard of Regensburg, 'Annales', p. 594; Walter of Guisborough, *Chronicle*, p. 228; *Chronicon de Lanercost*, p. 140; Giovanni Villani, 'Historia', cols. 337–8. See also Marino Sanudo Torsello, 'Liber Secretorum Fidelium Crucis', in *Gesta Dei per Francos*, ed. J. Bongars (Hannau, 1611; repr. Jerusalem, 1973), ii 231–2; Ludolph of Suchem, *De Itinere*, p. 47; E. Stickel, *Fall von Akkon: Untersuchungen zum Abklingen des Kreuzzugsgedankens am Ende des 13. Jahrhunderts* (Berne and Frankfurt-on-Main, 1975), 190–211; *Councils and Synods with other Documents relating to the English Church*, ed. F. M. Powicke and C. R. Cheney (Oxford, 1964), ii. 1109. For the origins of the concept and its popularity in the 12th and 13th cents. see Siberry, *Criticism*, pp. 69–108.

official papal apologetics of the crusades—proves perhaps that both P. A. Throop and E. Stickel exaggerated its inadequacy.[51]

Some contemporaries of the fall of Acre, perhaps the more thoughtful among the devout, found their consolation in the last resort, as happened during other periods of doom and adversity, in the belief that the tragedy was God's will (*Deus le vult*). Faced with the loss of the Holy Land, they resigned themselves to the inscrutability of God, as had Humbert of Romans a generation earlier when faced with the death of Louis IX. This attitude, which particularly characterized the German chroniclers, was perhaps intended to combat the defeatist opinions which verged on saying that Christ was unable to overpower Mohammed.[52]

Yet another explanation of the events, which could have justified and consoled, was that which inserted the loss of the Holy Land into the eschatological framework of the Last Days of the World. This again was a common German explanation. Some of the German accounts of the fall of Acre were accompanied by the 'Vision of Tripoli' and, with the apparently sole exception of the manuscript of St Jacques of Liège, the 'Vision' appeared in the early 1290s only in the German chronicles.[53] This goes a long way towards explaining the fact that when the idyllic vision of the peaceful return of the Holy Land to western Christianity seemed to materialize in 1300, the German response was the strongest in Europe.

The chronicles reflect on the whole a homogeneous attitude to the events and there were no meaningful 'class' or 'national' differences. The chronicles presented a basically uniform image of Acre, and Outremer. 'National' differences were, however, to be found in the reports of the Church councils convoked by papal request to formulate their advice regarding the means of the recovery of the Holy Land.

---

[51] P. A. Throop, 'Criticism of Papal Crusade Policy in Old French and Provençal', *Speculum*, 13 (1938), 379–80; Stickel, *Fall von Akkon*, pp. 214–15.

[52] John of Winterthur, 'Chronica', p. 42; Bohemond of Trier, 'Gesta', p. 475. See also Throop, 'Criticism of Papal Crusade Policy', p. 380 and n. 3.

[53] Above, n. 12; Erich Olaus, 'Chronica', in *Scriptores rerum Suecicarum medii aris*, ed. E. M. Fart (Uppsala, 1818–71), ii. 70, records for 1291 that the Dominicans of his city Uppsala, preached the nativity of Antichrist in Jerusalem. This perhaps points to the fact that the eschatological expectations following the loss of the Holy Land may rather be defined as a north European phenomenon.

Church Councils

The advice proffered by Church councils convoked late in 1291 or early in 1292 all over Christendom added a different dimension to the *de recuperatione* treatises as to what was considered to be the best means to recover the Holy Land. It is rather unfortunate that though such councils were held everywhere we know about the decisions of only a few of them. Yet seeing their general conformity it is possible to make some generalizations. As a whole they aimed at practical means and solutions for the conquest of the Holy Land and for its preservation once liberated. These voices came from prelates with no such particular interest in the cause of the crusade as that of the authors of *de recuperatione* memoirs.

The reports reflected a basically common concept of the future crusade. The insistence upon the general pacification of Europe including the Italian maritime powers,[54] upon the election of an emperor 'qui manu armata discordes ad pacis revocet unitatem',[55] upon the unification or merger of the military orders of the Holy Land,[56] and the nomination of a single leader of the future expedition[57] were regarded as necessary preliminaries for the launching of the crusade. Obviously, common to all was the concern for the necessary funds as well as for adequate manpower for the expedition and for the defence of

---

[54] William of Nangis, *Chronique*, i. 279 (apparently refers to the Council of Reims); Digard, *Philippe le Bel*, vol. ii, no. 13, pp. 281–2 (the decisions of the Council of Sens); *Councils and Synods*, ii. 1105 (the decisions of an unspecified English diocese), 1107, 1110–11 (the report of the Council of Canterbury).

[55] John of Thilrode, 'Chronicon', *MGH SS* xxv. 581; 'Annales Blandinienses', *MGH SS* v. 34 (both report the decisions of the Council of Reims); *Councils and Synods*, ii. 1105 (the decisions of the clergy of Norwich and of an unspecified English diocese), 1110 (the report of the Council of Canterbury); Digard, *Philippe le Bel*, vol. ii, p. 282 (the report of the prelates of Sens): Bartholomew Cotton, *Historia*, p. 211 (the report of the Council of Lyons).

[56] 'Chronicon Salisburgense', in *Scriptores rerum Austriacarum veteres*, ed. H. Pez (Leipzig, 1721–45), i. 390; Eberhard of Regensburg, 'Annales', p. 594; Weichard of Polhaim, (Continuatio', p. 813 (all sources refer to the Council of Salzburg); Mansi, xxiv. 1079 (the Council of Milan); *Councils and Synods*, ii. 1099, 1108, 1113 (the decisions of the the Council of Canterbury); Bartholomew Cotton, *Historia*, pp. 213 (the Council of Lyons), 214–5 (Council of Arles); John of Thilrode, 'Chronicon', p. 581 (Council of Reims).

[57] *Councils and Synods*, ii. 1104 (Council of Norwich), 1111 (Council of Canterbury); Digard, *Philippe le Bel*, vol. ii, pp. 281–2 (Council of Sens); John of Thilrode, 'Chronicon', p. 581 (Council of Reims).

the Holy Land when it had been recovered. The raising of funds, however, resulted in different proposals. The Fathers of the Council of Reims proposed that a force of knights and foot-soldiers should be permanently maintained in the recovered Holy Land. This should be financed from the properties of the Templars and the Hospitallers, but if their funds did not suffice, the whole of Christendom should contribute to the expenses.[58] An entirely different view was taken by the Council of Sens which proposed for this purpose a series of continuous and repeated expeditions as well as a severe cutting of all superfluous expenses. The same council also suggested that everybody—clergy, laity, and Jews—should be taxed.[59]

The proposals concerned with the maintenance of the Holy Land, when it had eventually been recovered, were sporadic before 1291, but became more frequent in the decades that followed the fall of Acre. The importance of Europe's earlier failure to supply the Latin Kingdom with material aid and adequate manpower now came to be generally understood. Yet both the French and the English prelates were reluctant to submit to additional taxation to finance the expedition. The Council of Canterbury which met at the New Temple in London and at Lambeth from 13 to 16 February 1292 was quite outspoken as to the sources of finance: they were to consist of the property of the Temple and of the Hospital, because they were originally endowed for the defence of the Holy Land; a general six years' tithe; and, finally, the resources of the nobility, knights, and other crusaders. Those willing to settle in the Holy Land, said the prelates of Canterbury, should rely on themselves and live of their own. The laity should be encouraged to contribute generously for the cause of the Holy Land and the funds collected by the papal envoys should be spent solely on its recovery.[60] The councils of Reims, Sens, and Arles advised on similar lines. At Reims it was suggested that all exemptions from the crusader *subsidium* should be cancelled, whereas at Sens and Arles it was insisted that the laity should be made to contribute to the crusade along with the clergy.[61]

[58] John of Thilrode, 'Chronicon', p. 581 (Council of Reims).
[59] Ibid. 582 (Council of Sens).   [60] *Councils and Synods*, ii. 1107–13.
[61] John of Thilrode, 'Chronicon', pp. 581–2; Bartholomew Cotton, *Historia*, pp. 214–15; Digard, *Philippe le Bel*, vol. ii, p. 283.

Despite the similarity of the councils' decisions, French and English councils reflect an adherence to their 'national' interests and to those of their monarchs. The recommendations of the Council of Sens, as G. Digard has demonstrated, were formulated in collaboration with the French court, and clearly reflected court opinion by recommending the organization of a crusade against Sicily and its allies. Philip IV, it is known, was in 1291–2 interested mainly in a crusade against Aragon, which should be launched before that to the Holy Land. Pope Nicholas IV, on the other hand, argued that the recovery of Sicily should not delay that of the Holy Land. The Council of Sens, as well as that of Lyons, obviously supported Philip against the pope. The prelates of Sens recommended the restoration of peace between Philip of France and Charles II of Anjou on the one hand and James II of Aragon on the other. Otherwise, they argued, the French knights would be unable to depart for the Holy Land and would be deprived of their natural leader. Peace must therefore be established, and if James of Aragon and his allies should persist in their conduct, it was against them that a crusade of all Christendom should be directed.[62]

Whereas the councils of Sens and Lyons insisted on the French king or on another Frenchman as the head of the expedition,[63] the English councils demanded this position for their own monarch Edward I. He should be assigned a tithe not only from England but from the whole of Christendom. Moreover, he should be granted the command over the Templars, Hospitallers, and other military orders which should be merged into one single order![64] On the whole, therefore, the opinions of the English prelates corresponded to those of Edward I as the opinions of the French prelates reflected the position of Philip IV.

[62] Digard, *Philippe le Bel*, vol. i, p. 139; vol. ii, pp. 281–2; above, ch. 2.
[63] Digard, *Philippe le Bel*, vol. ii, pp. 281–2; Bartholomew Cotton, *Historia*, pp. 210–12; It is worth mentioning that the appointment of the king of France as a leader of the crusade was recommended also by the Council of Milan. This suggestion was, it seems, the result of antagonism to the king of the Romans, but may also point to the reputation of Philip IV as *defensor fidei*. See Mansi, xxiv. 1079; Hefele and Leclerq, *Histoire des conciles*, vi. 327; below, ch. 5.
[64] *Councils and Synods*, ii. 1105 (the report of an unspecified English diocese), pp. 1112–13 (the report of the clergy of Canterbury).

Comparing the decisions of the Church councils with the *de recuperatione* treatises we find that they had much in common. There were, for example, certain preliminaries, such as the nomination of a single leader as the warrior-king, the pacification of Europe, the unification of the military orders of the Holy Land, and the provision of adequate financial aid and manpower. There were also points of similarity with other ideas which dominate the learned treatises. Thus at least one council, that of Milan, convoked by Archbishop Otto Visconti (26 November 1291), recommended an embargo on commerce with the Levant as a preliminary of the crusade.[65] Another, that of Sens, argued that, to ensure adequate provisions, the host should go through Byzantium and therefore recommended that 'Grecos ... ad plenam Romane Sedis obedientiam revocari'.[66] The Council of Canterbury, possibly under the influence of its Franciscan archbishop John Pecham, recommended dispatching to the East, on the eve of the crusade, Arabic-speaking 'spiritual warriors' to work there for the conversion of the infidels. This was much in tune with the ideas of Ramon Lull.[67] On one point the prelates were even more radical than the individual planners. Fidenzio of Padua and at that time Ramon Lull did not suggest in 1291 the merging of the military orders but this demand appeared in almost all the Church councils and it would reappear after 1292 in most of the crusade-plans. The councils, as we have seen, expressed relatively little or no concern for the 'technical' details of the crusade. This was, it seems, rather the result of the conviction that such matters should be left to experts than because of an unqualified agreement with the policy of the existing crusade.[68]

Conclusion

The abundance of references to the loss of the Holy Land shows that the event, though regarded as an ephemeral episode, a

---

[65] Mansi, xxiv. 1079.
[66] Digard, *Philippe le Bel*, vol. ii, p. 282.
[67] See *Councils and Synods*, ii. 1109–10; Kedar, *Crusade and Mission*, pp. 182–3, 201.
[68] See the report of the clergy of Reims in John of Thilrode, 'Chronica', p. 581. For the issue of the military orders, see A. J. Forey, 'Military Orders in the Crusading Proposals of the Late Thirteenth and Early Fourteenth Centuries', *Traditio*, 36 (1980), 321–3.

temporary setback, deeply stirred European society. As far as crusade-planning was concerned its effect was that of a catalyst which crystallized attitudes and solidified policies already voiced at the Second Council of Lyons, and sporadically, even earlier. The loss created variations on already-existing opinions rather than putting forward entirely new ideas. On the other hand, the emotional reaction reflected an intensification of a sentiment in favour of a crusade and the rise of eschatological expectations linked to a peaceful Christian return to the Holy Land. These eschatological expectations, perhaps linked with the general atmosphere of a *fin de siècle,* presented the loss of the last Latin outposts in the Levant as a part and parcel of the Last Days. On the whole the mood of contemporaries was pessimistic and it portrayed a certain concern, or even alarm and fear, regarding the outcome of a confrontation of Christianity with Islam. The all-prevailing feeling was—as the General Cortes of the crown of Aragon put it in November 1291—'que nuyl temps la Cristiandat no fo en maior peril que en est temps era'.[69] This awareness of danger strengthened the pro-crusader sentiments produced by the loss of the Holy Land. Pope Nicholas IV was personally aware of such sentiments and the urgent need to use them for a swift and effective action to recover the Holy Land. He died, however, even before the reports of the Church councils reached Rome. With him died the crusade-project for 1293.

[69] L. G. Anton, *Las Uniones aragonesas y las cortes del reino 1283–1301* (Saragossa, 1975), vol. ii, no. 308, pp. 450–1.

5
1292–1305
THE YEARS OF TRANSITION

The period which I have chosen to call the 'Years of Transition' was marked by the absence of any serious attempt on the part of the papacy, traditionally the main pillar of the crusade, to organize an expedition for the recovery of the Holy Land. The successors of Nicholas IV had to cope with the pressing needs and imperatives of European politics. Consequently, though the recovery of the Holy Land was not absent from the plans of Boniface VIII and his successor Benedict XI, neither came forward with a definite scheme for its realization. But symptomatically enough the two crusades launched by Boniface VIII were linked to the Holy Land; the Colonna were presented as 'enemies of the Church as well as of the Holy Land', whereas the crusade for Sicily was defined as aiding the cause of the Holy Land. During the pontificate of Benedict XI there was even the resurgence of a project, later adopted by Clement V, which linked the recovery of the Holy Land to that of Constantinople.

Yet though the period was a kind of intermission if judged in terms of tangible papal action, there was no such break in general enthusiasm for the crusade. The loss of the Holy Land kept alive and even triggered off a renewed zeal for its liberation. This mood bridged, so to speak, the time of papal inaction. It was during this period in 1300 that Europe was convulsed by an outbreak of crusader zeal, the result of what seemed like miraculous events in Syria, Palestine, and most of all, in Jerusalem. A more tangible outcome of these events was a series of attempted and even actual attacks upon the Mamluks in 1299–1302. The *spiritus movens* of these actions were the military orders the Templars and the Hospitallers, and the part played by them underlines their new and central position in the crusading movement following the loss of the Holy Land. During this period the military orders were already beginning to emerge as the most important single source of new crusading schemes.

The 'Angelic Pope' and the 'Most Christian King'

The four-month pontificate of Celestine V (August–December 1294) was too short to have a direct impact on the crusade. Yet it had some negative repercussions. Celestine, elected for his saintly reputation, was utterly unprepared for political life. Consecrated in Aquila, the old man (born c.1209–22) became a puppet in the hands of Charles II of Anjou. During the four months in his pontificate (until his abdication on 13 December 1294) he worked for Charles's interests, making systematic use of the name of the Holy Land and the funds destined for the crusade. Charles was granted (2 October 1294) two ecclesiastical tithes for the conquest of Sicily and the defence of his kingdom: one for four years in France—the kingdom of Arles, Lyons, Vienne, Besançon, Bourges, Aix; the other for one year in England. Celestine V saw the interest of the Holy Land as being intimately linked with that of Sicily. In his words the recovery of Sicily from the Aragonese would be of the greatest service to that of the Holy Land on account of its location, its fertility, and its possession of all the materials necessary for a crusade.[1] The expert character of this explanation coming from a saintly hermit suggests the direct influence of Charles II of Anjou. Yet this was in fact a policy held by all the popes (with the exception of Nicholas IV in 1291–2) after the Sicilian Vespers. Sicily, it should be remembered, enjoyed a principal role in the logistics of the crusade during the thirteenth century. Though in the writings of the theoreticians of the crusade after the loss of the Holy Land, it was replaced by Cyprus and later (c.1310) also by Rhodes, it never lost its standing as a supply-base for the Latin East.[2] This could only help to justify the papal insistence upon a crusade for the recovery of its own Sicilian fief before the crusade to the Holy Land. The linking of

---

[1] Potthast, no. 23985; *AE* ad an. 1294, no. 12; G. Digard, *Philippe le Bel et le Saint-Siège* (Paris, 1936), vol. i, pp. 172–88; R. Mols, 'Celestin V', DHGE xii (1953), 83–95.

[2] M. Mollat, 'Problèmes navals de l'histoire des croisades', *Cahiers de civilisation médiévale*, 10 (1967), 357–8.

Sicily with the Holy Land permitted the transfer of crusader money to Charles II of Anjou. Celestine V went so far as to threaten the Church in England with excommunication unless the collectors of the tithe granted by Nicholas IV to Edward I for the Holy Land transmitted such sums as were already in their hands to papal bankers. At the same time (13 October 1294) Boniface of Calamandrana, Hospitaller grand commander of Outremer, was ordered to transfer 15,000 gold florins of the sum assigned for the equipment of galleys for the aid of the Holy Land to the bankers of the papal Camera. The Hospital was also asked to give financial help to Charles.[3]

The pontificate of Peter of Morone, short as it was, was unfavourable to the crusade due to his policy of dispersing the means of the crusade and diverting them from the Holy Land. But the issue was more complex than that. The elevation of the saintly hermit to the papal throne inspired Joachimite expectations, thus fortifying a trend of thought hostile, or at least unfavourable, to the crusade. It is impossible to find out whether the election of Celestine was directly caused by Joachimite views and especially by their expectations of an 'Angelic Pope'. It is nevertheless clear that the election of the hermit-pope precipitated the crystallization of the image of the Angelic Pope in a clear and powerful form. The election of Celestine V caused intense joy among the Joachimites and particularly among the Spiritual Franciscans, and it intensified their expectations of the advent of the Age of the Spirit.[4] Celestine himself favoured the Spirituals. His indulgence of Santa Maria de Colledimezzo granted for the commemoration of his enthronement seemed actually to give papal consecration to the ideas developed by Peter John Olivi (who rejected the crusade to the Holy Land) on the necessity of multiplying the means of salvation on the eve of the End of the World, namely to extend

---

[3] John of Pontissara, *Register*, ed. C. Deeds (Canterbury and York Society, 19, 30; 1915–24), ii. 501–5; *Cartulaire général de l'ordre des Hospitaliers de St Jean de Jérusalem 1100–1310*, ed. J. Delaville le Roulx (Paris, 1894–1906)), no. 4260; Potthast, nos. 23966, 23997.

[4] The prophetic figure of the 'Angelic Pope' evolved as the embodiment of the Joachimite Third Age. Roger Bacon appears to be the first who, as early as 1267–8, clearly pointed ('Opus Tertium', in *Opera Inedita*, ed. J. Brewer (RS, London, 1857), 86–8) to a pope who would purge and reunite the Church. See M. Reeves, *Influence of Prophecy in the Later Middle Ages: A Study in Joachimism* (Oxford, 1969), 394–415; id., 'Joachimist Influences on the Idea of a Last World Emperor', *Traditio*, 17 (1961), 328.

the plenary indulgence like that of Portiuncula. The indulgence of Colledimezzo thus provided the Spirituals with yet another argument against the crusade.[5]

The Joachimite expectations inspired by the election of Celestine were further increased following his tragic end. The drama of Celestine V and Boniface VIII was seen in terms of the good and evil which characterized the Joachimite programme. Celestine became the prototype of the Angelic Pope and, as such, an integral element in the prophetic tradition. But the saintly reputation of Peter of Morone kindled all kinds of hopes and expectations. Among them, though it was one of the least important, was the hope of a crusade. In the English versions of his abdication speech (13 December 1294), Celestine allegedly referred to his inability to organize a crusade. The English linking of Celestine's abdication with the crusade shows that, at least in that country, the pope stimulated such hopes.[6]

The fact that Ramon Lull addressed a petition to Celestine V in which he combined the aim of missionary endeavour with that of armed action for the recovery of the Holy Land is of less significance with regard to contemporaries' expectations of Celestine V. Lull addressed his crusader treatises to all the popes from 1291 onwards, except for Benedict XI (1303–4). His 'Petitio Raymundi pro conversione infidelium' (Naples, 1294), addressed to Celestine V, summarized the ideas he had already expressed in his 'Tractatus'. Nevertheless, in contrast to the 'Tractatus', the 'Petitio' implied an armed crusade rather than stressed it. It is plausible to assume that Lull, along with his contemporaries, recognized the piety of Peter of Morone and believed that, as a friend of the Spiritual Franciscans, he would be more inclined to the policy of peaceful conversion than to

---

[5] Potthast, no. 23981; Digard, *Philippe le Bel*, vol. i, p. 196 n. 2. The date of the indulgence of Portiuncula is still questioned. The treatise defending it by Olivi indicates that it was accepted and disputed c.1279. See C. Brown, 'Portiuncula', *New Catholic Encyclopedia*, xi (New York, 1967), 602; C. Partee, 'Peter John Olivi: Historical and Doctrinal Study', *Franciscan Studies*, 20 (1960), 215–59.

[6] Bartholomew Cotton, *Historia Anglicana: necnon ejusdem Liber de archiepiscopis et episcopis Angliae*, ed. H. R. Luard (RS, London, 1859), 257; 'Annales prioratus de Dunstaplia', in *Annales monastici*, ed. H. R. Luard (RS, London, 1864–9), iii. 382–3; John of Halton, *Register*, ed. W. N. Thompson and T. F. Tout (Canterbury and York Society, 12–13; 1913), i. 29–30; *Historical Papers and Letters from the Northern Registers*, ed. J. Raine (RS, London, 1873), no. 68, p. 110; Digard, *Philippe le Bel*, vol. i, p. 203.

that of a crusade. He thus saw it as more suitable to address Celestine in terms of conversion than of crusade.[7] A few months later Lull, who either accompanied or followed Boniface VIII from Naples to Rome, presented him with virtually the same petition as he had presented to Celestine V. However, when addressing Boniface VIII, Lull referred as explicitly as in the 'Tractatus' to a 'crusade [*passagium*] for the recovery of the Holy Land'.[8] Lull himself spent over a year in Rome at the beginning of the pontificate of Boniface VIII. It is significant, however, that during the remaining six years of Boniface's reign Lull did not reappear in the Curia. According to his 'Vita' he came to the conclusion that no support for his plans could be expected from Boniface VIII.[9]

Lull's 'Petitio' to Boniface VIII is actually the only known crusader treatise composed during that pope's reign and dedicated to him. Another crusader plan once assigned to the reign of Pope Boniface, that of Galvano of Levanto, seems in fact to have been composed before the accession of Boniface to the papal throne (23 January 1295) and was dedicated to Philip IV. It would be only logical to assume that the decline of papal prestige and authority after the pontificate of Boniface VIII resulted in a certain decline in the position of the papacy as the leader of the crusading movement. However, already around 1295, some of the protagonists of the crusade were seeking support for their plans outside the Curia. This was not because they believed the pope to be incapable of organizing a crusade, but because they recognized the power of secular monarchs and above all that of the king of France.

---

[7] For the 'Petitio' addressed to Pope Celestine V see also A. S. Atiya, *Crusade in the Late Middle Ages* (London, 1938), app. 1, pp. 487–9. Atiya seems unaware that the treatise had already appeared in Golubovich, i. 373–5. In 1294 Lull stayed in Naples following his expulsion from Tunis in Jan. of that year. See J. N. Hillgarth, *Ramon Lull and Lullism in Fourteenth-Century France* (Oxford, 1971), p. xxiv and n. 11.

[8] Ramon Lull, 'Petitio Raymundi pro conversione infidelium' (Naples, 1294), in Golubovich, i. 373–5; id., 'Petitio Raymundi pro conversione infidelium' (Rome, 1295), ed. H. Wieruszowski, in Wieruszowski, 'Ramon Lull et l'idée de la Cité de Dieu', *Estudis Franciscans*, 47 (1935), 100–4. In both petitions, it is worth mentioning, Lull argued that the loss of the Holy Land was ascribed by the laity to the neglect of the *res publica*, *bonum publicum*, or *publica utilitas*. For him it was identical to the spiritual salvation of humanity and the expansion of the faith: see Wieruszowski, 'Ramon Lull et l'idée de la Cité de Dieu', pp. 96–9.

[9] 'Vita Beati Raymundi Lulli', ed. B. de Gaiffier, *Analecta Bollandiana*, 48 (1930), 165–6. See also Hillgarth, *Ramon Lull*, pp. 122–3 and n. 314.

Most of the plans invoking Philip of France as the guiding power of the crusade, appeared after 1305. Yet it was during the period of 1292–1305 that he had acquired this position. That he did so, was in this early period an attribute of his reputation as *defensor pacis*. Even before his elevation to the throne of France, Philip (1285–1314) had been regarded as an elect champion of Christ. In the 'De regimine principium' of Giles of Rome (Aegidius Romanus), composed *c.*1285, Philip appears as 'more than man, wholly divine' and 'quasi semideus'. The Council of Sens in 1292 addressed Philip as 'christianissimus'. The religious character of the French monarchy had never been so stressed as it was during the reign of Philip IV. This was certainly understood even by 'papists' like Galvano of Levanto.[10]

Galvano, a Genoese from the little town of Levanto near La Spezia, described himself as a physician. A Franciscan tertiary, he was aquainted with the Franciscan prior of the province of the Holy Land, Nicholas de Salix, who resided in Cyprus after the fall of Acre, and to whom he dedicated his 'Ars navigativa Spiritualis' (*c.*1300). It is doubtful if Galvano ever visited the Holy Land, although G. Golubovich assumed that he did on the basis of the map of the Holy Land which Galvano enclosed with his 'Liber sancti passagii Christicolarum contra Saracenos pro recuperatione Terrae Sanotae'. This map, as well as the ten last chapters of the 'Liber', have unfortunately been lost. The presence of the 'Liber' in an inventory of the papal library dating from 1295 assigns it to *c.*1291–5. The fact that no pope was ever mentioned by name suggests that it was composed during the papal vacancy (*c.*1292–4).[11] Something of a mystic and a physician, attached for some time to the papal court,

---

[10] Giles of Rome, *De regimine principium* (Rome, 1607), 19, 28; Hillgarth, *Ramon Lull*, pp. 110–11; Digard, *Philipp le Bel*, vol. ii, p. 281; J. R. Strayer, 'France the Holy Land, the Chosen People and the Most Christian King', in id., *Medieval Statescraft and the Perspectives of History* (Princeton, NJ, 1971), 300–14, esp. pp. 306–7 and n. 19; W. D. McCready, 'Papalists and Antipapalists: Aspects of the Church/State Controversy in the Later Middle Ages', *Viator*, 6 (1975), 241–73.

[11] Golubovich, i. 357–9; ed.'s introd. in Galvano of Levanto, 'Liber sancti passagii Christicolarum contra Saracenos pro recuperatione Terrae Sanctae', ed. C. Kohler, in 'Traité du recouvrement de la Terre Sainte adressé, vers l'an 1295, à Philippe le Bel par Galvano de Levanto, médécin génois', *ROL* 6 (1898), 343–57; J. Leclercq, 'Galvano di Levanto e l'Oriente', in A. Pertusi (ed.), *Venezia e l'Oriente fra tardo mdioevo e rinascimento* (Venice, 1966), 403–16.

Galvano dedicated some of his treatises to Boniface VIII, but his plan for the recovery of the Holy Land was dedicated to Philip the Fair. Galvano's project is mainly significant as inaugurating a flood of *de recuperatione* treatises presented to the French monarchs during the following two centuries. The use he made of the principles of the game of chess and the eulogies he lavished on Philip IV closely follow Giles of Rome. Galvano addressed Philip as 'christianissimus rex Francorum', 'principalis athleta Jesu Christi in orbe terrarum', and 'archipugil Jhesu Christi'. He dwelt at length on the example of Charlemagne, thus fortifying the Capetian claim to descent from the Carolingians. Far from implying the French right to the Holy Roman Empire as the French pamphleteers Pierre Dubois and William of Nogaret later did, Galvano appealed to Philip's piety rather than his political considerations. It was as 'prothopugnator et promotor fidei orthodoxe necnon et protector catholice et romane ecclesie' that Galvano exhorted Philip to launch a crusade. The recovery of the Holy Land was, for Galvano, above all an act of devotion. It was also primarily the recovery of the Holy Places. In his 'Thesaurus religiose paupertatis' he stated that he did not weep for the fall of Acre and Tyre and the other towns of Syria; he was not moved by the captivity of the vile multitude; but he deplored 'the loss of the Temple in which Jesus Christ had lived' and claimed that Christianity (*cultus Christianus*) should be restored to the Holy Land, to set it free from the Moslem sect.[12]

Galvano's appeal to the king of France (*c.* 1291–5) to take upon himself the leadership of a crusade, was not the only plea addressed to Philip during the 'Years of Transition'. As mentioned above, the Aragonese Arnold of Villanova appealed to Philip the Fair twice, in 1297 and 1301, to recover the Holy Land. In his tract 'De Adventu Antichristi' (1297) Arnold expected the coming of the Antichrist in 1335, at the time of the third Angelic Pope. He expected the latter to be accompanied by a king 'qui erit sibi filius et subditus, cum pace et crucibus

[12] Galvano of Levanto, 'Liber', pp. 348–50, 358–65; Leclercq, 'Galvano di Levanto', pp. 405–6; Pierre Dubois, *Summaria brevis*, ed. H. Kampf (Leipzig and Berlin, 1936), 129; M. J. Wilks, *Problem of Sovereignty in the Later Middle Ages* (Cambridge, 1963), 428–32; S. Lucas, 'The Low Countries and the Disputed Imperial Election of 1314', *Speculum*, 21 (1946), 74–5.

intrabit in Jherusalem et recipietur gratanter a Sarracenis'. It is impossible to ascertain whether Arnold had Philip in mind here. However, as the tract was written at the same time as he turned to Philip, this seems quite probable.[13]

Both Arnold's and Galvano's appeals to the French king seem almost fanciful as there was nothing in Philip's policy until 1305 to show his interest in a crusade. The appeals show how different were contemporaries' views of Philip from the opinion one gets from papal and French correspondence. Contemporaries were then already seeing in Philip a potential crusader. As 'defensor pacis', as 'protector catholice et romane ecclesie', Philip seemed destined to be a protector of the Holy Land. The appeals show that the ideal of St Louis was still very much alive, and they show also the extent to which the decline of the Holy Roman Empire and the rise of the French monarchy had influenced the European idea of the crusade. From the end of the thirteenth century onwards it was not the king of the Romans but the French monarch, as the descendant of Charlemagne, who was linked with the recovery of the Holy Land. In 1306 Pierre Dubois, in his *De Recuperatione Terrae Sanctae*, demanded an almost universal role for the French monarch, linking his claim with the descent from Charlemagne. From this source alone, he stated, the universal pence desired by all should flow, and therefore the imperial title must be given to the French royal house. The Holy Land recovered, it would be from Jerusalem that the French emperor would rule a federation of nations.[14] Such a plan, from which the papacy was conspicuously absent, could hardly have been formulated before the end of Boniface VIII's pontificate.

The Crusades of Boniface VIII 1295–1303

The sincerity of Boniface VIII's intentions for the recovery of the Holy Land has often been doubted; it has been argued that the crusade was used as a pretext, a goldmine to finance papal policy, and that the pope's proclamations were no more than a

---

[13] *Papsttum und Untergang des Templerordens*, ed. H. Finke (Münster, 1907), vol. i, pp. 220–1; Reeves, *Influence of Prophecy*, pp. 314–15.

[14] Pierre Dubois, *De Recuperatione Terre Sancte*, ed. C. V. Langlois (Paris, 1891), 98–9.

'façon de parler' or lip-service to current emotions.[15] The controversial character of the issue derives from the apparent dichotomy between *litterae* and *gesta*. Whereas Boniface's acts carry little conviction as to his interest in the Holy Land, his correspondence presents a pope deeply concerned with its fate. Moreover, the characteristics of the pope as immortalized by Dante, added to the obviously biased French propagandist treatises, the product of the French king's campaign against the pope which continued even after the pontiff's death. For Dante, Pope Boniface was the 'Prince of the modern Pharisees waging was near the Lateran'. Dante committed Boniface to Hell.[16] The charge that Boniface deliberately damaged the cause of the Holy Land was formulated by the propagandists of Philip IV, during the conflict of 1302–3 and was put into writing *c.*1303 –10 in the years immediately following the pope's death; this was part and parcel of the charges prepared for the trial against his memory instigated by Philip IV. The authors of the treatises did not invent their facts, but these were presented in a biased form and deliberately misinterpreted. There is some evidence to show that those who accused Boniface 'of caring nothing for the *negotium Terrae Sanctae*' held different opinions during the first years of his pontificate. So for example William of Nogaret, who inserted the above-quoted charge into his 'Allegationes excusatoriae super facto Bonifaciano, written *c.*1304, hail observed some ten years earlier in 1295 that 'the pope himself earnestly wished to undertake an expedition beyond the sea'. By 1310 the French propagandists had made Boniface into a devious almost anti-Christian monster who 'did more to induce believers to persecute the king of France and the French than the sultan of Egypt and the Saracens'.[17]

Actually, the accession of Benedetto Gaetani to the papal

---

[15] So Finke in his edn. *Aus Tagen Bonifaz VIII* (Münster, 1902), followed by F. Heidelberger, *Kreuzzugsversuche um die Wende des 13. Jahrhunderts* (Leipzig and Berlin, 1911), 10–23; Hillgarth, *Ramon Lull*, p. 75, as well as T. S. R. Boase, *Boniface VIII* (London, 1933), 222–7, *contra* H. K. Mann, *Lives of the Popes in the Middle Ages* (London, 1931–2), xviii. 14–27, and Digard, *Philippe le Bel*, vol. i, p. 206.

[16] Dante, *Inferno*, xxvii. 85–90; *Paradiso*, xxvii, *passim*; *Purgatorio*, xx, *passim*.

[17] *Histoire du differend d'entre le pape Boniface VIII et Philippe le Bel roy de France*, ed. P. Dupuy (Paris, 1655; repr. 1963), 213, 343; 'Allegationes excusatoriae super facto Bonificiano', in *Histoire du differend*, pp. 252–3; 'Formulaires de lettres du XIIe, du XIIIe et du XIVe siècle', ed. G. V. Langlois, *Notices et extraits des manuscrits de la Bibliothèque Nationale*, 34 (1891), 321; Digard, *Philippe le Bel*, vol. ii, p. 25 and n. 5.

throne as Boniface VIII (23 January 1295), seemed to herald the resumption of crusade-making. It was believed that the new pope, who as a cardinal under Nicholas IV has been much concerned with the Holy Land, would take up the cause of liberating the Holy Places. His first acts strengthened these expectations. On the day after his consecration (24 January 1295), he produced his first encyclical, stressing his concern for the Holy Land, which, in his words, could be recovered only by the united forces of the West.[18] Then in February the pope dispatched Bertrand de Got, bishop of Albano (later Clement V), together with Simon of Beaulieu, bishop of Palestrina, to England and France respectively, charged with the reconciliation of the warring kingdoms. The letter of introduction of the papal nuncios to Edward (19 February 1295), as well as another dispatch of 30 March 1295, repeatedly stressed that the war between the two kingdoms was delaying the *negotium Terrae Sanctae*.[19] English sources even knew that Pope Boniface had also written at that time to the kings of France, Germany, and Aragon appealing to them to conclude a general peace and to hasten the *passagium Terrae Sanctae*. It was also alleged that 'if the discords of kings could be ended the pope himself earnestly wishes to undertake an expedition beyond the sea' and that he intended to take part in it. Such a proclamation must have caused quite an impression in England where it was recorded by a number of chroniclers.[20] Four months later during the Congress of Anagni (12 June 1295) which reconciled Charles II of Anjou, Philip IV of France, and Charles of Valois with Frederick of Sicily and James II of Aragon, Boniface once again insisted upon launching a crusade.[21] Ratifying the peace (21 June 1295), Boniface explained his indulgence towards James II of Aragon, who had received papal absolution, by the needs of the Holy Land; the interests of the Holy Land were damaged

---

[18] *AE* ad an. 1295, no. 10.

[19] Boniface VIII, *Registres de Boniface VIII*, ed. G. Digard et al. (Paris, 1884–1939), no. 698; *Foedera, conventiones, litterae et cuiuscunque generis acta publica inter reges Angliae et alios quosvis imperatores, reges, pontifices, principes vel communitates 1101–1654*, ed. T. Rymer et al. (Hague, 1745), I. iii. 142–3.

[20] Bartholomew Cotton, *Historia*, p. 258; 'Annales de Dunstaplia', pp. 383–4; 'Annales prioratus de Wigornia', in *Annales monastici*, iv. 518–19; *Annales Angliae et Scotiae*, ed. H. T. Riley (RS, London, 1865), 400–1; *Historical Papers*, no. 68, p. 111; Boase, *Boniface VIII*, p. 224 and n. 3.

[21] Digard, *Philippe le Bel*, vol. i, p. 229 and n. 6.

by the dissent among the princes and it was the pope's intention to reconcile them.[22]

The crusade as envisaged by Boniface at that time was a gigantic all-European *passagium generale*. Expressing, in a sense, the unity of Christian Europe under the aegis of the Holy See, its realization was conditional upon the establishment of a complete and general peace in Europe. Boniface's promises to Lesser Armenia, his intention to build a fleet, and to maintain a sea-blockade against Egypt, make it plausible to suppose that he envisaged a crusade on the lines drawn up by Fidenzio of Padua, namely a simultaneous attack by sea and land. The route to be taken by the land-force was also that recommended by Fidenzio: a landing in Armenia and then a march into Syria.[23]

Peace in Europe and the unity of Christendom were always considered by the papacy to be the essential and preliminary conditions for the launching of a general crusade; during the pontificate of Gregory X, this was also one of the basic ideas brought forward by expert advice. The loss of the Holy Land was in Boniface's opinion caused by European discords; as the Catholic princes were occupied by wars among themselves there was no one to fight for the Holy Land. In his linking of the idea of peace with that of the recovery of the Holy Land, Boniface was following a long line of popes and especially Innocent III.[24] Like his great predecessor, Boniface saw himself charged by the 'King of Kings' with the task of maintaining peace among the Christian princes to enable them to render service to the Celestial King in the form of the *negotium Terrae Sanctae*. The crusade was an expression of the unity of Christendom, a means of appealing for peace, and at the same time an end in its own right.[25]

Thus during the first two years of his pontificate Boniface worked systematically on both complementary levels: the pacification of Europe and the organization of the crusade. A

---

[22] Boniface VIII, *Registres*, no. 184, see also no. 168.
[23] Ibid., nos. 2310, 2654.
[24] See e.g. ibid., no. 868. See also Digard, *Philippe le Bel*, vol. i, p. 206; J. Leclercq, *Idée de la royauté du Christ au Moyen Âge* (Paris, 1959), 56–64; J. Riley-Smith, *What were the Crusades?* (London, 1977), 41–2; cf. Boase, *Boniface VIII*, p. 224.
[25] Boniface VIII, *Registres*, no. 184.

long list of actions and proclamations bore witness to his preoccupation. Shortly after his consecration on 24 April 1295, he renewed the indulgences granted by Nicholas IV to the potential participants of the crusade planned for 1293.[26] Similarly he tried to strengthen the blockade of Egypt; on 12 May 1295 he renewed the ban on trade with the Saracens; the decrees against all traffic to the sultanate were henceforth renewed every six months throughout the whole of Boniface's pontificate, and inquisitors were ordered to take special cognizance of the offence. At the same time Boniface authorized the king of Aragon to collect fines and property confiscated in the illegal trade with the Saracens in his kingdom; moreover, absolutions were to be given to offenders willing to donate a quarter if male, and a fifth if female, of their gains from the illegal trade with the Saracens for charitable or pious purposes.[27]

Though it was not explicitly stated, it seems plausible to suppose that the blockade of Egypt was among the tasks which the pope intended for the fleet he had decided to build. In 1294 James II of Aragon had already suggested that a fleet of sixty galleys should be built at the expense of the Church in his own dockyards. The decision to build a fleet, taken on 20 January 1296, was apparently prompted by the need for a more or less independent naval power for the war against Sicily and the eventual expedition for the recovery of the Holy Land. The tasks of the fleet were, in the pope's words, the aid of the Holy Land and the war against the enemies of the Church. The appointment of James II of Aragon as the 'general Standard-Bearer, Captain and Admiral of the Church' was as much due to the awareness of the rising power of Aragon as to the papal interests in Sicily; moreover, there were the repeated declarations of James about his intended departure for the Holy Land.

---

[26] Ibid., no. 124.
[27] Ibid., nos. 778, 848, 1591, 1654, 2339, 3109, 3111, 3354, 3421, 4420, 5015, 5020, 5346; *Cerdeña y la expansión mediterránea de la Corona de Aragón*, ed. V. Salavert y Roca (Madrid, 1956), no. 19, p. 19. See W. Heyd, *Histoire du commerce du Levant au Moyen Âge*, trans. F. Raynaud (Leipzig, 1936), ii. 30–1. From 1302 the volume of trade of Aragon with Egypt declined rapidly; in 1302 it was over 107,342ss whereas in 1312 it was just over 35,708ss. See J. Trenchs Odena '"De Alexandrinis": El Comercio prohibido con los Musulmanes y el papado de Aviñón durante la primera mitad del siglo XIV', *Anuario del estudio medievales*, 10 (1980), 274–9; J. N. Hillgarth, 'Problem of a Catalan Mediterranean Empire 1229–1327', *EHR* suppl. 8 (London, 1975), 7, 41.

James, however, cautious as he was, stipulated that he would not be bound to take part in the expedition if its participants included one of his enemies. However, if he were to participate he would be granted Corsica and Sardinia, as well as a triennial ecclesiastical tithe of Aragon. For providing ten galleys or more, he and his kingdom would be taken under the protection of the Holy See.[28]

Financial arrangements connected with the envisaged expedition bore witness to the pope's interest even in the minutiae of its preparation. Boniface transferred the crusader tithes imposed and collected under his predecessors directly to the papal bankers, in order to be ready to finance the crusade. On 14 July 1295 he renewed the order of Celestine V that money from the Scottish tithe for the Holy Land (some £10,000) should be paid to certain papal bankers; the Scottish bishops were bidden to denounce as excommunicated those who had not yet paid the tithe.[29] In 1296 Boniface reopened the matter of the collection of the tithe imposed by Nicholas IV (16 March 1291) on the clergy of the British Isles for a period of six years. The order for the collection, which had been interrupted by the wars with France, Wales, and Scotland, was repeated and on 5 February 1300 Boniface requested the collection of the remaining three years of Nicholas's tithe.[30] A year later, however (10 February 1301), Boniface granted Edward I the proceeds of the first three years of Nicholas's tithe which were already in the king's possession. This was probably done on the assumption that Edward would depart on a crusade.[31] On 5 December 1301 Boniface imposed a new triennial tithe on English ecclesiastical benefices, of which the annual income was above 7 *livres tournois*. This was justified by the need to maintain what the Mongols had won in the Holy Land.[32] For France, Boniface ordered on 8 June 1296 that the tithe collected for the Holy Land, as well as sums from the redemption of vows and legacies, should be

---

[28] *Cerdeña*, pp. 115–17, no. 19, pp. 17–20; *AA* vol. i, no. 10, pp. 14–15.
[29] *Historical Papers*, no. 71, pp. 114–16, and above, n. 3.
[30] Boniface VIII, *Registres*, nos. 1129, 3165, 3538–42; W. E. Lunt, 'Collectors' Accounts for the Clerical Tenth levied in England by Order of Nicholas IV', *EHR* 31 (1916), 102–3.
[31] Boniface VIII, *Registres*, no. 3925; *Foedera*, I. iv. 6–7; W. E. Lunt, 'Papal Taxation in England is the Reign of Edward I', *EHR* 30 (1915), 416–17.
[32] John of Halton, *Register*, i. 146–9.

transferred to his bankers.³³ He also ordered the collection of the tithe imposed by the Second Council of Lyons. Specific orders were directed to Dalmatia and the diocese of Ferrara as well as to Germany, Hungary, Bohemia, and Poland.³⁴

An interesting phenomenon of the period was the diminishing number of privileges commuting and redeeming crusader vows; moreover, there was also a tendency from the time of Nicholas IV to the close of the pontificate of Boniface VIII to make their attainment far more difficult.³⁵ Papal legates, dispatched to France to request an aid of Philip IV to Charles II of Anjou, were granted (12 February 1297) a limited faculty to dispense from vows; dispensations were granted only to the old and to the disabled. The price of redeeming vows was considerable: 300 *livres tournois*, a sum equivalent to the cost of arming a warrior for the Holy Land. Guy of Flanders, for instance, was ordered to leave in his will the sum of 10,000 *livres tournois* for the redemption of his vow.³⁶ What actually came to the papal treasury is difficult to gauge. By 1301 Boniface was complaining about the grave conditions of his finances due to the ruinous expenses of the Sicilian war. Whether there was any truth in the statement of Walter of Guisborough that the conspirators of Anagni aimed at the treasure Boniface intended for the *negotium Terrae Sanctae* cannot be proven.³⁷

Another aspect of Boniface VIII's interest in the crusade was his preoccupation with the Hospitallers and the Templars. His attitude proved that their position had changed after 1291. Following the loss of the Holy Land the orders emerged as the single most important factor in any future crusade. Nicholas IV, as well as the *de recuperatione Terrae Sanctae* treatises, assigned to the orders a leading role both in the recovery of the Holy Land and in the future Christian maintenance of the reconquest. At the same time, as already seen, European public opinion was openly hostile to the orders and they were made the scapegoat for the failure of Outremer. Thus a contradictory

---

[33] Boniface VIII, *Registres*, no. 1096.   [34] Ibid., nos. 1305, 3543–4, 4408–10.
[35] N. Paulus, *Geschichte des Ablasses in Mittelalter von Ursprunge bis zur Mitte des 14. Jahrhunderts* (Paderborn, 1922–3), ii. 33–9; J. A. Brundage, *Medieval Canon Law and the Crusader* (London, 1969), 133.
[36] Boniface VIII, *Registres*, nos. 86, 450, 452–5, 833, 2316, see also no. 2315.
[37] *Foedera*, I. iv. 7; Walter of Guisborough, *Chronicle*, ed. H. Rothwell (London, 1957), 356.

situation was created. Deprived of their immediate *raison d'être*, even their position in their refuge in Cyprus was difficult. Due to restrictions imposed on them by King Henry II of Lusignan to prevent them from becoming as powerful in Cyprus as they had been in Syria, the orders found it difficult to support the forces they were trying to maintain in Cyprus. They had to cut down their forces on the island and to concentrate upon building up fighting fleets and developing naval warfare. This policy was, as recently remarked by A. Luttrell, the natural result of the new circumstances of being island-based and dependent on a monarch suspicious of their growing power. It is likely, however, that no less decisive in their transformation from a force of land-based knights to a sea-power was the policy of Nicholas IV and the recommendation of the crusade-planners. The latter, as has been said, advised maritime warfare and the transfer to the orders of the responsibility for the maintenance of the naval blockade against Egypt. What happened in Cyprus was thus encouraged and perhaps even initiated by Pope Nicholas IV and much in tune with the advice of crusade theorists.[38]

The development of the naval forces of the orders was rather slow. In 1300 the Hospitallers, Templars, and the king of Cyprus were able to raise among them only sixteen galleys.[39] Although the pope referred in 1297 to the use the Hospitallers were making of their navy in the prosecution of the war against Islam,[40] there is no evidence that the matter was discussed during the visit of William of Villaret, then prior of St Gilles (future master-general), and Boniface of Calamandrana, the commander general of Outremer, to Rome in 1295. On the other hand there was a clear indication, referred to by James of Molay the master-general of the Templars, that Pope Boniface actually considered at that time the union of the orders and only after much consideration rejected the scheme.[41] Instead he became involved with the reform of the Hospital. Anxious about the state of the order, Pope Boniface was presented in

[38] J. Riley-Smith, *Knights of St John in Jerusalem and Cyprus c.1050–1310* (London, 1967), 201–5; A. Luttrell, 'Hospitallers in Cyprus after 1291', *Acts of the First International Congress of Cypriot Studies* (Nicosia, 1972), 162–3; above, ch. 3.

[39] Luttrell, 'Hospitallers in Cyprus', p. 163 and n. 3, *contra* Riley-Smith, *Knights of St John*, pp. 200, 209, 329–30. [40] *Cartulaire*, no. 4336.

[41] James of Molay, in *Dossier de l'affaire des Templiers*, ed. G. Lizerand (Paris, 1964), 4.

1295 by Boniface of Calamandrana and by William of Villaret with plans to reform the structure of the Hospital. He even rebuked Odo des Pins the master-general (1294–6) for not observing the statutes and customs of the Hospital. Supporting the reformers, Pope Boniface later confirmed their decision to call the chapter-general of 1300 to Cyprus instead of to Avignon, where the new master-general William of Villaret (1296–1305) wished to convene it. Thus William of Villaret's attempt to concentrate his rule in Europe was thwarted.[42] In 1295 Boniface VIII also received the visit of the master-general of the Templars James of Molay. Following this visit, Boniface granted the orders the same privileges in Cyprus as in the Holy Land. He also addressed Edward I of England and asked Philip IV of France and Adolf of Germany to allow the master general and brothers of the Temple to export from their lands supplies needed for the sustenance of Cyprus. During his tour of England and France (1295–6), James of Molay presumably collected money for the Holy Land from the Templars' estates. The main purpose of his visit was to make up for the losses of the order in Acre and perhaps also to stimulate enthusiasm in the West for a crusade. With Boniface he must have discussed the future of the crusade as well as the question of a possible merger of the orders. The only tangible result of Molay's efforts, however, was the papal intercession on behalf of the order at the courts of Europe and Cyprus.[43]

Boniface tried also to placate the factions in Cyprus. On 19 March 1298 he wrote to King Henry II of Cyprus and, reminding him of the great sufferings of the orders in the Holy Land and of their faithful service of his kingdom, he asked him to give them favourable treatment. Simultaneously Boniface wrote to James of Molay. It was the king of Cyprus, he wrote, who had granted refuge to the order after the expulsion of the Christians from Syria; the Templars and the Hospitallers should therefore make peace with the king and unite against the

---

[42] *Cartulaire*, nos. 4293, 4461–4, 4468; Riley-Smith, *Knights of St John*, pp. 205–6; 208, 298–303.

[43] Boniface VIII, *Registres*, nos. 487–9; *Foedera*, I. iii. 147; M. Barber, 'James of Molay, the Last Grand Master of the Temple', *Studia Monastica*, 4 (1972), 94–7; M. L. Bulst-Thiele, *Sacrae Domus Militiae Templi Hierosolymitani Magistri* (Göttingen, 1974), 305–8.

enemies of Christianity. In the course of the following years Boniface's intervention continued. On 10 June 1299, for instance, he rebuked the king for not observing the conditions of arbitration that had been set up to deal with his disputes with the ecclesiastics of the kingdom, including the military orders. The pope's efforts to placate the factions in Cyprus were only partly successful. In 1300–1 both the king of Cyprus and the orders were able to take part in expeditions to Syria and Egypt, but the friction between the king and the orders continued.[44]

At the same time, as already stated, Boniface was trying to pacify Europe. The impression created by his constant appeals to the secular heads of Europe is that the sole aim of the pope's sustained efforts was the recovery of the Holy Land.[45] He succeeded in reconciling France with England and Venice with Genoa until 1299.[46] The war for the domination of Sicily, however, still went on. From 1296, when Boniface gave up hope of reaching a settlement with Frederick III who was crowned king of Sicily at Palermo on 3 May 1296, the issue of the recovery of the papal fief of Sicily became inseparably linked with that of the Holy Land. The recovery of Sicily became, in the papal view, an essential pre-condition for that of the Holy Land.[47] This was reflected in Boniface's response to the appeal of King Sempad of Lesser Armenia. The pope exhorted the king (5 October 1298) to resist until help could reach him. He assured him of his attention and urged the kings of France and England, whom he hoped he had by now reconciled, to go to his help. Writing to the patriarch of Lesser Armenia (26 October 1298), Boniface assured him that once the Sicilian affair, a matter which caused him 'sleepless nights', was over, a crusade would be set on foot. Meanwhile Lesser Armenia must remember that the Lord is always near those who are in tribulation. 'From the very beginning of our ascension to the Apostolic

---

[44] Boniface VIII, *Registres*, nos. 2438–9, 3060–2, 3114; G. Hill, *History of Cyprus* (Cambridge, 1948–52), ii. 198–203; Riley-Smith, *Knights of St John*, p. 204; below, nn. 61, 64–5.

[45] See e.g. Boniface VIII, *Registres*, nos. 698, 783, 853, 868–70, 2310, 2321; *Foedera* I. iii. 189, 193, 203, 215.

[46] A. E. Laiou, *Constantinople and the Latins: The Foreign Policy of Andronicus II 1282–1328* (Cambridge, Mass., 1972), 101–14; Boase, *Boniface VIII*, pp. 203–7.

[47] Boniface VIII, *Registres*, no. 2886; Laiou, *Constantinople*, pp. 128–9; E. G. Léonard, *Angevins de Naples* (Paris, 1954), 186–91.

dignity', he continued, 'we have devoted anxious thoughts and study to the work of maintaining peace between the western kings and princes, especially among those who have been most eager to fight readily and usefully for the Holy Land'. Now by God's grace peace reigned between Edward I of England and Philip IV of France and the co-operation of James II of Aragon for the recovery of Sicily had been secured. The patriarch had therefore to bear up till help came, which by God's mercy 'will be sooner than is generally supposed'.[48]

The Sicilian war, which was proclaimed as a crusade, resulted in the diverting of crusader privileges and funds from the Holy Land. Philip IV of France was thus granted (8 August 1297) half of the French tithes for his help to Charles II of Anjou. At the same time he was ordered to return crusader money he had kept in France.[49] James II of Aragon, Charles II of Anjou, Charles of Valois, and their followers fighting for the papal cause were granted the same indulgences and privileges as those which had been bestowed on the crusades in the Holy Land. The revenues of ecclesiastical tithes in various parts of Europe, and even sums kept by the Holy See in Italian banks for the Holy Land, were now diverted and bestowed upon them. Actually, Boniface was the first to abandon completely the careful division kept by his predecessors between the Lyons' tenth and that levied for the Aragonese crusade, and he used the Lyons' tenth as payments to Charles II of Anjou and James II of Aragon. In 1299, when James II of Aragon showed signs of withdrawing from the Sicilian War, the bank of the Clarenti of Pisa was ordered to hand him over the sum of 32,000 gold florins.[50]

The entrance of Charles of Valois into the Sicilian war following the withdrawal of James II of Aragon in 1299, inaugurated a plan which linked the recovery of Sicily by the combined papal–Valois–Angevin forces to the recovery of the Latin Empire of Constantinople. Charles of Valois was to

---

[48] Boniface VIII, *Registres*, nos. 2653–4, 2663; *Foedera*, I. iii. 204; S. Der Nersessian, 'Kingdom of Cilician Armenia', in *Setton's History*, iii. 657.
[49] Boniface VIII, *Registres*, nos. 2361–7.
[50] Ibid., nos. 1575, 2340, 3001, 3072, 3917, 4395–7, 4484–9, 4625; N. J. Housley, *Italian Crusades: The Papal–Angevin Alliance and the Crusades against Christian Lay Powers 1254–1343* (Oxford, 1982), 103–4.

restore the pope's right in Sicily and then conquer the Byzantine Empire.[51] The plan was not entirely new. The conquest of Constantinople was *post facto* sanctioned by Innocent III (1204). The argument was that the domination of Constantinople was necessary as a base for a land-route to the Holy Land. Innocent III had also granted the same indulgences to those going to the aid of the Latin Empire as to those who aided the Holy Land. Urban IV (1261-4), who declared in 1261 the recovery of Constantinople as a crusade, justified it on the ground 'of the Greeks who again had fallen away from Rome'; moreover 'if the Greeks seize all Romania, the way to Jerusalem will be barred'. Since 1261 overt attempts were made in the West to restore the Latin Empire. Only in 1331, with the death of Philip of Taranto grandson of Charles I of Anjou and heir to his aspirations, was the intention to reconquer Constantinople replaced by a different plan: a coalition with the Byzantine Empire against a common enemy, the Ottoman Turks.

The plan to capture Constantinople and restore the Latin Empire, however, was not a policy which the papacy consistently followed between 1262-1331. Gregory X (1271-6) was convinced that not only would good relations with Byzantium benefit Christendom, but, more important, that without Greek support Jerusalem could neither be retaken nor be maintained under Christian domination. His negotiations with Emperor Michael VIII Palaeologus (1258-82) culminated in the Second Council of Lyons where an agreement on a religious union was signed. Now that the two Churches were united, Gregory expected that both would join in a great crusade to restore Jerusalem. Though the union existed only on paper, it acted as a powerful brake on the aspirations of Charles I of Anjou. With the Greeks apparently reconciled, no expedition against Byzantium could have been regarded as a crusade. The successors of Gregory X (Innocent V (1276), John XXI (1276-7), and Nicholas III (1276-80)) were equally unfavourable to the projects of Charles I of Anjou. It was not until the elevation of Martin IV (1281-5), that the papacy, without whose sanction no expedition against Constantinople could be launched, turned a favourable ear to Charles's plan. Repudiat-

---

[51] Laiou, *Constantinople*, pp. 129-30.

ing the union of the Second Council of Lyons, Martin IV excommunicated Emperor Michael and urged Charles to lead a crusade against 'the Greek schismatics'. The popes of the late thirteenth century and of the early fourteenth century continued (with the exception of Pope Nicholas IV), the policy of Martin IV. Instead of a precarious *entente* with the Greeks, they preferred a military effort to restore the Latin Empire.[52]

From 1293, when the Infante Frederick of Aragon, later Frederick III of Sicily, made his first appearance as a potential claimant to the Byzantine throne, until the peace of Caltabelotta in 1302, the plan for the recovery of the Byzantine Empire appeared time and again in the various schemes to end the war over Sicily. Boniface, reverting to Martin IV's aggressive policy towards Byzantium, made it a pawn in the Sicilian game. At first in 1295 he tried to wed Catherine of Courtenay, titular empress of Constantinople, to Frederick of Sicily. Frederick was promised 60,000 gold ounces and an additional 30,000 ounces annually for three years to help conquer the empire of Romania and get out of Sicily. Later (30 July 1299) Boniface tried to marry off Frederick to Maria the daughter of Charles II of Anjou. The kingdom of Jerusalem would form the dowry of the princess, and additionally, Frederick would hold Corsica and Sardinia as well as the island of Rhodes as papal fiefs. All this failed, as Frederick, understandably enough, preferred to hold Sicily than to scheme for the possession of Constantinople, Jerusalem, or Rhodes, which he might never see.[53]

Catherine of Courtenay was married to Charles of Valois in 1301. It is impossible to agree with D. Geanakopolos that this marriage followed the suggestion of Pierre Dubois's *Summaria brevis* (1300). Actually Philip IV of France was eager to bring Catherine under his influence as early as 1294 and he opposed the negotiations of Boniface to marry off Catherine to Frederick of Sicily. The Courtenay–Valois marriage meant the transfer of the claims to the eastern heritage and the Angevin ambitions to the French royal family. Obviously Boniface VIII was interested in recovering Sicily; he wanted to draw Charles of

---

[52] Geanakoplos, 'Byzantium', pp. 27–43, and above, chs. 1–2.
[53] Boniface VIII, *Registres*, nos. 809, 874, 3398; Laiou, *Constantinople*, pp. 53–5, 128–31, 200–1; Luttrell, 'Hospitallers in Cyprus', p. 165.

Valois into Italy on the assumption that Charles would be eager to end the war there, in order to pursue his eastern ambitions. In 1302 Charles promised Boniface VIII and Philip IV that he would not lead an expedition to Constantinople before waging war in Sicily. His Sicilian expedition was the pre-condition of any help that pope and king might give Charles for the recovery of Constantinople.[54]

The peace concluded at Caltabelotta (31 August 1302) between Charles II of Anjou and Frederick of Sicily, which regulated the Sicilian problem after twenty years of strife, could have marked a turning-point in the preparations for a crusade for the Holy Land, but in the meantime papal plans had been overshadowed by the recurrent strife of 'regnum et sacerdotium'. Boniface, completely involved in his dispute with Philip of France, was in no position to pursue plans for a crusade either to the Holy Land or to Constantinople. The latter plan was in fact pursued by Charles of Valois. Linked to the recovery of the Holy Land, the plan for the conquest of Byzantine was to become an important feature of the policies of Boniface's successors Benedict XI (1303–4) and Clement V (1305–14). Boniface himself never went so far as to link the conquest of Constantinople with that of the Holy Land.[55]

It was yet another crusade, that against the Colonna, which won Boniface the ill-famed reputation of 'the Prince of the modern Pharisees'. Dante put these words into the mouth of the great Ghibelline leader Count Guy of Montefeltro (1223–98). A penitent rebel, Guy took in 1298 the Franciscan habit for which Boniface had himself sent the cloth. In the very month of the fall of the fortress of Palestrina (June 1298), Guy died in Ancona as he was making a pilgrimage to the Holy Land.[56] The crusade against the Colonna (1297–9) as well as that for the recovery of Sicily, both of which are beyond the scope of this study, show that in the complex machinery of papal spiritual authority, the crusade was a feature of the utmost importance.

[54] Geanakoplos, 'Byzantium', p. 43 n. 45, who is mistaken in assuming that the memoir is still unpublished. See Pierre Dubois, *Summaria brevis*. See also Brandt in his trans. of Pierre Dubois, *Recovery*, pp. 156 n. 7, 211; Laiou, *Constantinople*, pp. 52–4, 129–30.
[55] Laiou, *Constantinople*, pp. 128–34, and below, ch. 6.
[56] John Elemosina, 'Chronicon', in Golubovicah, ii. 128–9; Boase, *Boniface VIII*, pp. 82, 179.

The papal appeal to take the Cross against the Colonna, 'the enemies of Christendom and the Holy Land'—in the words of the pope—met with tremendous success in Italy. The sanctioning of the expedition as a crusade partly solved the problem of finance as it made possible the use of the tithe collected for the Holy Land. Even the military orders were asked for assistance; the Templars and the Hospitallers were asked (23 February 1298) for 12,000 gold florins each, and the Teutonic Knights for 5,000 'to revenge the losses and injuries caused to the Church by the schismatic Cardinals James and Peter Colonna. Boniface was later accused of seizing the goods of the Templars for his own wars. But when in 1310 the turn of the Templars came, it was more convenient to present them as supporters of the pope.[57]

Whereas the crusade against the Colonna points to the importance and efficiency of the crusade as an instrument of papal policy, the French allegations against the pope during the conflict of 1302–3 demonstrate the prominent part the Holy Land could play in international politics. Though the Holy Land was incorporated only in about 1305 into French plans for expansion, French propaganda some three years earlier was already making use of the equation of 'war for France' with 'war for the Holy Land'. A sermon issued at the request of Philip IV during the war against Flanders after the battle of Courtrai (11 July 1302), included the following: 'He who carries war against the king [of France] works against the entire Church, against the Catholic doctrine, against holiness and justice and against the Holy Land.'[58] This sounds like a royal counterpart of the links constantly made by Boniface between the cause of the Holy Land on the one hand, and the crusades against Sicily and the Colonna, as well as his struggle against Philip IV, on the other. The king was in fact accused by the pope (on 13 April

---

[57] Boniface VIII, *Registres*, nos. 2273, 2352, 2375–83, 2386–8, 2426–30; *Histoire du differend*, pp. 2617, 2623, 2878; *Cartulaire*, nos. 4407–8, pp. 342, 529; Barber, 'James of Molay', pp. 101–2; D. P. Waley, 'Papal Armies in the Thirteenth Century', *EHR* 72 (1957), 26; id., *The Papal State in the Thirteenth Century* (London, 1961), 243–9.

[58] J. Leclerq, 'Un sermon prononcé pendant la guerre de Flandre sous Philippe le Bel', *Revue du Moyen Âge latin*, 1 (1945), 164–72, esp. p. 170; E. H. Kantorowich, '*Pro patria mori* in Medieval Political Thought', *AHR* 56 (1951), 484; G. M. Spiegel, '"Defense of the Realm": Evolution of a Capetian Propaganda Slogan', *Journal of Medieval History*, 3 (1977), 115–33.

1303) of injuring 'public welfare, the growth of the Catholic faith, the preservation of the freedom of the Church, and the aid of the Holy Land'.[59]

To sum up, though during the pontificate of Boniface VIII little was done to see in motion a crusade for the recovery of the Holy Land, the two terms 'crusade' and the 'Holy Land' were often used throughout Christendom as very effective slogans and battle-cries. Often combined, those two terms never disappeared and thus kept awareness of the problem alive. It was partly this frame of mind which explains how one of the last great performances of the medieval papacy, the papal jubilee of 1300, was associated with the recovery of the Holy Land. It was precisely this focal year 1300 which brought to the fore emotions and currents of thought which would have otherwise passed undetected. Their almost dramatic appearance during Boniface's pontificate complement the picture of the pope and of European public opinion in relation to the *negotium Terrae Sanctae*.

The Papal Jubilee of 1300 and the Holy Land

The emotions of the Europeans and their attitudes both towards the Holy Land and the crusade during the 'Years of Transition' are well reflected by their reaction to what came during the year of the papal Jubilee of 1300 to be regarded as the Mongol recovery of the Holy Land—for Christendom.

In December 1299 the Mongol il-khan of Persia Ghazan (1295–1304) invaded Syria. Assisted by King Hetoum II of Lesser Armenia he occupied Aleppo, defeated the Egyptian army before Homs on 23 December, and on 6 January 1300 (Epiphany) entered Damascus.[60] On the march, probably somewhere between Aleppo and Damascus, he wrote (21 October 1299) to the king of Cyprus and to the master-generals

---

[59] Boniface VIII, *Registres*, no. 5342; see also ibid., nos. 2886, 4127, 4395, 4425; *MGH: Legum Sectio IV, Constitutiones*, vol. iii, no. 546, pp. 515–16.

[60] R. Röhricht, 'Études sur les derniers temps de royaume de Jérusalem', *AOL* i (1881), 646–8; J. Somogyi, 'Adh-Dhabai's Record of the Destruction of Damascus by the Mongols in 699–700/1299–1301', in S. Löwinger and J. Somogyi (eds.), *Ignace Goldziher Memorial Volume*, i (Budapest, 1948), 353–86.

of the Temple and Hospital, inviting them to join the expedition. The letter was received in Cyprus on 3 November. King Henry II and the master-generals, however, could not agree on a common plan of action. Although Ghazan sent a second message by the end of November 1299, no action was taken.[61] Meanwhile the Mongol and Armenian troops raided the country as far south as Gaza. According to an Armenian source confirmed by Arab chroniclers, Hetoum II with a small force reached the outskirts of Cairo and then spent some fifteen days in Jerusalem visiting the Holy Places.[62] Ghazan himself returned to Persia in February 1300, leaving behind him in Syria a part of his troops under the Emir Mulai. The weakening of the Mongol forces enabled the Mamluks to reoccupy Syria. Thus, from the beginning of January till the end of May 1300, there were no Mamluk forces left in Syria. For that brief period the Mongol il-khan was *de facto* lord of the Holy Land.[63]

The Mongol interlude showed the Christians how easily the Holy Land could be conquered. Christendom reacted immediately by a series of attempts to achieve that conquest. Guy of Ibelin, lord of Beirut and Jaffa, and John of Antioch set out late in 1299 for Byblos to join the king of Lesser Armenia, who stayed there with Ghazan. On their arrival they discovered that Ghazan had already left Syria and they returned to Cyprus. In the following summer (1300) a fleet of sixteen galleys with some smaller vessels was equipped by King Henry II of Cyprus, the Temple, and the Hospital. With the king was his brother Amalric of Tyre, the master-general of the Temple James of

---

[61] Florio Bustron, *Chronique de l'île de Chypre*, ed. M. L. de Mas-Latrie (Paris, 1886), 129–30; Francesco Amadi, *Chronique, Chroniques d'Amadi et de Strambaldi* ed. M. L. de Mas-Latrie (Paris, 1891), 234–5. See also a letter of Pietro Gradenigo doge of Venice to Pope Boniface in *RIS* NS xiii. 396–8.
[62] Moufazzal Ibn Abil Fazail, 'Histoire des Sultans Mamluks', ed. E. Blochet, *Patrologia Orientalis*, 14 (1920), 622–72, and 20 (1929), 17–94; An Nuwairi in D. P. Little, *Introduction to Mamluk Historiography* (Wiesbaden, 1970), 24–7; Makrisi, *Histoire des Sultans Mamluks*, ed. and trans, M. Quatremère (Paris, 1837–45), 134–72; Ptolemy of Lucca, 'Annalen', ed. B. Schmeidler, *MGH SRF* viii (Berlin, 1955), 236; 'Chronique de royaume de la petite Armenie', *RHC Hist. arm.* i. 659–60. The conquest of Jerusalem by the Mongols was confirmed by Niccolò of Poggibonsi who noted (*Libro d'Oltramare 1346–1350*, ed. P. B. Bagatti (Jerusalem, 1945), 53, 92) that the Mongols removed a gate from the Dome of the Rock and had it transferred to Damascus.
[63] Hayton, 'Flos historiarum Terre Orientis', *RHC Hist. arm.* ii. 319–21; D. Sinor, 'Mongols and Western Europe', in *Setton's History*, iii. 535.

Molay, the master-general of the Hospital, and Isol le Pisan (Zolus Bofeti), Ghazan's nuncio to Henry II. The expedition left Famagusta on 20 July 1300 and, after briefly raiding Alexandria, Acre, Tortosa, and Maraclea, returned to Cyprus.[64] At that time a messenger from Ghazan arrived in Cyprus with the news that the khan was again preparing for a campaign in Syria and invited the Cypriots to meet him in Lesser Armenia. Amalric of Tyre raised a force of about 300 men, which was joined by the Templars and Hospitallers with a similar contingent. They sailed to the island of Ruad, opposite Tortosa, and captured that city, but retired a few days later to Ruad, as their allies did not appear. The Mongols failed to arrive until the following February (1301), delayed by torrential rains. Accompanied by the king of Lesser Armenia, the Mongols, commanded by one of Ghazan's emirs, raided Syria as far as Homs and withdrew. By this time Amalric's and most of the Hospitallers' forces had returned to Cyprus.[65] In Ruad there now remained a predominantly Templar garrison under the command of Bartholomew, the marshal of the order. It consisted of 120 knights, 500 archers, and 400 servants, men and women. James of Molay apparently attempted to organize Ruad as a base for attacks on the Saracens. The garrison left in Ruad, however, was in a precarious position. It lacked both ability and resources. When in 1302 the sultan of Egypt sent against them one of his emirs with a fleet, the Templars had no galleys to oppose it. King Henry II and the masters general of the orders contemplated sending a relief expedition and had

[64] 'Templier de Tyr', in *Gestes des Chiprois*, ed. G. Raynaud (Geneva, 1887), 302–3; Marino Sanudo Torsello, 'Liber Secretorum Fidelium Crucis', in *Gesta Dei per Francos*, ed. J. Bongars (Hannau, 1611; repr. Jerusalem, 1973), 241–2; Florio Bustron, *Chronique*, pp. 130–4; Francesco Amadi, *Chronique*, pp. 235–9; J. Richard, 'Isol le Pisan: Un aventurier franc gouverneur d'une province mongole?', *Central Asiatic Journal*, 14 (1970), 186–94. Among the participants of this expedition was a Raymond Visconti who was a castellan of Tyre. Another participant was Baldwin de Pinquenin, a Cypriot admiral: see 'Templier de Tyr', p. 303; J. Richard, 'Un partage de seigneurie entre Francs et Mamelouks: Les "Casaux de Syr"', *Syria*, 39 (1953), 73 n. 2. On 25 Feb. 1300 the Templar preceptor Petrus de Vares hired in Famagusta a vessel from the Genoese Petrus Rubeus 'for a departure to Syria, Tortosa, Tripoli, Tyre and Acre': 'Actes passées à Famagouste de 1299 à 1301 par le notaire génois Lamberto di Sambuceto', ed. C. Desimoni, *AOL* 2 (1884), 42.

[65] Florio Bustron, *Chronique*, pp. 132–4; Francesco Amadi, *Chronique*, pp. 237–9; 'Templier de Tyr', pp. 305–6; *Papsttum*, ed. Finke, vol. ii, pp. 3–5; Bulst-Thiele, *Domus*, pp. 310 n. 73, 366.

assembled ships in Famagusta. They were too late. In October 1302, after a siege that lasted almost a year, the Templars surrendered and the Moslems seized the islands. The survivors, about 280, were taken as prisoners to Cairo.[66] Though the unfortunate expedition to Ruad was recorded by one European chronicler only, John of Paris, this incident must have been known and certainly remembered. It became almost a *cause célèbre* when, during the trial of the Templars, Gerard of Villiers, preceptor of the Temple in France, was accused with another knight of being responsible for the loss of the island and for the deaths of the brothers stationed there.[67]

Two other attempts were a 'Genoese crusade' and that of Roger Lauria. Following the arrival of the news about the Mongol victories in Syria, Genoa witnessed an outbreak of crusader zeal: several noble ladies, among them a Spinola, a Doria, and a Grimaldi, sold their jewellery to finance a fleet intended for Syria. This expedition, described by Boniface as *passagium quasi particulare*, for which he granted papal permission (9 August 1301), was to be led by Benedetto Zaccaria and assisted by Franciscans of the province of Genoa. A son of one of the city's leading families, Benedetto had gained the reputation of being among the ablest negotiators and most daring seamen of his time. His involvement in the crusader East went back to 1288, when he was sent to Tripoli with full powers of the republic to negotiate beyond the seas. After negotiating with the claimants to the crown of the 'kingdom of Acre', he performed the more useful action of transporting inhabitants of Tripoli, shortly before its fall (1289), to Cyprus. In 1300 Benedetto, with the aid of the ladies of his native city, equipped several vessels for the expedition to the Holy Land. The participants of the expedition were granted full crusader indulgences. Although the Genoese enthusiasm for their own crusade was immense, the whole project was dropped when the Genoese learned about a papal prohibition of the conquest and

---

[66] Francesco Amadi, *Chronique*, p. 239; Marino Sanudo Torsello, 'Liber', p. 242; Bulst-Thiele, *Domus*, p. 311; Hill, *Cyprus*, ii. 215–16; Barber, 'James of Molay', pp. 98–9; id., *Trial of the Templars* (Cambridge, 1978), 13–14. Later, in 1306, Henry of Cyrpus was accused by the Cypriot barons of deliberately causing the fall of Ruad; see 'Documents chypriotes', ed. C. Kohler, *ROL* ii (1905/8), 447–8.

[67] John of Paris, 'Memoriale historiarum', *RHGF* xxi. 640; Bulst-Thiele, *Domus*, p. 311 n. 77; J. Michelet, *Procès des Templiers* (Paris, 1841), i. 38–9; Barber, *Trial*, p. 128.

the fortification of towns in Syria, and especially of Tripoli, in the interest of any particular individual or commune. Benedetto Zaccaria who was apparently interested in winning Tripoli for Genoa, proceeded thereupon to occupy the island of Chios instead (1304).[68] Later in 1303 Roger Lauria the 'admiral of the kingdoms of Sicily and Aragon', who on behalf of James II commanded the operations in Sicily, offered to maintain permanently ten armed vessels against the enemies of the Church, the Saracens, and those who broke the papal ban on trade with Egypt and Alexandria. After being granted special privileges from the Holy See (20 April 1303), Lauria made abortive preparations for a descent on Syria, which terminated in an amalgamation of his forces with those of the Catalan Company.[69]

The series of attempts to conquer the Holy Land was only a part of the reaction of Europeans. Narratives as well as numerous letters prove that between February and September 1300 many Christians in the West, including Pope Boniface VIII, laboured under the impression that the Holy Land, including Jerusalem with the Holy Sepulchre, had been conquered by the Mongol Ghazan from the Moslems and was to be handed over to the Christians. Contemporary European chroniclers present for that year a detailed and uniform description of this recovery. The majority of accounts ascribed the conquest to a Mongol il-khan, joined by the Christian kings of Lesser Armenia and Georgia, and in some versions by the Lusignan king of Cyprus. 'Cassanus', as the Mongol il-khan was referred to by the chroniclers, was regarded by some as the divine instrument to bring back the Christians into the Holy Land.[70] The pagan Ghazan appears even in the official version of the events, namely in the letters of Pope Boniface VIII. Writing to Edward

---

[68] Boniface VIII, *Registres*, nos. 4380–6; *AE* ad an. 1301, nos. 33–4; Golubovich, iii. 29–36; Röhricht, 'Études', *AOL* i. 648–51; Y. Renouard, *Hommes d'affaires italiens au Moyen Âge* (Paris, 1949), 99–100; G. Caro, *Genua und die Mächte am Mittelmeer 1237–1311: Ein Beitrag zur Geschichte des 13. Jahrhunderts* (Halle, 1895–9), ii. 316–19; Digard, *Philippe le Bel*, vol. ii, p. 129 n. 2.

[69] Boniface VIII, *Registres*, nos. 5168–71; R. I. Burns, 'The Catalan Company and the European Powers 1305–1311', *Speculum*, 29 (1954), 755.

[70] See e.g. 'Continuationes Anglicae Fratrum Minorum', *MGH SS* xxiv. 258; 'Chronicon Lemovicense Brevissimum', *RHGF* xxi. 807; 'Annales Frisacenses', *MGH SS* xxiv. 67; Bohemond of Trier, 'Gesta', *MGH SS* xxiv. 481–2; S. Schein, '*Gesta Dei per Margolos 1300*: The Genesis of a Non-Event', *EHR* 95 (1979), 805–8.

I (on 7 April 1300), Boniface announced the good tidings from the East. The Holy Land, he said, had been recovered from the Saracens by the acts of 'viri magnifici gentis Tartaricae dominantis, qui non renatus fonte baptismatis, nondum orthodoxae fidei lumini illustratus'.[71] Boniface produced an even more detailed version of the events a year later (on 26 February 1301). God, Boniface affirmed, wanted to put an end to the misery of the deserted Holy Land; as there was not even one among the sons of the Church who wished to help, he chose a Mongol ruler, a pagan, to liberate it from the hands of the Saracens.[72]

Like Pope Boniface himself, the chroniclers who reported the recovery of the Holy Land often returned to its loss in 1291. Thus the recovery was seen as God's vengeance. English and German chroniclers linked the recovery with the liberation of Christians captured in Acre. The *Annales Regis Edwardi Primi* had the story of a certain Galfridus de Semari who was seized during the capture of Acre when on pilgrimage. He had been a captive for nine years when he was set free by Ghazan after his conquest of Babylon on 11 August 1300. The German Cistercian annals of Kaiserheim mentioned Christian captives in Damascus who after their liberation by Ghazan went to Cyprus. The linking of Ghazan's recovery of the Holy Land with the deliverance of the captives of Acre was not necessarily a piece of wishful thinking. In the 1300s some of the Christian prisoners held in Egypt were actually given their freedom following the intervention of James II of Aragon with the sultan.[73]

According to the chroniclers, the Mongol il-khan conquered not only the Holy Land but also Egypt. On 6 January 1300 (Epiphany) he sang mass, together with the king of Lesser Armenia, at the Holy Sepulchre. It was from Jerusalem that the

---

[71] For this letter see *Annales Regis Edwardi Primi*, ed. H. T. Riley (RS, London, 1865), 465–70; *Foedera*, 1. iv. 1–2.
[72] *Foedera*, 1. iv. 6–7.
[73] Pietro Cantinello, 'Chronicon', *RIS* NS xxviii. 93; Ptolemy of Lucca, 'Annalen', p. 236; *Annales Regis Edwardi Primi*, p. 442; 'Annales Caesarienses', ed. G. Leidinger, as 'Annales Caesarienses seu Kaisheimer Jahrbucher', in *Sitzungsberichte der Königlich Bayerischen Akademie der Wissenschaften*, 7 (1910), 36; A. S. Atiya, *Egypt and Aragon: Embassies and Diplomatic Correspondence between 1300 and 1330 AD* (Leipzig, 1938), 21 and n. 2; Golubovich, iii. 73, 82.

il-khan wrote to Boniface VIII asking for bishops, clergy, settlers, and peasants to be sent to the Holy Land.[74] One of the chroniclers remarked explicitly that the il-khan asked for soldiers, needed for defence only. The khan, it was also alleged, wrote to the masters-general of the Hospital and the Temple to come to the Holy Land and recover their castles.[75] Some of the chroniclers claimed to know about the actual departure to the East of clergy and contingents of the orders.[76] The return of the Christians to the Holy Land in 1300 was presented, especially by German and Italian chroniclers, as a kind of peaceful migration to settle the land under Mongol protection. Characteristically enough, a chronicler terminated his narrative by saying: 'Item rex Hermini cantavit missam in Jerusalem in die Ephifanie Domini, et Tartari bapticati sunt in flumine Gordano, omnes Saracenos quos inveniebant interficientes preter rusticos quos conservant ad colendum terras.'[77] These accounts gave the impression that the Holy Land was again in Christian hands as in the golden days of Godfrey of Bouillon, but this time without a crusade. The question thus arises as to how the news about a Mongol victory in northern Syria, accompanied by futile attempts of contingents of Cypriots and military orders to get a foothold in Syria, was transformed in 1300 into the story of the recovery of the Holy Land for Christendom. The accounts of the recovery of the Holy Land were based, it seems, either on flying rumours or on correspondence from Cyprus which transformed the Mongol–Saracen confrontation which actually took place into a Mongol recovery of the Holy Land. However, the collective image of what could be best described as a kind of 'Gesta Dei per Mongolos' derived from a more complex set of circumstances which accumulated in the year 1300. It can be said that had the news from Syria arrived in Europe at any other time, it would have never caused such an uproar.

[74] 'Annales Frisacenses', p. 67; Bohemond of Trier, 'Gesta', p. 483; Pietro Cantinello, 'Chronicon', p. 94; 'Continuationes Anglicae Fratrum Minorum', p. 258; Eberhard of Regensburg, 'Annales', *MGH SS* xvii. 599; 'Annales Forolivienses', *RIS* xxii. 176; William of Nangis, *Chronique latine de Guillaume de Nargis de 1113 à 1300 avec les continuations de cette chronique de 1300 à 1368*, ed. H. Géraud, i (Paris, 1843), 308.
[75] Bohemond of Trier, 'Gesta', p. 483.
[76] 'Annales Frisacenses', p. 67; 'Cronica Reinhardsbrunnensis', *MGH SS* xxx. 642.
[77] Pietro Cantinello, 'Chronicon', p. 94; 'Annales Caesenates', *RIS* xiv. 1120.

AD 1300 was the year of the jubilee in Rome. On 23 February of that year Pope Boniface issued the bull *Antiquorum Relatio* which promised 'the fullest pardon of their sins' to all those 'truly penitent and confessed' who made a pilgrimage to the basilica of the Holy Apostles. As the jubilee indulgences offered by Boniface had been popularly regarded as *poena et culpa*, the response was overwhelming and the influx of pilgrims from all over Europe was beyond all expectations. This influx together with the jubilee celebrations which were echoed all over Europe, played a major role in transforming an ephemeral event in the Levant into one of allegedly major significance in Christian–Moslem–Mongol relations in the East. This metamorphosis can be also explained by the positive image western Europe had at that time of the Mongols, which reinforced the credibility of their recovery of the Holy Land. The Mongols, still regarded in the early 1260s as soldiers of Antichrist or the apocalyptic Gog and Magog came to be regarded during the next decade as the glorious champions of Christianity against Islam and, moreover, a powerful ally against the Saracens. The numerous embassies dispatched to the West by the il-khans of Persia had aimed, since 1262, at a joint campaign in Syria. Moreover, in their letters to popes as well as secular rulers the il-khans, e.g. Hulagu in 1262 and Arghun in May 1289, promised that after their victory over the Mamluks they would freely hand over either Jerusalem or the entire Holy Land to Christendom.[78]

It was, however, the fortuitous coincidence that the news from the East reached Europe in 1300 that transformed the Mongol victory into a kind of 'Gesta Dei per Mongolos', an expression of God's will: 'A Domino factum est istud et est mirabile in oculis nostris' (Ps. 117: 23).[79] The mystery of a new century loosened the imagination of many Europeans and also produced a religious turmoil which found its outlet in eschatological expectations. The agitation at the turn of the century

---

[78] See above, ch. 3 n. 23.
[79] W. E. Lunt, *Papal Revenues in the Middle Ages* (Columbia University Records of Civilization, 19; New York, 1934), i. 121, ii. 452; J. Sumption, *Pilgrimage: An Image of Medieval Religion* (London, 1975), 231–6; Digard, *Philippe le Bel*, vol. ii, pp. 24–5; R. Folz, *Concept of Empire in Western Europe from the Fifth to the Fourteenth Century* (New York, 1969), 146; Bohemond of Trier, 'Gesta', p. 483.

was so high that in October 1299 the Council of Béziers condemned, in one of its canons, those who announced the end of the world and the advent, if not already the presence, of Antichrist.[80] In 1300 Fra Dolcino, the leader of the Apostolic Brethren, proclaimed in a manifesto, that in the Last Days God would send his own *congregatio* to preach the salvation of souls and lead the Church into the Fourth Period, when it would become *bona et pauper*. Dolcino and his followers expected that soon after 1300 the outward transfer of authority from the Roman Church, no longer the Church of God, but 'illa Babylon meretrix magna', to the Apostolic one, upon whom had descended the power of St Peter, would be accomplished by the violence of a revolution. Pope Boniface VIII, the cardinals, prelates, clergy, and all military and monastic orders would be exterminated by the sword of God wielded by a new emperor. Dolcino believed that this emperor would be Frederick III, king of Sicily. After the holocaust a new and holy pope would be chosen by God, and under his obedience would be placed the Apostolic Order, with all the remnants saved from the sword by divine grace and chosen to become Apostolic Brethren. This would be followed by a new outpouring of the Holy Spirit.[81] The so-called 'Prophecy of John of Parma', written in Italy after 1292, allegedly the result of a vision seen while John was lamenting the fall of Acre,[82] predicted *ex eventu* the fall of Tripoli and Acre. However, it also predicted the election of an Angelic Pope who together with an emperor, probably a German ruler, would go to Jerusalem 'cum pace et crucibus' and receive it gratuitously from the Saracens; then Antichrist would appear.[83] Thus it combined two themes: that of the Last Emperor, which already had a long history behind it, and that of the Joachimites on the Angelic Pope. Moreover, one extremely popular eschato-

[80] *Thesaurus novus anecdotorum*, ed. E. Martène and U. Durand (Paris, 1717), iv. 225–8; Digard, *Philippe le Bel*, vol. ii, pp. 15–16.

[81] Reeves, *Joachim of Fiore*, pp. 47–8; R. E. Lerner, 'An "Angel of Philadelphia" in the Reign of Philip the Fair: The Case of Guiard-Cressonessant', in W. C. Jordan *et al.* (eds.), *Order and Innovation in the Middle Ages: Essays in Honour of J. R. Strayer* (Princeton, NJ, 1976), 354–5.

[82] 'Vision seu prophetia fratris Johannis', ed. E. Donckel, in *Römische Quartalschrift*, 40 (1932), 367–74; Lerner, 'Medieval Prophecy', pp. 11, 15.

[83] *Aus Tagen Boniface VIII*, i. 216–23; 'Notices et extraits de documents inédits relatifs à l'histoire de France sous Philippe le Bel', ed. E. P. Boutarie, *Novices et extraits des manuscrits de la Bibliothèqie Nationale*, 20 (1862), 235–7.

logical jingle predicted the birth of Antichrist for 1300.[84] It had been stated also that this year, mystically interpreted, was to mark the final disappearance of Islam.[85] Thus the year 1300, more than any other, was propitious for miracles. When in the spring news from Syria began to arrive in Europe, the city of Rome played a major part in its diffusion. The pope, it was said, had ordered solemn processions to celebrate the recovery of the Holy Land, and offered full crusader indulgences to those who would go there as pilgrims or settlers.[86] It is possible, however, that some of these so-called celebrations of the recovery of the Holy Land were in fact celebrations of the jubilee.

The sources reveal active crusader propaganda in Brabant in 1300.[87] At the same time John of Brittany took the Cross.[88] Already on 18 May 1300 James II of Aragon had written from Lerida to Ghazan as 'king of the kings of all the Levant' and as one 'elected by the Omnipotent to take revenge on his enemies and recover the Holy Land'. James offered the supposed conqueror ships, supplies, and men for a fifth of the regained Holy Land and for free access to the Holy Places and to the Holy Sepulchre.[89] Ramon Lull, having heard in Majorca about Ghazan's victory, hurried to Syria to meet the il-khan. He sailed early in 1301 and progressed as far as Cyprus, where he found out that the news was false. It seems that Lull's disappointment with Ghazan affected his firm belief in the conversion of the Mongols.[90]

In fact Ghazan's career as the il-khan of Persia, turned into a fable by the westerners, seems to have caused embarrassment to the 'eastern' historians like Hayton, the 'Templar of Tyre', and later on Marino Sanudo Torsello. It was precisely Ghazan's accession to the throne of Persia in 1295 that marked

---

[84] Lerner, 'Refreshment', p. 138.
[85] Röhricht, 'Études', *AOL* 1. 649 n. 75; Throop, *Criticism of the Crusade*, pp. 134–5.
[86] 'Annales Frisacenses', p. 67, Bohemond of Trier, 'Gesta', p. 483.
[87] 'Martini Continuatio Brabantina', *MGH SS* xxiv. 261.
[88] Boniface VIII, *Registres*, nos. 3719–20. John was killed in an accident which occurred in Lyons during the coronation of Clement V on 14 Nov. 1305: see J. Petit, *Charles de Valois 1270–1325* (Paris, 1900), 104–5.
[89] *AA*, vol. iii., no. 42, pp. 91–3, see also no. 43, p. 95.
[90] Ramon Lull, 'Vita Coaetanea', ed. H. Harada, *CCCM* xxxiv (1980), 294–5; J. M. Victor, 'Charles de Bovelles and Nicholas Le Pax: Two Sixteenth Century Biographies of Ramon Lull', *Traditio*, 32 (1976), 327–8; Ramon Lull, 'Liber de fine', ed. A. Madre, *CCCM* xxxv (1981), 267.

the definite triumph of Islam among the Mongols. Ghazan abandoned the traditional policy of tolerance and, abjuring Buddhism a few days after his succession, adopted sunni Islam. It was also in the first years of his reign that the Christians in Iran suffered a great decline of fortune. Rashid ad-Din asserted that Ghazan invaded Syria in 1299 in a mood of religious zeal; he had been outraged by a group of Egyptians who captured Mardin and drank wine in the mosques during the Holy Fast of Ramadan.[91] According to Hayton's opinion, which is corroborated by other sources, Ghazan's pro-Moslem attitude changed and gave way, as early as 1298, to plans for an alliance with the Christian world and he became more tolerant towards the Christians in his kingdom.[92] Ghazan, it seems, did not let his Moslem religious feelings cloud his political judgement. He recognized the Mamluks as his principal enemy and it was to propose a joint expedition against the sultan of Egypt that he sent his two embassies of 1300–1 and 1302–3 to Europe. Ghazan's letter to Boniface VIII dated 12 April 1302 suggests that, having received an encouraging message from the pope, he was counting on Christian participation in his expedition to Syria in 1303 and that his sole aim was the 'great work' i.e. the war against the Mamluks.[93] In the spring of that year he sent a new army into Syria. Led by a certain Gotholosa and the king of Lesser Armenia, this army was defeated by the Mamluks in Marj-ab-Suffar, near Damascus (20 April 1303). Ghazan died in the following year (11 May 1304) when preparing for yet another expedition to Syria.[94] The negotiations between the Christian West and the Mongols of Persia continued in the

[91] Rashid-ad-Din, *Jami-al-tabarikh*, ed. Alizade (Baku, 1957), 332–3; J. Richard, 'The Mongols and the Franks', *Journal of Asian History*, 3 (1969), 55 n. 42; Sinor, 'Mongols', pp. 535–7.

[92] Hayton, 'Flos', p. 316; 'Templier de Tyr', p. 297; Marino Sanudo Torsello, 'Liber', p. 239; Paulinus of Venice, 'Chronologia Magna', in Golubovich, ii. 97.

[93] Sinor, 'Mongols', pp. 536–7; L. Petech, 'Marchands italiens dans l'Empire mongol', *Journal asiatique*, 250 (1962), 564–6; A. Mostaert and F. W. Cleaves, 'Trois documents mongols des archives secrètes vaticanes', *Harvard Journal of Asiatic Studies*, 15 (1952), 467–71; L. Lockhart, 'Relations between Edward I and Edward II of England and the Mongol Il-Khans of Persia', *Iran: Journal of the British Institute of Persian Studies*, 6 (1968), 29; J. A. Boyle, 'The Il-Khans of Persia and the Prince of Europe', *Central Asiatic Journal*, 20 (1976), 37–8.

[94] Hayton, 'Flos', pp. 321–4, 330; J. B. Chabot, 'Histoire du patriarche Mar Jabalaha et du moine Rabban Cauma', *ROL* 2 (1894), 252–3; Röhricht, 'Études', *AOL* 1. 648.

years which followed. During the pontificate of Clement V a military alliance was often recommended in the *de recuperatione Terrae Sanctae* treatises of that period. However, the events of 1300 were not without effect upon the attitude of the Europeans to the Mongols. Disappointment such as that evidenced by Ramon Lull must have been shared by many others. After 1300 not only did hope in the conversion of the Mongols diminish, but also belief in the possibility of co-operating with them against the Moslems.[95]

The reaction of Europe to the news from the Holy Land in 1300, the avalanche of correspondence and chronicler accounts, reflect an immense interest taken in the Holy Land by the West in general, and by Pope Boniface in particular. The pope received the news about the recovery of the Holy Land shortly after 19 March 1300. An unsigned letter of this date, which was addressed to him, came from Venice. Its contents suggest that it was written by the Doge Pietro Grandenigo (1289–1311). The correspondent from Venice drew his information from Venetian merchants, who had just arrived from Cyprus, which they left on 3 February. His report is a more or less accurate description of the situation in Syria just before Ghazan's return to Persia (6 February 1300). Assuming that the Mongols were in full possession of the Holy Land, the doge terminated his letter by congratulating Pope Boniface as the pontiff elected by God to have the Holy Land recovered during his pontificate. He also expressed a pious wish to follow the tradition of his predecessors, 'doges and Venetians, in their devoted efforts to aid the Holy Land'. Offering Venetian assistance, the doge asked the pope for instructions.[96] The pope reacted almost immediately. On 7 April 1300 he informed Edward I and possibly other monarchs of Europe of the 'great news ... joyful news to be celebrated with special rejoicing', namely the conquest of the Holy Land by the pagan Ghazan and his offer to hand it over to the Christians. The pope went on to exhort the rulers of the West to unite their forces with those of the conquerors for the aid of the Holy Land (*subsidium ac succursum*) as well as its future prosperity. The believers were exhorted to depart for the newly recovered Holy Land. As a

[95] See below, ch. 6.   [96] *RIS* NS xiii. 396–8.

passagium generale was not expected, the pope wrote that all those who had already taken the Cross should depart and thus fulfil their oath. Edward I was requested to encourage his subjects to visit the Holy Places; the prospective pilgrims were promised full crusade indulgences. All of the former Outremer prelates were requested to return to their posts.[97] In the *Processus adversus Fredericum et Siculos rebellos* published on the same day (7 April 1300), Boniface did not omit to mention 'the miraculous liberation of the Holy Land from the hands of the Saracens'.[98] Writing on 20 September 1300 to the archbishop of Nicosia, the pope referred to the preservation of the recently recovered Holy Land. Isol le Pisan was referred to as 'vicarius Siriae ac Terrae Sanctae a Casano imperatore Tartarorum institutus'. This title, used by the papal chancery, points to an office ascribed by the pope to Isol; he was regarded as Ghazan's vicar nominated to supervise on his behalf the re-establishment of the Franks in the territories returned to them. This information may have derived from the Florentine Guiscard Bustari, an ambassador of the il-khan, who arrived in Rome in the summer of 1300. His embassy was mentioned by several chroniclers and in some letters. It was said that the envoy of the il-khan was accompanied by a retinue of one hundred men, all clad in Tartar garments. According to Florentine accounts, Boniface VIII received at the time of the jubilee a dozen ambassadors, all of them Florentines, dispatched to him by various kings and princes. Among them was Guiscard Bustari and his hundred companions.[99] As to the pope, he was no doubt very proud of the fact that the Holy Land was to be returned to Christendom during his pontificate, yet with the Sicilian war on he was in no position to do much for the maintenance of this conquest.[100]

The European accounts of the events were, on the whole, uniform. The image is in fact a function of the correspondence which referred to the events in Syria. The number of surviving letters is strikingly large as compared with a single letter, that of John of Villiers, which had described the fall of Acre only nine

[97] See above, n. 71.
[98] Boniface VIII, *Registres*, no. 3879.
[99] Richard, 'Isol le Pisan', pp. 188–90, 193; Petech, 'Marchands', pp. 563–6. In a document drawn up on 25 May 1301 in Famagusta Isol appears as 'misaticus . . . Casani imperatoris Tartarorum', *Notai genovesi*, vol. iii, no. 381, p. 457.
[100] Boniface VIII, *Registres*, nos. 3879, 3917.

years earlier. A comparison between the chroniclers and the letters points to contacts between the two on a regional level. The letters of 1300 became the main sources from which the chroniclers drew their information for their fabulous accounts of the events. The contents of those letters vary in details only, which explains how the uniform, collective image came into being.[101] Nevertheless there were some regional differences between the chroniclers' accounts. The Italians and Germans emphasized and developed the legendary aspects. The recovery of the Holy Land became for them a part of eschatological expectations, which were usually stronger in Germany and Italy than in England and France. Ghazan thus became a tool in the almost idyllically peaceful return of the Christians to the Holy Land. The English accounts were meagre, especially when compared with their accounts for 1291–2. This is surprising, as by 1295 the crusade was being actively preached in England.[102] The French accounts, which were the most realistic, pointed to the increase of French involvement in the affairs of the Holy Land.

The years 1300, which could have marked a turning-point as far as the crusade was concerned, passed by without any tangible result. As soon as the Europeans, including Pope Boniface, learned about the actual state of affairs in the Middle East, the rejoicing was over; the excitement and uproar died as quickly as they arose. Only few chroniclers reported for 1301 the Mamluk reoccupation of the Holy Land.[103]

The Scheme of Benedict XI: A General Crusade to the Holy Land through Byzantium

The pontificate of Boniface's successor the Dominican general Nicholas Boccasini, Benedict XI (23 October 1303–7 July 1304), was important mainly as a transitional period between

[101] *Hagnaby's Chronicle*, fos. 474–84; 'Annales Caesarienses', pp. 3–37; 'Bericht über die Schlacht bei Hems am 23 Dec. 1299', ed. W. Wattenbach, in *Neues Archiv*, 4 (1879), 207–8; Heidelberger, *Kreuzzugsversuche*, pp. 81–4.
[102] John of Pontissara, *Register*, pp. 191–3; Robert Winchelsey, *Register*, ed. K. Graham (Canterbury and York Society, 51–2; 1952–6), i. 26–30.
[103] See e.g. *Grandes Chroniques de France*, viii. 191–2; William of Nangis, *Chronique*, i. 311.

the pontificates of Boniface VIII and Clement V. Benedict XI's policy was aimed at the restoration of papal prestige. The organization of a crusade was for him one of the main means to this end. However Benedict XI, like his predecessor Boniface VIII, never came close to a real attempt to organize a crusade. In this aspect his pontificate marks the continuation as well as the end of the period that began with the death of Nicholas IV in 1292. At the same time, it was Benedict XI who, to some extent, prepared the ground for the resumption of the attempts to launch a crusade by his successor Clement V. Benedict's policy, after eight years of conflict, reconciled France with the papacy. When papal crusader activity was resumed, Philip IV, now an ally as well as host of the pope, could have been considered, if not as the potential leader of the planned *passagium generale*, at least as an active co-operator in the pope's projects. It was also Benedict XI who, supporting the plan of Charles of Valois for the reconquest of Constantinople, initiated one of the crusader projects of Clement V. This plan, which combined the reconquest of Constantinople with the recovery of the Holy Land, was to dominate Clement V's crusader policy during the first years of his pontificate and was not abandoned even when other projects were adopted.[104]

With the peace between France and the Holy See restored, Charles of Valois requested the aid of Benedict XI for the recovery of Constantinople. Like Charles I of Anjou before him, Charles of Valois was fully aware of the fact that for such a plan the help of the papacy was absolutely necessary. If the enterprise was declared a crusade its leader would not only obtain money from crusading subsidies, but his followers would receive all the privileges and indulgences attached to the crusade. Early in 1304 he had therefore requested Pope Benedict to allow men who took the Cross to fulfil their obligations by following him to Constantinople. The position taken by Benedict was the traditional one. He originally refused to order the collection of a tithe for the expedition unless it was to form part of a general crusade. For this pope, Byzantium was a secondary objective; the primary aim was the recovery of Jerusalem.

---

[104] Digard, *Philippe le Bel*, vol. ii, pp. 186–209; Brandt, in his trans. of Pierre Dubois, *Recovery*, pp. 25–8; below, ch. 6.

Later, however, Benedict XI's position changed. He agreed to grant those who would join Charles of Valois crusading privileges and indulgences. He urged believers to help him, bestowing upon them crusader indulgences unless a general crusade preceded his expedition. Under the same condition he granted Charles the income from the commutation of crusader vows in France, as well as the profits from crusader revenues and redemptions, except the tithes accorded for the Holy Land. Writing on 20 June 1304 to the bishop of Senlis, Benedict referred, as other popes had done during the reign of Michael VIII Palaeologus (1261–82), to the difficult position of the Byzantine Empire, which could not oppose the Turks for long. It was, he felt, the duty of good Christians to bring the schismatics back to the Catholic Church, and at the same time prevent the conquest of a large and ancient part of Christendom by the Turks. Moreover, the reconquest of the Byzantine Empire by the combined forces of the West could aid the recovery of the Holy Land.[105]

It is probable that Benedict XI referred here to expert advice which stressed the importance of the Byzantine Empire in relation to the Holy Land. Advice to that effect was produced by Frederick III of Sicily early in 1304. At that time both Frederick and James II of Aragon, involved in the affairs of the Catalan Company, began to make plans for the conquest of Constantinople, on their own account. Already in 1304 Frederick had informed the Holy See of his intention to conquer Constantinople. Writing to Benedict XI (7 July 1304), Frederick requested permission to send to Byzantium a fleet of ten ships under the command of his half-brother Sancho of Aragon, stressing how useful these lands would be as a base against Egypt. The occupation of the Greek islands might be a first blow against Egypt; if the Catholics held these islands, they could undermine the economic and political strength of the Egyptians by a boycott of their trade.[106] Frederick was, it seems, familiar with the ideas advocated by Marino Sanudo Torsello. The two probably met when the latter stayed in 1300 at the court of Palermo. He had already referred to the necessity

---

[105] Benedict XI, *Registres de Benoît XI*, ed. C. Grandjean (Paris, 1905), nos. 1006–8; Petit, *Charles de Valois*, p. 91; Laiou, *Constantinople*, pp. 138, 202–3.
[106] *AA*, vol. ii, no. 431, pp. 681–4; Burns, 'Catalan Company', pp. 755–6.

of the conquest of the Greek islands in the first draft of his 'Liber Secretorum Fidelium Crucis', the so-called 'Conditiones Terrae Sanctae' composed between 1306 and 1309.[107]

It follows from the overtures of men like Frederick III of Sicily, James II of Aragon, Charles of Valois, and later Philip of Taranto, that in order to receive the support of Benedict XI, they had to link their plans for an attack on Byzantium to that for a crusade to the Holy Land, and that crusade in turn with a plan for the naval blockade of Egypt. With respect to this last-named objective, Benedict XI was following in the footsteps of his predecessors. During the nine months of his pontificate he not only twice renewed their prohibitions of trade with Egypt, but added to their number.[108] On 18 November 1303 he renewed the decrees of the Second Council of Lyons and of Nicholas IV, and extended the penalties to exports of food, wine, and oil. This renewal and extension were repeated on 26 March 1304. Eight days later he appealed to the council and commune of Venice that they should observe the prohibitions for the sake of the Holy Land. At the same time, however, he declared that the Venetians might export to Egypt all products not specifically mentioned in his bulls, such as cloth and clothing.[109] He also granted an absolution to three merchants of Ragusa who in 1304 were caught returning from Egypt, as well as to a large number of Genoese and Venetians, who were excommunicated for the same offence of trading with the Saracens.[110]

The relaxation of the blockade of Egypt, especially as far as Venice was concerned, was possibly connected with the plans of Charles of Valois and aimed at winning over the maritime republics. As a policy of appeasement it met with some success. Despite the Venetian–Byzantine truce of 1302, the Venetians concluded (in 1306) some sort of alliance with Charles of Valois, and thus solved the problem of transport. At the same time, however, the new papal policy encouraged Venice to

[107] J. Prawer's foreword to Marino Sanudo Torsello, 'Liber', pp. vii–viii. The still unedited 'Conditiones' became later the first book of the 'Liber': see ibid. 29–30.

[108] *AA*, vol. ii, no. 431, p. 682; Burns, 'Catalan Company', p. 756 n. 18.

[109] Benedict XI, *Registres*, nos. 1101, 1315; *Diplomatarium Veneto-Levantinum*, ed. G. M. Thomas (Venice, 1880–99), v. 19–21; Heyd, *Histoire du commerce*, ii. 26.

[110] Benedict XI, *Registres*, nos. 86, 90, 351, 769, 819, 861; Heyd, *Histoire du commerce*, ii. 50 and n. 3.

pursue its traditional interests in the Levant. The commune had already renewed its formal relations with the sultan in 1302, when its ambassador Francesco de Canali was installed as consul in Alexandria. It was then granted a charter by Qalawun in which he confirmed the concessions accorded to it previously. In 1304, the Venetians re-established a foothold in Syria. In the summer of that year the commune dispatched two envoys to the emir of Safed and was granted free access to the Holy Sepulchre as well as assurances for the well-being of its people and their property in Syria. Genoa, which only few years earlier, in 1301, had witnessed an outbreak of crusader zeal, preferred in 1304 to become the only western friend of the Byzantine Empire.[111]

Thus in the autumn of 1304 the eyes of Europe were turned to Constantinople. The Catalans were already there; their famous company had appeared in the East only a few months earlier (1303). Deprived of employment by the signing of the peace of Caltabelotta (31 August 1302), this powerful group of adventurers from Catalonia and Aragon made its way to the East, where it offered its services to Andronicus II Palaeologus (1282–1328). Most of the provinces of Byzantine Asia Minor were already overrun by the advancing Turk tribes and the remainder of Asia Minor was in grave peril. In September 1303 Andronicus therefore accepted the offer of Roger de Flor, the leader of the Catalan Company against the Turks in Bithynia. The Catalans defeated the Turks in several campaigns in Asia Minor. However, soon emboldened by their success and disgruntled by the irregularity of their pay, the Catalans began to pillage Byzantine territory around Constantinople. Relations between them and the Greeks grew increasingly tense until 30 April 1305, when Roger was suddenly assassinated in the palace of the imperial prince Michael IX.[112]

In 1305 Sancho of Aragon, who had received the moral backing of the Holy Father, launched his expedition. In April of that year he arrived with his fleet in Romania. As he could not agree on the terms of co-operation with the Catalans, Sancho

---

[111] *Diplomatarium*, v. 30–2; M. L. de Mas-Latrie, 'Traité des vénitiens avec l'émir d'Acre en 1304', *AOL* I (1881), 406–8; Heyd, *Histoire du commerce*, ii. 37–41; Laiou, *Constantinople*, pp. 147–57.

[112] Geanakoplos, 'Byzantium', pp. 45–7.

was reduced to attacking the Greek islands on his own. He stayed for some time around Gallipoli to give naval protection to the Catalans, but after the end of May 1305 he left for Sicily. The Catalans shut themselves up in the fort of Gallipoli and raised on its walls the standard of St Peter, proclaiming themselves crusaders fighting for the Church. Meanwhile a great fleet was being prepared in Sicily. By spring 1305 most of the European powers, France, Venice, Aragon, Naples, and Sicily, were openly committed to the winning of Constantinople. At that moment a new pope, Clement V, was elected (5 June 1305). His main aim was the recovery of the Holy Land. The plan for the conquest of Constantinople had to be incorporated into his own.[113]

[113] *AA*, vol. ii, no. 431, p. 682; Burns, 'Catalan Company', pp. 755-6 and n. 18; Laiou, *Constantinople*, pp. 138-45.

6

1305–1308
IN SEARCH OF A PROJECT

The elevation of Clement V to the papal throne (14 November 1305) marked the beginning of serious and intensive efforts to launch a crusade. It was during this pontificate that many of the treatises of the genre *de recuperatione Terrae Sanctae* were composed. No less than fifteen treatises are extant; no year passed without the proposal of at least one project, and attempts were made to implement them. It is not easy to say why the nine years of the pontificate of Clement V were a period of intensive and unrelenting crusade efforts. Obviously this was due to no one single factor but rather to a complex set of interacting factors: the more-than-favourable disposition of the leaders of Europe, the survival or persistence of attachment to the memories of the Holy Land, and, last but not least, the propitious political circumstances, that is, the relative pacification of Europe. Still, if there was one factor which was more important than the others, it was the double sponsorship of the crusade by the two heads of Christendom, the pope and the king of France. It seems somehow ominous (but was it a coincidence only?) that Bertrand de Got, archbishop of Bordeaux, took at his consecration the name of Clement. His namesake was Clement IV, the staunch friend of St Louis. The alliance between pope and king renewed an old pattern of co-operation, enhancing thereby the chances of success. And it is not out of context to remember that it was during the pontificate of Clement in 1309 that John of Joinville finished his life of Louis IX, the crusading king of France, who had been officially recognized since 1298 as a saint. The epilogue of his 'Vita' surely points the finger at Philip IV: 'It has brought moreover, great honour to those of the good king's line who are like him in doing well, and equal dishonour to those

descendants of his who will not follow him in good works'.[1]
This was a call to a crusade from *outre-tombe*.

The *Crux Transmarina* in European Politics 1305–1308

Clement V launched himself into fervent crusading activities immediately after his election to the papal throne (5 June 1305). Long familiar with crusade-politics, having served under Celestine V and especially under Boniface VIII, he was from the outset determined to embark upon a systematic approach to the crusade. In this aspect he vividly recalls Gregory X.[2]

Clement announced his intention to organize a crusade in the encyclical which proclaimed his coronation (1305).[3] It was, however, not until three years later, at the beginning of 1308, that he formulated his plan of action. The interim three years, including almost a year of illness (1306), were devoted to preparations, deliberations, the search for candidates to lead an expedition, and the choice of a suitable plan of action.[4] The question was taken up as early as 23 December 1305 at Lyons, when the newly elected pontiff met with Philip IV. He urged the king of France to take the Cross. Philip agreed (29 December) when granted the right to choose and fix the time. By then Philip IV's project of the crusade was already formulated: a grandiose plan to establish French hegemony in the East, it

---

[1] John of Joinville, 'Life of Saint Louis', in *Chronicles of the Crusade*, trans. M. R. B. Shaw (London, 1963), 351; R. H. Bautier, 'Diplomatique et histoire politique: Ce que la critique diplomatique nous apprend sur la personnalité de Philippe le Bel', *Revue historique*, 259 (1978), 93; S. Schein, 'Philip IV and the Crusade: A Reconsideration', in P. W. Edbury (ed.), *Crusade and Settlement* (Cardiff, 1985), 121–6.

[2] C. V. Langlois, 'Documents relatifs à Bertrand de Got (Clément V)', *Revue historique*, 40 (1889), 48–50; E. Renan, 'Bertrand de Got: Pape sous le nom de Clément V' *Études sur la politique religieuse du règne de Philippe le Bel* (Paris, 1899), 387–9; G. Mollat, 'Clément V', *DHGE* xii (1953), 1115–27; K. Wenck, *Clemens V und Heinrich VII* (Halle, 1882), 29–37; F. Heidelberger, *Kreuzzugsversuche um die Wende des 13. Jahrhunderts* (Leipzig and Berlin, 1911), 24–5; L. Thier, *Kreuzzugsbemühungen unter Papst Clemens V 1305–1314* (Düsseldorf, 1973), 106–8; See also *Papsttum und Untergang des Templerordens*, ed. H. Finke (Münster, 1907), vol. i, pp. 104–7.

[3] See e.g. letter to archbishop of Trier, 16 Nov. 1305, in *Nova Alammaniae*, ed. E. E. Stengel (Berlin, 1921), i. 18–19.

[4] *Papsttum*, vol. ii, nos. 13–18, pp. 15–25.

included the conquest of the Byzantine Empire by Charles of Valois.[5] At Lyons (November 1305–January 1306) Philip was obviously anxious to win papal support for the plans of his brother. Clement V succumbed to the French approaches, but he did not see their plans within the context of a general crusade. He opposed any deviation of the crusade or its being conditioned in any way by the conquest of Constantinople. True, the capture of Constantinople would prove helpful in the liberation of the Holy Land but this was not a *sine qua non* for launching a crusade.[6] Yet Clement, like his predecessor, could not disregard the grave danger menacing the Byzantine Empire. Once conquered by 'Turks and other Saracens and Infidels' it would be extremely difficult to recover. This would certainly hardly help the crusader effort. In short, Clement supported Charles of Valois to please his royal brother and because this seemed to serve strategically and politically the cause of the Holy Land. The conquest of Constantinople was to remain more popular in the circle of Philip than in that of the pope. The French plan expounded by Pierre Dubois was absent from Marino Sanudo Torsello's 'Conditiones Terrae Sanctae' (1309), from the plan of the Armenian Hayton (1307), and from the memoirs of both masters general James of Molay (*c.*1306) and Fulk of Villaret (*c.*1305), as well as from Ramon Lull's 'Liber de fine' (1305).[7]

Charles of Valois stayed in Lyons after the coronation of Clement (14 November 1305) until the end of January 1306, as he was gravely wounded during the coronation ceremony. On that occasion Clement granted him a crusading subsidy for the next two years in France, and a three-year subsidy in the kingdom of Sicily.[8] The projected expedition against Constantinople was proclaimed as a crusade, and its participants were

---

[5] K. Wenck, 'Aus den Tagen der Zusammenkunft Papst Klemens V und König Philipps des Schönen zu Lyon: November 1305 bis Januar 1306', *Zeitschrift fur Kirchengeschichte*, 27 (1906), 189–203.
[6] Cf. A. E. Laiou, *Constantinople and the Latins: The Foreign Policy of Andronicus II 1282–1328* (Cambridge, Mass., 1972), 203; R. I. Burns, 'Catalan Company and the European Powers 1305–1311', *Speculum*, 29 (1954), 758–9.
[7] Clement V, *Regestum Clementis Papae V*, ed. monks of the Order of St Benedict (Rome, 1885–92), no. 243; J. Petit, *Charles de Valois 1270–1325* (Paris, 1900), 103–4.
[8] Clement V, *Regestum*, nos. 243–4; Petit, *Charles de Valois*, pp. 104–6; Laiou, *Constantinople*, p. 204.

accordingly granted the same rights as the crusaders for the Holy Land. The maritime republics Venice and Genoa were urged (14 January 1306) to help Charles as 'the recovery of the empire would facilitate that of the Holy Land'. The republics, should they participate in this enterprise, would be offered the same privileges as those for crusaders in the Holy Land. Genoa, which had vested interests in Byzantium, identified herself with the empire and was consequently unwilling to support Charles of Valois. Venice was more eager to co-operate and became Charles's staunchest supporter. An alliance between Charles of Valois and Venice was concluded on 22 June 1306, and a treaty was signed by the Doge Pietro Gradenigo and the French plenipotentiaries on 19 December 1306. With few changes it reaffirmed the treaty of July 1281 between Charles I of Anjou and Venice. The alliance allowed Charles of Valois to demand all the ships he might need as long as he did so before the following Easter. Venice promised to lend her ships at current prices and even constructed a number of ships at Charles's demand.[9]

The expedition was to leave from Brindisi between March 1307 and March 1308. In the spring of 1307 the hopes for the success of this 'crusade' ran high. On 3 June 1307 Clement excommunicated Emperor Andronicus II Palaeologus and threatened his Catholic allies with excommunication and the forfeiture of their property.[10] In May 1307 Thibaut of Chepoy, Philip IV's 'grand maître des arbalétriers', whose services were lent now to Charles of Valois, embarked from Brindisi, as planned, with a joint Veneto-Valois force of ten galleys commanded by a Venetian captain, to attack imperial lands and waters. This was a force envisaged as the first wave of the long-heralded crusade. In the summer, Chepoy appeared at Negroponte and advanced on Macedonia. By August he had concluded his negotiations with the Catalan Company. The Catalans, who now called themselves 'crusaders', took an oath of fealty to the 'emperor of Constantinople', that is Charles of

[9] Clement V, *Regestum*, no. 248. On the motivation of the republics see Marino Sanudo Torsello, 'Letters', ed. F. Kunstmann, in 'Studien über Marino Sanudo den Alteren', *Abhandlungen der Bayerischen Akademie der Wissenschaften (Historische Klasse)*, 7 (1853), 774–6; Laiou, *Constantinople*, pp. 205–8.
[10] Clement V, *Regestum*, no. 1759.

Valois, and to his representative Chepoy as their new leader. James II of Aragon gave his formal sanction to the treaty with the Catalan Company, when ordering the company (10 May 1308) to obey Charles and his emissaries. This gave Chepoy control, at least a nominal one, over the most efficient army in Romania.[11]

In 1308 Chepoy and the Catalans, based now at the stronghold of Cassandreia, launched several attacks against the fortified monasteries of Mount Athos and on Thessalonica. Their siege of the city in spring 1308 came to nothing, but it seems that they succeeded in gaining control of the entire area around the capital. During the next two years, until his return to Europe, Chepoy busied himself mainly with diplomatic activities. He negotiated treaties with Athens and Thessaly, made contact with Lesser Armenia, with the Levantine Venetians, and in all probability with his allies from among Byzantine nobles and Serbs. He also helped to finance and provision the company. But after the years of agitation, the whole net of alliances and preparations came to a dead end. The ambitions of the company as well as events in Constantinople made the whole enterprise less and less feasible. At the time when Chepoy was preparing the expedition in the East and when his fleet was roaming the Eastern Mediterranean, Charles of Valois was contracting a complicated system of alliances in the West. By the beginning of 1308 his anti-Byzantine league included the pope, the kings of France, Aragon, Sicily, Lesser Armenia, Naples, and Serbia, as well as the Catalans, a number of Greek princes (among them the Latin duke of Athens), and the republic of Venice. The kings of Lesser Armenia and Serbia promised to adopt Catholicism. The enterprise for the reconquest of Constantinople was thus,

[11] The attempts of Charles of Valois and of the Catalan Company to recover Constantinople have become the subject of a large number of studies: see G. Schlumberger, *Expedition des 'Almugavares' ou routiers catalans en Orient* (Paris, 1902), *passim*; J. Petit, 'Un capitaine du règne de Philippe le Bel, Thibaut de Chepoy', *Moyen Âge*, 2 (1891), 224–39; id., *Charles de Valois*, pp. 106–7; Burns, 'Catalan Company', pp. 751–71; H. Moranvillé, 'Projets de Charles de Valois sur l'Empire de Constantinople', *BEC* 51 (1890), 61–86; J. N. Hillgarth, 'Problem of a Catalan Meditteranean Empire 1229–127', *EHR* suppl. 8 (London 1975), 7–54; A. Löwe, *Catalan Vengeance* (London, 1972), 105–6. See also Laiou, *Constantinople*, pp. 200–42. For the papal efforts to bring about the conversion of the Armenians and the Serbians see e.g. Clement V, *Regestum*, nos. 748, 1368.

in theory at least, supported by all the major Latin Mediterranean powers except for Genoa. Unexpectedly the plan suffered a serious setback. With the death of Charles's wife Catherine of Courtenay (October 1307), the Courtenay claim to Constantinople now bypassed Charles and went to his daughter Catherine of Valois. Additionally, because of the tension on the frontiers of Flanders, neither Charles nor any of the great nobles of France could leave the country in 1307 on a lengthy expedition. Moreover, Charles could not levy the tithes granted to him in France, because his brother Philip was collecting them and intended to use this privilege for the next two years. Venice, disturbed by Charles's inability to fulfil his part of the bargain, broke the Valois alliance and signed (11 November 1310) a truce with Emperor Andronicus. The Catalan Company proved utterly uncontrollable and at the beginning of 1310 Chepoy was back in France with little to show for his efforts.[12]

This changed situation made Charles of Valois a sad and frustrated ruler. Disappointed by his unsuccessful attempts to gain the crown of the empire following the assassination of Albert of Austria (1 May 1308), he failed again trying to gain Constantinople. He remained 'fils de roi, frère de roi, père de roi, et jamais roi'. Yet despite the failure of Charles's plan, the West was reluctant to abandon the theoretically brilliant combination of legal claims and actual strength available, so to speak, *in situ*: during the Council of Vienne, the papal vice-chancellor Cardinal Arnold Novelli still urged the use of the Catalan Company for a crusade against Byzantium and the Turks. The plan to attack Constantinople as a preliminary to the conquest of the Holy Land was continuously urged by Philip IV. It was his initiative to marry off Catherine of Valois, the issue of the Courtenay marriage, to Philip of Taranto (6 April 1313), thus uniting the two legally strongest claims to the Byzantine Empire. By then, however, the outlook in the empire became far less promising than it had been in 1308. The Catalan Company captured the Latin duchy of Athens, depriving the West of a convenient base for the crusade against Constantinople. The Catalans then renewed, according to

---

[12] See above, n. 11.

Marino Sanudo Torsello reluctantly, their subordination to the House of Aragon represented by Frederick III of Sicily, who appointed his second son, for five-year-old Infante Manfred, as duke of the newly conquered duchy. The Catalans, whom Clement had favoured only a few years earlier when they were fighting under the banners of Charles of Valois, and whom he had even been ready to recognize as crusaders, had become by 1312 'filii pravitatis', blinded by the wiles of the devil. Horrified by their ravages and crimes, Clement threatened (2 May 1312) these 'sons of perdition' with excommunication unless they abandoned 'certain conventions and pacts with the enemies of the faith' against Philip of Taranto. He warned them that he was directing Fulk of Villaret to help Philip of Taranto to expel them from Greece. On the same day the pope wrote to Villaret that if the company did not desist, in obedience to his commands, from its enmity to the prince of Taranto, the Order of St John should aid the prince. But the Hospital, engaged in Rhodes, declined, preferring to involve Catalan and Aragonese strength in the Levant by supporting James II of Aragon's marriage to Maria of Lusignan in 1315. The Catalan duchy of Athens was to last about three-quarters of a century, drawing its dukes from the Catalan houses of Sicily and Aragon, consistently opposed by all the popes of Avignon.[13]

It is clear that Clement V, though he supported the plan of Philip IV for the recovery of Constantinople, did not forgo his own schemes. Reluctant to rely on Philip's promise to lead a crusade, he looked around for other candidates for the prestigious task. In 1305–6 Clement was active in negotiations with Edward I, Albert I of Germany, and James II of Aragon—a fact somehow overlooked by historians. If one believes his reiterated statements, Edward did not give up his intention to depart for a crusade until his death (7 July 1307). Clement V, a Gascon and a former royal clerk, intuitively relied on the king of England, especially as by 1305 Edward's difficulties in Scotland appeared to be over and Edward's concern for the recovery of the Holy Land was still renowned. But if Clement

---

[13] Clement V, *Regestum*, nos. 7890–1, 10167–8; P. Topping, 'Morea 1311–1364', in *Setton's History*, iii. 104–10; K. M. Setton, 'Catalans in Greece 1311–1380', in *Setton's History*, iii. 167–73; A. Luttrell, 'Crusade in the Fourteenth Century', in J. R. Hale *et al.* (eds.), *Europe in the Late Middle Ages* (London, 1965), 132–3.

put any hopes in the candidature of Edward as the leader of the crusade, he was soon disillusioned. The war in Scotland broke out again in 1306. Only a few days after the peace treaty with France was finally signed by his emissaries at Poitiers, Edward I died during his march to the Scottish border.[14]

Albert I of Habsburg had already stated his wish to set out for a crusade in 1300 when his emissaries at the Curia were seeking recognition for him by Boniface VIII. It was not until three years later on 30 April 1303 that Boniface accepted Albert as king of the Romans. A delegation dispatched by Albert to the newly elected Clement to seek the imperial crown announced that he had taken the Cross. Writing to Albert on 13 October 1305 Clement invited him to come in person or send an embassy to discuss the issue of the *passagium*. Referring to the imperial coronation, the pope refused Albert's demand on the grounds that the time had not yet come. When the latter was murdered on 1 May 1308 the negotiations for his imperial coronation were still going on. It was obvious that Albert would insist upon getting the imperial crown before binding himself to lead a crusade. It was the same insistence on coronation that characterized the crusader policy of the next king of the Romans, Henry VII of Luxembourg.[15]

The third of the potential candidates for the leadership of the crusade was James II of Aragon. As already mentioned, James had stated his intention to set out on a crusade before the loss of the Holy Land c.1289–91. Two years after the fall of Acre (11 November 1293) he wrote to the khan of Persia and to the kings of Cyprus and Lesser Armenia, once again repeating his intention of going to the aid of the Holy Land. At the same time he

---

[14] See Edward I, in M. L. Bulst-Thiele, *Sacrae Domus Militiae Templi Hierosolymitani Magistri* (Göttingen, 1974), 366–7; see also *Foedera, conventiones litterae et cuiuscunque generis acta publica inter reges Angliae et alios quosvis imperatores, reges, pontifices, principes vel communitates 1101–1654*, ed. T. Rymer et al. (Hague, 1745), I. iv. 41, 49, 56–7; Pierre Dubois, *Recovery of the Holy Land by Pierre Dubois*, trans. and ed. W. I. Brandt (New York, 1956), 69–70 n. 4. Towards the close of the 13th cent., authors of model letters were using subjects suggested by Edward's zeal for the crusade. See C. V. Langlois in his edn. of Pierre Dubois, *De Recuperatione Terre Sancte* (Paris, 1891), 2 n. 1. See also M. Prestwich, *War, Politics and Finance under Edward I* (London, 1972), 37, 191, 269.

[15] *MGH: Legum Sectio IV, Constitutiones*, ed. J. Schwalm (Hanover, 1904–6), vol. iv, no. 173, p. 143; no. 204, p. 174; *AA*, vol. iii, no. 64, p. 144; C. J. Hefele and H. Leclercq, *Histoire des conciles*, vi (Paris, 1914–15), 539 n. 3; T. S. R. Boase, *Boniface VIII* (London, 1933), 237–8, 327–30; below, ch. 7.

dispatched an embassy to the East to negotiate for military alliances with the Mongols of Persia and the kings of Cyprus and Lesser Armenia.[16] These proclamations of James were made in order to secure his release from the sentence of excommunication pronounced on him in 1291 by Nicholas IV. They immediately followed the mission of Boniface of Calamandrana (November 1293), who had persuaded James to accept the conditions offered by Charles II of Anjou in order to put an end to the Sicilian war. Anxious to make his peace with the papacy, James agreed (7 July 1294) that the Church should arm a fleet in Aragon for the aid of the Holy Land under the command of Boniface of Calamandrana or Roger Lauria. On the eve of his reconciliation with the Church in 1295 James promised to take the Cross and save the Holy Land by armed force, and thus emerged as the only European monarch who seemed to be prepared to depart for the crusade. His position as a potential leader of a crusade was given an official character when Boniface adopted in 1296 the plan to build a fleet and nominated James 'the general standard-bearer', 'captain and admiral of the Church', 'for the aid of the Holy Land and against the enemies of the Church'.[17] As already mentioned, the fleet as well as James's services were used by Boniface solely for the Sicilian war. But was James actually prepared to undertake a crusade? It is doubtful how strongly he was attracted to the crusading ideal. As rightly pointed out by J. N. Hillgarth, James underwent sporadic periods of enthusiasm during which he seemed about to intervene in the East. This occurred on the eve of the loss of the Holy Land. In 1296 he sought to establish a protectorate over oriental Christians and the Holy Places in Palestine; in 1300 he hoped to recover the Holy Land for Aragon by means of an alliance with the Mongols; later, together with his brother Frederick III of Sicily, he toyed with the idea of recovering the Holy Land via Constantinople. Nevertheless, he steadily refused to be drawn into the papal plans for a crusade in the East. Like his elder

[16] *AA*, vol. ii, no. 459, pp. 740–3; N. Iorga, *Philippe de Mézières (1327–1405) et la croisade au XIVe siècle* (Paris, 1896), 34 n. 2. The il-khan of Persia was at that time Arghun's brother Kaichatu (1291–5); see L. Lemmens, 'Die Heidenmissionen des Spatmittelalters', *Franziskanische Studien*, 5 (1919), 38.

[17] *AA*, vol. i, no. 10, pp. 14–15; *Cerdeña y la expansión mediterránea de la Corona de Aragón*, ed. V. Salavert y Roca (Madrid, 1956), no. 19, pp. 17–20; above, ch. 4.

brother Alfonso II, James was prepared, when in difficulties as in 1290 and 1292, to ally himself with Egypt against a possible crusade. In the later part of his reign, when reconciled with the papacy, he lent a polite ear to papal crusading plans, but he preferred in practice to negotiate diplomatically with Egypt and to confine his crusading to the Iberian Peninsula. From 1305 James insisted that the war against the Saracens of Granada should be completed before the war against the Saracens of the East. Though the plans he presented to the Curia tended to link his plan for a crusade against Granada with the papal plan for the recovery of the Holy Land, in 1308–9 he did his best to ruin the papal–Hospitaller crusade. His eastern policy became confined to maintaining peaceful relations with the sultans of Egypt and gaining the crown of Jerusalem for Aragon.[18]

What is most striking about James's diplomatic relations with Egypt is that some of his embassies were dispatched precisely at the time when he seemed most interested in the Holy Land and in the crusade. It follows from a letter (6 April 1300) of al-Malik an-Nasir to James that the latter dispatched his first embassy to Egypt in the course of 1300, precisely when he was negotiating with the il-khan of Persia for a part of the Holy Land, which James then believed had been reconquered by the Mongols. In 1305 James sent a third embassy to the sultan. A month later James made an unconditional offer of himself, his lands, and treasure for a crusade. This was done at the conference of Montpellier (7–11 October 1305). Ramon Lull, who was present at one of the meetings between James II of Aragon, his uncle James II of Majorca, and Clement V, was deeply impressed by James's zeal for the crusade and presented him with the 'Liber de fine'. This offer was hardly the right background for the terms of a tentative treaty contemplated at that time between Aragon and Egypt. The terms were the following: a permanent peace between Egypt and Aragon, and Naples and Sicily; the king of Aragon would impede any movements in the West for a *passagium generale* not only by using his influence but also by arming 100 galleys to intercept it; Catalans and other Christians would be allowed to carry

---

[18] Hillgarth 'Problem', pp. 40–1; below, ch. 7.

timber and iron to Egypt. However, an unexpected episode, which occurred in the course of the visit of this Aragonese embassy in Egypt, led to an order from Cairo for the arrest of all Aragonese subjects and the confiscation of their property in Alexandria, as well as the suspension of Aragon's diplomatic relations with Egypt for eight years. As the Aragonese ambassador Neymerich Dusay, accompanied by the sultan's envoy to James, the Emir Fakhr-al-Din, laden with valuable presents and a company of freed prisoners, was about to sail in 1306 from Alexandria, the sultan was informed that one of the prisoners (captured in Ruad) was the son of a powerful monarch for whose liberty he could exact a heavy ransom. The sultan at once recaptured his former prisoner, and Dusay retaliated by seizing Fakhr al-Din's possessions and placing him on a barge which drifted back to Alexandria. The diplomatic relations between Aragon and Egypt were not altogether without advantages to James himself as well as to some Christians. James's influence at the sultan's court became so well known in the West that some westerners used his name in Egypt under false pretences. James II probably received a grant from the sultan in 1293 in return for the promise that his navy would not join any new crusade. The sultan granted passage to pilgrims bearing letters from James. More grudgingly he freed a number of captives who were Aragonese subjects. In 1303 he set at liberty four Aragonese Franciscans captured in Tripoli and Acre. Following the third embassy dispatched by James, the sultan freed twelve captives. In 1303 two churches were reopened in Egypt, at James's request. In 1306 one of the Templars captured in Ruad was released.[19]

At the same time, as shown by his diplomatic activities and correspondence, James aspired to the crown of Jerusalem. On the eve of the fall of Acre James tried to achieve this aim as a pre-condition of his departure for the aid of the Holy Land; in 1300 he tried to gain the kingdom from the il-khan of Persia. Later, between 1309 and 1311, Arnold of Villanova was engaged on behalf of James II of Aragon and Frederick III of Sicily in persuading Robert of Naples to renounce his title to the

---

[19] J. N. Hillgarth, *Ramon Lull and Lullism in Fourteenth-Century France* (Oxford, 1971), 65 n. 58; A. S. Atiya, *Egypt and Aragon: Embassies and Diplomatic Correspondence between 1300 and 1330 AD* (Leipzig, 1938), 17–19, 21–8, 33, 35–41.

crown of Jerusalem in favour of Frederick, who was interested in the crusade. James himself was negotiating (1311–15) his second marriage with a princess of Cyprus in order to acquire the succession to the Cypriot throne and to the title of Jerusalem claimed by the rulers of Cyprus.[20]

James's attitude to the crusade was on the whole characterized by the same traits that recur in the attitude of other maritime Mediterranean powers like the Italian communes: the trade and the diplomatic relations with Egypt were too profitable to be abandoned. On the other hand, there were advantages in crusading. It was possible to obtain crusader rights and privileges without having even the intention of going on crusade. The gaining of the Holy Land would be most welcomed but crusading had to be subordinated to 'national' politico-economic interests. Yet in 1305 Clement was no less impressed than Ramon Lull by James's seemingly unconditional offer of himself, his lands, and treasure for a crusade. Shortly after the conference at Montpellier Clement granted to James the revenues of an ecclesiastical tithe of Aragon for the next four years (17 October 1305). These revenues were to enable him to take possession of Sardinia and Corsica, and once the domination of the islands had been achieved he would be able to devote himself to the cause of the Holy Land. Clement confirmed also the grant made to James by Boniface VIII of the revenues from fines on the contraband trade with Egypt. It soon became obvious, however, that James's offer had strings attached to it, and, more important, it envisaged Granada rather than the Holy Land as its main objective. On 28 May 1306 his representatives at the Curia were ordered to demand from the pope the aid of the Templars and Hospitallers, together with knights, foot-soldiers, and money, for Sardinia and Corsica. This expedition had to be recognized as a crusade; the revenues from fines on trade with Egypt were to be extended to all the papal territories and to Genoa. In 1307 James began active preparations for an attack on Granada. His aim was now that his crusade should be granted the same privileges as those bestowed on the participants of the Hospitaller *passagium particulare*. In the face of James's demands Clement vacillated.

[20] *AA*, vol. ii, no. 440, pp. 701–2; Hillgarth, *Ramon Lull*, p. 66 n. 60.

Voicing the cause of the Holy Land, he refused to surrender to James's request until November 1309 and even then he granted to James only part of his demands.[21]

Obviously in 1305 Clement V could not have known that in 1308 the only candidate for the leadership of the crusade would be either Philip IV or another scion of the French royal family. However, even in 1305 it was clear that it would be impossible to organize a general crusade in the immediate future. The enforcement of papal prohibitions against trade with Egypt and the defence of Cyprus and Lesser Armenia, were all urgent problems which needed solution, before any plan for a crusade could be effected.

These problems, including the plan of Charles of Valois, were discussed by the pope and Philip IV during their conference at Lyons. Here it was decided to enforce the papal prohibitions on the trade with Egypt. Clement V was certainly aware of the ineffectiveness of the papal legislation. Yet he did no more than renew the prohibitions of Nicholas IV,[22] convinced perhaps that a maritime blockade would be more effective in putting an end to the illegal trade. The enforcement of the ban had been attempted since the fall of Acre by the fleets of the Templars and Hospitallers, as well as by the king of Cyprus Henry II of Cyprus. Maritime patrols based on Cyprus were successful in capturing a large number of vessels on their way to or from Egypt. By 1305, however, it was seen that these measures, taken as a whole, created more difficulties than they solved. They brought about constant conflicts between the maritime powers (Genoa, Venice, the Catalans), the orders,

---

[21] Clement V, *Regestum*, nos. 223, 225, 357; *Cerdeña*, no. 129, pp. 167–8; no. 158, pp. 203–4; no. 323, pp. 403–4; no. 332, p. 413–14; no. 340, pp. 425–7; no. 345, p. 431; *Papsttum*, vol. ii, no. 101, pp. 182–4; *AA*, vol. iii, no. 91, pp. 198–9; Hillgarth, *Ramon Lull*, pp. 67–71; below, ch. 7 nn. 7–30. At the beginning of 1306 James demanded from Fulk of Villaret a loan of 10,000 marks. The master general refused, complaining of a lack of money due to the expenses of the expeditions to Ruad and Lesser Armenia: see *AA*, vol. iii, no. 65, pp. 145–6; A. Luttrell, 'The Aragonese Crown and the Knights Hospitaller of Rhodes 1291–1350', *EHR* 75 (1961), 1–19, esp. p. 5.

[22] Clement V, *Regestum*, nos. 2994–5; *Ordonnances des roys de France de la troisième race*, ed. M. de Laurière (Paris, 1723), i. 505 (28 Aug. 1312). This is the sole ordinance of Philip IV forbidding trade with the infidels. It follows from the document that some French of the Midi had betrayed to the sultan the secret decisions made at Lyons in view of a crusade in order to sell to Egypt more war-materials and slaves: *Ordonnances*, i. 505–6. See also *Papsttum*, vol. ii, no. 7, pp. 7–8; Thier, *Kreuzzugsbemühungen*, pp. 33–9.

and the king of Cyprus, without being able to stop effectively the traffic with Egypt.[23]

By the beginning of 1306 the issue of the defence of Cyprus, and especially that of Lesser Armenia, had become acute. The Saracen pirate raid against the Ibelin castle in Cyprus in 1302, for instance, made clear how vulnerable the island was. On 26 April 1306 Amalric, brother of the king and nominal lord of Tyre, led a rebellion with the support of a large party of Cypriot nobles, on the grounds that King Henry II was neither offering the kind of leadership needed to defend the island nor showing interest in a crusade. The king was forced to turn over the government to his rebellious brother, to approve the latter's election as 'rector, governor, and administrator of the kingdom', and to retire to his country estate. A declaration and a charter drawn up (26 April 1306) by Amalric and his party emphasized the king's failure to take active measures to protect the realm against the Genoese and against an expected attack by the sultan of Egypt. Fortifications and provisions were lacking; the king, it proclaimed, had failed to help Lesser Armenia against the incursions of the Egyptians or to acquire allies. The kingdom was in the grip of famine, which became worse from day to day, forcing many of the king's subjects to flee overseas, even to Saracen lands, where they would not starve. The list of complaints was long, culminating in the accusation that during the expedition under Amalric to Ruad (1300) the king had not allowed it to be provisioned in its hour of need and had forced the members of the expedition to obtain supplies at their own cost. The Temple and Hospital, as well as all the clergy, had greatly suffered from the

---

[23] Henry II of Lusignan, 'Consilium', ed. M. L. de Mas-Latrie in Mas-Latrie, *Histoire de l'île de Chypre sous la règne des princes de la maison de Lusignan* (Paris, 1852–61), 121–5; Marino Sanudo Torsello, 'Liber Secretorum Fidelium Crucis', in *Gesta Dei per Francos*, ed. J. Bongars (Hannau, 1611; repr. Jerusalem, 1973), ii. 31 n; W. Heyd, *Histoire du commerce du Levant au Moyen Âge*, trans. F. Raynaud (Leipzig, 1936), ii. 29–30. See also Clement V, *Regestum*, nos. 752, 1034–6, 1247–8, 1250; *Cartulaire général de l'ordre des Hospitaliers de St Jean de Jérusalem 1100–1310*, ed. J. Delaville le Roulx (Paris, 1894–1906), nos. 4727–8. Marino Sanudo was certainly mistaken in claiming that King Henry II of Cyprus was engaged with the enforcement of the ban 'non requisitus'; see Heyd, *Histoire du commerce*, ii. 30 and n. 1.

king's ill will, which had made him reluctant to administer justice.[24]

The length at which the above-mentioned document dealt with Armenia can be explained by the influence of Hayton 'the Monk', the lord of Gorighos. Hayton, having fallen out with King Hetoum II, retired from Armenia to Cyprus in 1305 and entered the Premonstratensian Order at Bellapaïs. In the quarrel between King Henry and the later's brother he sided with Amalric. In 1306 he went to the papal court in Poitiers and there, besides pursuing the cause of Armenia, he represented that of Amalric. Made prior of a convent of his order near Poitiers, he wrote there in 1307, at the request of Pope Clement V, his 'Flos historiarum Terre Orientis', advocating a *passagium* by the way of Cyprus and Armenia. It was presented to the pope in August 1307.[25]

Meanwhile, following its appeal for aid, Lesser Armenia moved into the forefront of the oriental problems. Lesser Armenia's appeal to the West was prompted mainly by the deterioration of its relations with its allies the Mongols of Persia, as well as by the emergence of a unionist party in the kingdom. This party, headed by the king and Patriarch Gregory, was inclined to a union with Rome. This was demonstrated on 19 March 1307 by the decision of the Council of Sis which made concessions to the Roman creed.[26] After the death of Ghazan (17 May 1304), the Egyptians renewed their raids against Armenia. Between the years 1300 and 1304, William of Villaret the master-general of the Hospital twice brought aid to Lesser Armenia. In the summer of 1305 an Egyptian army of 3,000 men advanced as far as Tarsus, but was defeated and cut to pieces by King Hetoum near Ayas (17 July 1305). Thereupon the sultan hastened to conclude a truce with Armenia.[27]

---

[24] See C. Kohler in his edn. 'Documents Chypriotes du début de XIVe siècle', *ROL* 11 (1905/1908), 444–56; G. Hill, *History of Cyprus* (Cambridge, 1948–52); 216–18; *AA*, vol. ii, no. 463, pp. 745–6.

[25] Hayton, 'Flos historiarum Terre Orientis', *RHC Hist. arm.* ii. 255; Hill, *Cyprus*, ii. 226 and n. 5. Hayton dicated his treatise in French to Nicholas Falcon, who translated it into Latin. Later (1351), John of Ipres translated the Latin version into French. See Kohler in his introd. to Haton, 'Flos', pp. xxxv, lxxxv.

[26] Thier, *Kreuzzugsbemühungen*, pp. 77–82 and n. 3.

[27] *AA*, vol. iii, no. 65, p. 146; Hayton, 'Flos', pp. 330–3; Geoffroy of Paris, *Chronique metrique attribuée à Geoffroy de Paris*, ed. A. Diverrès (Strasburg, 1956), p. 154, ll. 3275–95, dated this battle mistakenly as 1306.

Though it remained under permanent threat of a Mamluk invasion, the situation of Armenia in 1306 was apparently not worse than before. Yet in the spring of 1306 a joint Armenian and Cypriot delegation arrived at the Curia in Bordeaux, asking for aid. The king of Lesser Armenia Leo IV (1305-7) stressed in his letters to the pope, as well as in those addressed to Edward I of England, the difficult position of his kingdom, the treachery of the Mongols, and the need for a crusade. As the king of Armenia was aware that a general crusade was as yet unlikely, he asked the pope for a contingent of 300 knights and 500 foot-soldiers. Clement, as it follows from his letter (2 July 1306) to Leo, decided to organize immediate aid for Armenia, appealing to Genoa and to Arthur of Brittany. Genoa was actually the only maritime power not involved in the 'conquest of Constantinople'. Besides, it had its own vested trade interests in Lesser Armenia, and its eventual intervention was facilitated by the truce it enjoyed with Cyprus. Arthur of Brittany, on the other hand, was called to the aid of Armenia on account of the will of John of Brittany, his father. John, a *crucesignatus* since 1300, who was accidentally killed during Clement's coronation, bequeathed a sum of money for the Holy Land. Clement intended to use this legacy, to which he added a sum from the papal Camera, for the aid of Armenia, if Arthur would agree to lead a military contingent to that country. Arthur's reply seems to have been negative and Genoa was not more enthusiastic. Genoa suspected that Clement was using Armenia as a pretext only, and that he actually intended to use their fleet for Charles of Valois's expedition against Constantinople. Around 1308 Oschin, king of Lesser Armenia (1307-20) again appealed for aid, complaining about the treachery of the il-khan of the Mongols. Clement thereupon appealed to the newly elected Henry VII of Luxembourg (27 November 1308). The king of the Romans wrote to the king of Armenia c. 1309, promising aid in men and ships, but actually no such aid was sent to the East.[28]

Conspicuously, Clement V did not appeal to the military

---

[28] Clement V, *Regestum*, nos. 748, 750-1, 1033; *Acta Imperii Angliae et Franciae 1267-1313*, ed. F. Kern (Tübingen, 1911), no. 164, pp. 108-9; no. 190, p. 128; 'Lettres inédites concernant les croisades 1275-1307', ed. C. Kohler and C. V. Langlois, *BEC* 3 (1891), 61-2; Boniface VIII, *Registres de Boniface VIII*, ed. G. Digard et al. (Paris, 1884-1939), nos. 3713, 3719-20; Hill, *Cyprus*, ii. 227-8; Heyd, *Histoire du commerce*, ii. 83-4; Thier, *Kreuzzugsbemühungen*, pp. 79-82 and n. 4.

orders for direct aid to Lesser Armenia as had been done by his predecessors. Evidently the pope intended to prevent the orders from sending their forces to Armenia, so that they could be used mainly for the Holy Land. It was 'ad deliberandum et consulendum de recuperatione Terrae Sanctae', that the masters-general James of Molay and Fulk of Villaret were summoned (on 6 June 1306) to Poitiers. The pope, as indicated in the summons, wanted their advice on such a plan of action for the Holy Land, that would somehow, coincidentally, provide also immediate aid to the kingdoms of Cyprus and Lesser Armenia.[29] Yet the main object of the meeting was to hear the opinion of the masters on the proposals for the merging of their orders. By May 1307 it was rumoured at the Curia that Clement V intended to convince the masters-general that their orders must be united.[30] Actually, nothing definite is known about Clement's opinion at that time. His relations with the Hospital and the Temple, and their masters-general, were cordial, but clearly Clement was already hard pressed by Philip IV to get involved in the proposals for the reform of the orders. It is known from his memoir devoted to this subject (c.1306) that the master-general of the Temple James of Molay regarded the plan for the merger as scandalous and dishonourable. In his opinion the competition between these ancient and venerable orders stimulated zeal and brought results, whereas the existing jealousies were too strong to allow a union. Unfortunately the Hospitaller counterpart of the Templar memoir, namely that composed by Fulk of Villaret, does not survive. It seems, however, from other sources, that it was more favourable to the plan of union. On the whole around 1306 the plan for the unification of the two military orders, backed by Philip and advocated in most of the crusader plans of the period, was more popular than ever. Ramon Lull even went so far as to condemn to Hell those who opposed it.[31]

---

[29] Clement V, *Regestum*, no. 1033: *Cartulaire*, no. 4720; *Vitae Paparum Avoniensium*, ed. G. Mollat (Paris, 1914–27), vol. i, p. 6; *Grandes chroniques de France*, ed. J. Viard, viii (Paris, 1834), 257.

[30] *Papsttum*, vol. ii, no. 23, p. 36.

[31] James of Molay, 'Concilium super negotio Terre Sancte', *Vitae Paparum*, vol. iii, pp. 150–4; Ramon Lull, 'Liber de Fine', ed. A. Madre, *CCCM* xxxv 1(1981), 74; *Papsttum*, vol. ii, no. 75, p. 118; G. Lizerand, *Clément V et Philippe IV le Bel* (Paris, 1910), 85–6; M. Barber, 'James of Molay, the Last Grand Master of the Temple', *Studia Monastica*, 14 (1972), 105–6.

From 1305 onwards Philip IV outspokenly advocated the unification of all the military orders. The king of France would become the ruler of the orders on renouncing his kingdom and accepting that of Jerusalem. His successor as head of the orders would be either one of his sons, or, failing this, a person he would nominate. The resemblance between Philip's ideas and those of Lull's 'Liber de fine' suggested to Hillgarth that Philip may have been inspired, if not by the 'Liber de fine' itself, then by his earlier contacts with Lull or alternatively by the 'Tractatus de modo convertendi infideles seu Lo Passatge' (c.1291–2), preserved in a French manuscript of the fourteenth century.[32] As a matter of fact, most of Philip's ideas correspond very closely to the plan of Charles II of Anjou (c.1292–4). The similarities between the ideas of Charles and those of Lull were, as already stated, so close as to suggest mutual influences. In all probability it was Charles who was first influenced by Lull, but the relationship was reversed later on. Philip could therefore have been inspired by either.

It is clear that Philip's plan aimed at the full control of the military orders, and above all, at the complete control of the crusading movement. As to his plan of action, Philip's support of his brother Charles of Valois meant that the recovery of the Holy Land should be preceded by the conquest of Constantinople. Yet it is to do less than justice to Philip to impute that he tried to exploit the crusade for the aggrandizement of France, or in order to put pressure on Clement V.[33] J. N. Hillgarth rightly remarked that the constant efforts made by France during the years 1305–14 to gain control of the crusading movement would have been pointless had Philip really decided never to launch a crusade.[34] The king of France as *rex christianissimus* regarded the crusade as something which involved the honour of France. His idea of his own mission, based on his deep piety, was reinforced by his victory over Boniface VIII, and, as he confessed, by the death of his wife, Joan of Navarre (2 April

---

[32] For Philip IV's plan see *Papsttum*, vol. ii, no. 75, pp. 118–19; Hillgarth, *Ramon Lull*, pp. 72–3.
[33] For such opinions see e.g. Wenck, *Philip*, pp. 61–2; Thier, *Kreuzzugsbemühungen*, pp. 49–50.
[34] Hillgarth, *Ramon Lull*, pp. 76–7. Hillgarth's study has replaced Lizerand's *Clément V* as far as Philip's attitude to the crusade is concerned.

1305). Finally at peace at home and with his neighbours, he was after 1305 prepared to follow in the footsteps of St Louis. It is probable that Philip had presented his plan to the pope in December 1305 at Lyons. It was certainly discussed at their next meeting in May 1307 at Poitiers, as from that time it became common knowledge. It was also widely known that Clement had rejected Philip's plan for the military orders. It was one of the few issues on which Philip could never win the pope to his point of view. Clement V, though he was to capitulate to Philip over the Templars, contrived to resist his plan for the merging of the military orders under French royal command and later, for his attack against the Order of St John. Thus the common contemporary opinion, which had it that Philip 'faciet quidquid vult', was not altogether correct.[35]

The masters-general of the military orders were summoned to appear in Poitiers on All Saints' Day 1306. Their meeting with the pope was, however, delayed by the latter's illness (August 1306–January 1307). The master-general of the Temple James of Molay arrived from Cyprus late in 1306 or early in 1307, and came at the end of May to Poitiers. Fulk of Villaret, involved since April 1306 in the conquest of Rhodes, was in Poitiers by 31 August 1307. A few months later the entire issue of the orders received a wholly different dimension. With the arrest of the Templars in France on 13 October 1307 the plan for the fusion of the two orders lost its meaning, and the trial of the Templars became the *cause célèbre* of the century. Henceforth crusader plans would reckon with the Hospital only.[36]

---

[35] *Papsttum*, vol. ii, no. 7, p. 9; no. 34, p. 51; no. 75, p. 118; no. 101, p. 183; R. Fawtier, *Europe occidentale de 1270 à 1380*, i (*Histoire du Moyen Âge*, vi *Histoire générale publiée sous la direction de G. Glotz*; Paris, 1940), 413; Hillgarth, *Ramon Lull*, pp. 77–8; S. Schein, 'Philip IV the Fair and the Crusade: A Reconsideration', in P. W. Edbury (ed.), *Crusade and Settlement* (Cardiff, 1985), 121–6.

[36] *Papsttum*, vol. ii, no. 23, p. 36; no. 28, pp. 43–4; *Vitae Paparum*, vol. ii, pp. 46–8. Fulk of Villaret stayed again at the Curia from June to Aug. 1308: *Cartulaire*, nos. 4749, 4786, 4792, 4800–1, 4812, 4816–7; *Vitae Paparum*, vol. ii, p. 48; A. Luttrell, 'Hospitallers in Cyprus after 1291', *Acts of the First International Congress of Cypriot Studies* (Nicosia, 1972), 166 n. 4.

The *de recuperatione Terrae Sanctae* Treatises of Clement V's Pontificate

The opinions of the masters-general regarding the crusade and their plans of the unification of the orders reached Clement (*c*.December 1306) in the form of treatises.[37] They were added to the large number of such documents which multiplied from the first years of Clement's pontificate. Pamphlets on the theme of *de recuperatione Terrae Sanctae* now came to dominate the literary scene. Summing up their characteristics, the author of the latest study of the subject refers to the predominance of national interests and to the prominent place given in the plans to organization, finance, and routes for the crusade. The recovery of the Holy Land appears in the plans as a traditional duty only. It was economic, political, and colonizing aims that were, according to this scholar, in the foreground. This was stressed by the place assigned to naval forces as a preliminary to the crusade. And it is symptomatic that in all the plans except that of Ramon Lull, Jerusalem somehow remained in the background and no place was left for a peaceful dialogue with the Saracens.[38]

It is possible to come to very different conclusions. First of all the plans were not innovations but essentially variations on earlier themes. The points enumerated could be applied to crusade-planning since 1291 and partly even since the Second Council of Lyons, and thus are hardly peculiar to the pontificate of Clement V. This was certainly true, as already pointed out, of the place of naval–economic warfare. Again, the thesis that the nature of the individual proposals was determined by the national origin of their writers is unconvincing. With the exception of Pierre Dubois and William of Nogaret, nationality was hardly a determinant and moreover, their major interest was still in a crusade. As the recovery of the Holy Sepulchre was

---

[37] Finke, in *Papsttum*, i. 125 n. 3, dates these treatises as 1306. The treatise of Fulk of Villaret was, as he himself testifies, written *c*.1305. It was certainly composed before the revolt of Amalric of Lusignan (26 Apr. 1306). See Fulk of Villaret, 'Hec est informatio et instructio nostri magistri Hospitalis super faciendo generali passagio pro recuperatione Terre Sancte', ed. J. Petit, as 'Memoire de Foulques de Villaret sur la croisade', *BEC* 9 (1899), 602–3. The treatise of James of Molay was composed apparently *c*.1305–6. See Mollat in *Vitae Paparum*, vol. iii, p. 145 n. 1; Thier, *Kreuzzugsbemühungen*, pp. 50–1. [38] Thier, *Kreuzzugsbemühungen*, pp. 75–6.

equivalent to that of the crusade to the Holy Land, it need not be specifically mentioned. The treatises were written not as *excitatoria* for a crusade but as tangible plans for its implementation. This explains the insistence on the economic and political advantages, and those for the process of colonization. These, additionally, have to be seen in the spirit of the period. As to the absence of a peaceful dialogue, this, it seems, was the typical attitude of the West to Islam since the fall of Acre. Moreover, the plans under discussion were for the recovery of the Holy Land and not for missionary conversion. On the whole the plans composed during the period 1305–12 were far less original than it is commonly accepted, or than they appear when individual plans are discussed in isolation from other contemporary treatises on the same subject. The famous 'De Recuperatione Terre Sancte' of Pierre Dubois, the 'Flos Historiarum Terre Orientis', of Hayton, and even the memoir of Marino Sanudo Toisello, all developed already-prevailing opinions. Moreover, there was very little in the crusade plans of Clement V's pontificate which differed from those which immediately followed the loss of the Holy Land. With the exception of James of Molay, they focused on a maritime blockade of Egypt.

The master-general of the Temple not only rejected the idea of a limited expedition (*passagium parvum*) to Lesser Armenia, but actually denounced the plan to launch such an expedition for the enforcement of the maritime blockade of Egypt. Relying upon his own experience at Ruad, he argued that, as there are no bases to work from, such limited expeditions would be useless. Moreover, he argued that as the main strength of Egypt is in its land-forces, it is against them that the crusade should be directed. His general crusade was an expedition in the most traditional sense. It should be realized on a large scale, with 12,000 to 15,000 mounted knights and 5,000 foot-soldiers. Transport ships should be built rather than galleys and this for reasons of economy. The troops should be raised from the kingdoms of France, England, Germany, Sicily, Aragon, and Castile, as well as other kingdoms whose monarchs would come to an agreement with the pope. He recommended Cyprus as the point of disembarkation. As to the place of landing on the Asian shore, this should be kept a secret. He offered, therefore, to give his advice on this point to the pope and the king of France in private. In the meantime some military aid agreed with Cyprus should be dispatched to the island and a fleet of ten armed galleys should be prepared for the defence of the island as well as to prevent trade with the Saracens. The captain recommended was Rogeronus, the son of the famous Roger Lauria (d.1305), a man 'qui non dubitet perdere temporalia bona per potentiam civitatum maritimarum'. Neither Templars nor Hospitallers should be in the fleet, since it might have to fight the Genoese and Venetians and would thereby damage the orders' standing in these cities.[39]

The plan of James of Molay favouring the traditional general crusade was more in tune with the opinions current before the loss of the Holy Land than with the other *de recuperatione* treatises inspired by the fervent crusading activities undertaken by Clement V. The critical approach to the traditional model of the crusade already evident at the Second Council of Lyons, now became even more pronounced than in 1291–2. The strategy advocated was that of preliminary expeditions as a preamble to the major crusade, the *passagium generale*. Opinions varied as to details of such expeditions. Ramon Lull advocated in the so-called 'Liber de fine' or 'De expugnatione Terrae Sanctae' (April 1305), as well as in the 'Liber de acquisitione Terrae Sanctae' (March 1309), a six-year blockade by a fleet based in Rhodes and Malta and placed under the command of an admiral appointed by the warrior-king.[40] Fulk of Villaret specified the number of vessels needed at twenty-five galleys to be manned and equipped by the king of Cyprus.[41] Pierre Dubois insisted in his 'Oppinio cujusdam' (1308) on a 100 or more galleys and a suitable contingent of warriors of a new military order which would replace all the orders founded for the benefit of the Holy Land.[42] Hayton's 'Flos historiarum Terre Orientis' (August 1307) urged that a fleet of ten galleys, carrying 1,000 knights and 3,000 foot-soldiers, under the com-

[39] James of Molay, 'Concilium', pp. 147–9.
[40] Ramon Lull, 'Liber de fine', pp. 270–82; id., 'Liber de acquisitione Terrae Sanctae', ed. E. Kamar, in *Studia Orientalia Christiana Collectanea*, 6 (1961), 106–8.
[41] Fulk of Villaret, 'Hec est informatio', p. 606.
[42] Pierre Dubois, 'Oppinio cujusdam suadentis regi Franciae ut regnum Jerosolimitanum et Cipri acquireret pro altero filiorum suorum ac de invasione Egipti', in id., *Recovery*, trans. Brandt, pp. 154–5.

mand of a captain and a papal legate, should wage war against the sultan at sea, imposing a naval blockade on Egypt and ravaging its coast. At the same time, and this is the most original part of Hayton's plan, the Mongols would intercept eastern trade with the Mamluk territories. They would deplete the enemy's garrisons in northern Syria by attacking the provinces situated on the border of the western Mongol Empire. The Christian maritime force should at that time conquer the island of Ruad and then, in conjunction with 40,000 oriental Christian archers from Lebanon who would undoubtedly join them, occupy Tripoli, a convenient port for the disembarkation of the general crusade. The Mongols might then conquer Syria and consequently the Holy Land. Moreover, from what was known about the conditions and the ways of the Mongols, they would probably had over the custody of these lands freely to the Christians.[43]

This preliminary device of weakening Egypt by a maritime blockade would be advocated later at the Council of Vienne by crusade-theoreticians, such as Ramon Lull and William Durant the younger, bishop of Mende (1296–1330), as well as by monarchs, such as James II of Aragon and Henry II of Cyprus. The latter, a man of action whose kingdom was within the easy reach of the Saracens, the Turks, and the *mali christiani*, opposed categorically the immediate launching of a general crusade. He was ready to supply himself a fleet of fifteen to twenty galleys to be manned mainly by crossbowmen. This fleet would be independent of the maritime powers who had interests in the trade with Egypt and whose loyalty to the Christian cause was therefore doubtful. Relying upon repeated papal prohibitions against those trading with the Saracens, the fleet should secure their strict observance for two or three years; it could also cause considerable damage to the coastal towns of the enemy and strike terror in their hearts until the time was ripe for the general crusade.[44]

Both Fulk of Villaret and Marino Sanudo Torsello insisted upon yet another preparatory expedition to ensure the security of the subsequent general crusade. According to Fulk of

[43] Hayton, 'Flos', pp. 355–7.
[44] P. Viollet, 'Guillaume Durant le Jeune, évêque de Mende', *HL* xxxv (1921), 129–34; Henry II of Lusignan, 'Consilium', pp. 121–2; below, ch. 8.

Villaret, a fleet of fifty or sixty galleys, able to carry forty to fifty horses each, should be sent to the East a year before the departure of the crusade. If it proved successful, the general crusade would arrive when the Saracens were in no position to resist it.[45] Marino Sanudo began his 'Conditiones Terrae Sanctae', intended for Clement V, in March 1306. Completed in 1309, it became later the first book of his *opus magnum* the 'Liber Secretorum Fidelium Crucis' finished only in 1321 and submitted to John XXII and Charles IV of France (in 1323). For many ages to come, his plan enjoyed fame far beyond its originality. Its real value, seen in the context of crusade-planning after the fall of Acre, has to be reconsidered. Against the background of plans which have already been analysed and the papal–Hospitaller 'particular expedition' proclaimed on 11 August 1308, the plans advocated in the 'Liber Secretorum' were far less original than has generally been assumed. As a three-stage enterprise, the plan resembled that of Fulk of Villaret with whom Marino Sanudo was well acquainted. Like Fulk of Villaret and Henry II of Cyprus, Marino Sanudo held the view that the expedition should be directed straight against Egypt. The first stage would be a maritime blockade by ten galleys manned by 250 men to enforce the papal prohibitions on trade and to interrupt the commercial relations of Egypt with Byzantium and the Mongols of the Crimea. It would be the pope's task to bring together a fleet and to finance it, whereas the galleys could be raised without difficulty: one from the Zaccarias of Genoa, the lords of Chios; one from the Venetians, namely from William Sanudo from Naxos together with the Chisi, the lords of Tinos and Myconos; a third from the patriarch of Constantinople; two from the Hospitallers; four from the king of Cyprus; and one from the archbishop of Crete. Recommending this organization, Marino explicitly referred to the papal–Hospitaller expedition of 1309. Two or three years of blockade would pave the way for the first, preliminary land expedition which Marino Sanudo described in the second book

---

[45] Fulk of Villaret, 'Hec est informatio', pp. 606–7. Atiya, *Crusade in the Late Middle Ages* (London, 1938), 56–7, confuses the Fulk of Villaret's plan, which he apparently did not know, with his letter to Philip IV, in J. Delaville le Roulx, *France en Orient au XIVe siècle: Expeditions du maréchal Boucicaut* (Paris, 1885–6), ii. 3–6. This letter should be dated 27 Jan. 1309 and not as in Delaville le Roulx, AD 1311.

of the 'Liber Secretorum', written in 1312. This expedition should be commanded by a Venetian, who would have under him, 15,000 foot, 300 knights, and fifty galleys, all financed by the Church. The fleet would be manned mainly by Venetians, but some Baltic and Scandinavian countries could also furnish suitable sailors. This force would be sufficient for the occupation of Egypt. Hostilities must start on the soil of Egypt and, following the 'exemplum Venetae nationis', the crusaders should establish a foothold on the coast and then gradually extend their colonization operations inland. Between April and October a force of 5,000 foot and 1,500 knights would attack the powers which traded with the sultan, namely Tunis, Turkey, and Constantinople. The operation would be accomplished within four or five years. His basic argument was 'quod subiugata terra Aegypti, Terra Promissionis se tenere non poterit'. Thus, only when the second expedition was concluded, would the general crusade begin. It would mobilize 50,000 foot and 2,000 horse. In contrast to the foregoing expeditions manned by professional *stipendiarii*, the general crusade would be made up of *crucesignati* who would have responded to the papal call. The expedition would conquer the island of Rosetta and thence proceed to the Holy Land. Other probable consequences of the subjection of Egypt and Syria would be the surrender of Constantinople, Barbery, the whole North African coast, and even 'the island of India'.[46]

The most striking features of Marino Sanudo's plan were the detailed descriptions of the various aspects of the enterprise. As one who considered himself, and probably was, an expert on the affairs of the Levant, he furnished the most detailed description of the coasts and the nations beyond the sea. A Venetian, born into a family of merchant princes, and at the same time attached to the Curia, Marino Sanudo was also able to supply details on shipping, navigation, military equipment, warfare, manning, provision, and other relevant matters. The best

---

[46] Marino Sanudo Torsello, 'Liber', pp. 20, 30–1, 34–6, 39, 44, 48, 81–95; A. Magnocavallo, *Marin Sanudo il vecchio o il suo progetto di crociata* (Bergamo, 1901), 54–5, 61, 81–4. The 'Conditiones', as already stated, is still unedited. Incorporated in the 'Liber', it is thus transmitted in its last version. Magnocavallo often quotes the original version. See also C. J. Tyerman, 'Marino Sanudo Torsello and the Lost Crusade: Lobbying in the Fourteenth Century', *TRHS* 32 (1982), 61, 67.

constructed and informed was his plan of economic warfare. We know now that the plan to undermine the economic power of Egypt, a major factor of its expected downfall, was not, as has been commonly held, Marino's Sanudo's invention.[47] It was advocated by all the theoreticians of the crusade and had been part of official papal crusader policy as far back as 1215. Nor was Marino Sanudo's suggested strategy original. What was really original was his economic reasoning which brought this aspect of warfare to its most far-reaching logical conclusions. This, unfortunately, has been misconstrued by some historians, who have regarded Marino Sanudo's plan as an expression of the trade interests of Venice.[48] If indeed he was representing commercial interest, it was that of all the maritime republics. His was an attempt to reconcile their particular interests and create a common denominator both for vested trade interests and the crusade. Too much of a realist, it was his conviction that Italian maritime co-operation was indispensable for the success of a crusade. His politico-economic theory of the crusade was as a matter of fact one of the best examples of the new orientation. The crusade was still the Holy War, a devoted *passagium generale*, but at the same time it was a carefully planned economic and military enterprise.

The plan to send the general crusade by sea directly to Egypt rather than to the Holy Land was hardly common to all; opinion on this matter was far more divided than it was about the preliminary activities. This was, it seems, the result of the particular inclinations of the writers and the vested interests they represented. Sometimes a plan was simply conditioned by the plan of action adopted by the pope or the monarchs of France or Aragon. This explains the fluctuations in the plans of Ramon Lull and Pierre Dubois. Whereas Lull recommended in

---

[47] Using William of Tyre and Jacques de Vitry, Marino Sanudo's 'Liber' forms a sort of summary and concludes a chapter in crusade-thinking. Besides the plans of Fulk of Villaret and Hayton, Marino Sanudo may have been familiar with the plan of William Adam. The latter offered ideas remarkably similar to those of Hayton and the anonymous 'Directorium ad passagium faciendum'. For Marino Sanudo's connections with William of Nogaret, Pierre Dubois, and Ramon Lull, see Tyerman, 'Marino Sanudo Torsello', pp. 59–60.

[48] Tyerman, 'Marino Sanudo Torsello', pp. 70–3, 276–7, 280–2, *contra* Delaville le Roulx, *France*, i. 35–9; Atiya, *Crusade*, pp. 114–27; Thier, *Kreuzzugsbemühungen*, pp. 57–63.

the 'Tractatus' of 1292 a general crusade by land through Byzantium and Armenia, the 'Liber de fine' of 1305 dismissed this route as dangerous, long, and expensive. For largely similar reasons it dismissed three other alternatives: the invasion of Egypt by the way of Rosetta, the sea-route via Cyprus and Armenia, and the attack on Tunis. To his new way of thinking the most suitable line of attack was through Spain, namely Andalusia. This seems an excessively indirect approach, but Lull explained that the kingdom of Granada was surrounded partly by sea and partly by the Christian kingdom of Aragon and Castile and thus would be cut off from Saracen assistance. Spain was not only fertile and healthy, but had plenty of horses and above all was close at hand and therefore easily accessible. The warrior-king could swiftly accomplish the conquest of Granada with a small army. Crossing over to Barbary, to pass through the emirate of Ceuta would be an easy affair. From here the road would lead to Tunis. What would follow, Lull concluded, would be the capture of Egypt and the Holy Land. Lull came out with this plan in 1305. Possibly by then he doubted the ability of the Holy See to organize a general crusade. The 'Liber de fine' proved his rejection of an all-European enterprise in favour of a sort of 'national' expedition. It could appeal only to a Spanish monarch and more precisely to James II of Aragon. That the plan appealed to the latter, is known from the fact that James hurried in Montpellier to present the 'Liber de fine' to Clement. In April 1308, as in 1305, Lull was still interested in a Spanish crusade.[49] In March 1309, however, he completed a new plan for a crusade, the 'Liber de acquisitione Terrae Sanctae'. Here the crusade against Granada was strongly advocated, yet no longer as the only land crusade against Islam. A simultaneous expedition to the East by way of Constantinople should take place. Thus in 1309 Lull returned to his thesis of 1292. The same proposal was also to be made in his petition to the Council of Vienne in 1311. The reason for his change of mind between April 1305 and March 1309, was as specified by Lull in his preface to the 'Liber de acquisitione', the affair of the Templars. He also hinted at

---

[49] Ramon Lull, 'Liber de fine', pp. 276–7, 280–2; Kamar, in his edn. of Ramon Lull, 'Liber de acquisitione', pp. 87–92; Hillgarth, *Ramon Lull*, pp. 66–9 and n. 61.

another reason: 'in order to acquire the Holy Land it is necessary to bring about a concord between both empires [East and West], so that the city of Constantinople may be obedient to the Roman Church . . . and that the Greek schism may disappear'. This could be accomplished if it was 'carefully planned and by the use of force, with the aid of the Venerable Charles (de Valois) and the Reverend Master of the Hospital and this will be easily done with the goods of the Church'. Lull thus explained his new and certainly more optimistic plan by the trial of the Templars and the plans of Charles of Valois for the recovery of Constantinople. The support of the French crusade against Constantinople actually brought his plan into line with other French royal projects, such as the removal of the main obstacle to an eventual effort to recover the Holy Land. This was a tacit agreement to destroy the Templars in favour of a new order. Ultimately it meant to endorse French policy rather than that of Clement. Yet one can agree with J. N. Hillgarth that Lull's relations with the French court did not preclude similiar ones with the king of Aragon. In 1305 and again in 1308 and 1309 Lull appeared to be closely connected with James II.[50]

The French plans were also propagated, mostly in an exaggerated form, by Pierre Dubois. From our particular point of view, the theories of the celebrated Dubois are of little interest. Hardly original, the practical opinions be presented were precisely those he plagiarized. As E. Power remarked, 'Dubois had a very imperfect sense of what could and what could not be accomplished in actual fact'.[51] Moreover, he had very exalted notions of the military resources of France. In the *De Recuperatione Terre Sancte* (1306–7) he suggested an expedition by land because warriors and their mounts were normally weakened by a sea-voyage. Vessels capable of transporting such a large number of people were not available; neither was there any port

---

[50] Ramon Lull, 'Liber de acquisitione', pp. 104, 108–9, 115; id., 'Petitio Raymundi in concilio generali ad adquirendam Terram Sanctam' (1311), ed. H. Wieruszowski, in Wieruszowski, 'Ramon Lull et l'idée de la Cité de Dieu', *Estudis Franciscans*, 47 (1935), 104–9. For Lull's support of the French king see also his 'Liber natalis pueri parvuli Christi Jesu' (Paris, 1311), ed. H. Harada, *CCCM* xxxii (1975), 70–1; Hillgarth, *Ramon Lull*, pp. 67–129.

[51] E. Power, 'Pierre de Bois and the Domination of France', in F. G. C. Hearnshaw (ed.), *Political Ideas of Some Great Medieval Thinkers* (London, 1923), 161.

where they could embark, nor would they be able to disembark at any single port at the same time. Under these circumstances the few arriving would be cut to pieces by the enemy, helped by the wicked angels, who opposed this expedition. Pierre Dubois, however, was hardly consistent. He accepted without question the legend of Charlemagne's crusade, and advised that the great emperor's example be followed, as well as those set by Godfrey of Bouillon and Frederick I Barbarossa. Accordingly this project should be executed by four armies, three of which would go by sea. The fourth and largest should go by land. The land-route should be followed by the Germans, Hungarians, Greeks, and all living to the north of them. However, the Germans and the Spaniards, Dubois argued, though renowned as warriors, had on account of the incessant wars of their kings long since ceased to come to the aid of the Holy Land, nor would they be able to do so in the future. The French, English, and Italian contingents which do not fear the sea might journey by the maritime route to the Holy Land. Following a successful crusade, and leaving in the Holy Land a force sufficient for its defence, the victorious army might return by the way of Constantinople, where it would enable the king's brother Charles of Valois to take over the Greek Empire. This would involve a military attack unless 'the unjust usurper Palaeologus', that is Andronicus II, should be willing to withdraw. Additionally, this effort would be made only if Charles formally bound himself to bring aid to the defence of the Holy Land whenever the need might arise, since after gaining the Greek Empire he would be nearer the Holy Land than any other princes.[52] Thus Dubois, like the French monarchy, arrived after 1305 at a combination of the plan for the conquest of Constantinople, advocated in his *Summaria brevis* composed in 1300, with the plan for the recovery of the Holy Land. In the *Summaria*,[53] it is worth noting, the Holy Land was not even once

---

[52] Pierre Dubois, *Recovery*, trans. Brandt, pp. 86–7, 71, 156, 182–4 n. 50, 196; R. Fawtier, *Europe occidentale de 1270 à 1380*, i. 178–86.

[53] Composed between July 1300 and the date of the marriage of Charles of Valois and Catherine of Courtenay on 28 Jan. 1301. See H. Kämpf in his edn. of Pierre Dubois, *Summaria brevis* (Leipzig and Berlin, 1936), p. ii. The *Summaria* proposed that Philip IV should obtain for Charles of Valois the hand of Catherine, the heiress of the Latin Empire and that by a preliminary agreement he should be recognized as lord of that empire in exchange for the assistance he would furnish in recovering it.

mentioned. The *De Recuperatione*, to sum it up, provided a plan for the establishment of French hegemony over the west and the east, through a crusade. In Dubois's opinion: 'expedirect totum mundum subiectum esse regno Francorum'.[54]

In the course of 1308 Dubois devoted to the crusade two treatises, 'Oppinio cujusdam suadentis regi Franciae ut regnum Jerosolimitanum et Cipri acquireret pro altero filiorum suorum ac de invasione Egipti', a kind of postscript to be added to the 'De Recuperatione' and the so-called 'Pro facto Terre Sancte'. As the 'Oppinio cujusdam' was composed after Ascension Day (23 May 1308), the two treatises were written almost simultaneously. It seems that Dubois was stimulated to compose the 'Oppinio cujusdam', by Clement's proclamation on 11 August 1308 of the papal–Hospitaller *passagium particulare*, whereas the death of Albert I of Austria prompted the composition of 'Pro facto Terre Sancte', of which the main interest was in the French succession to the western Empire. It suggested an expedition by land through Lombardy, Genoa, and Venice, in the footsteps of Charlemagne and Frederick I. Dubois argued again that embarkation difficulties prescribed the land-route. This plan thus enlarged the establishment of French hegemony over Lombardy, Genoa, Venice, Tuscany, and Hungary.[55] The aim of the 'Oppinio cujusdam' mirrored the persistent rumours circulated at that time, that Philip IV planned to bestow the crowns of Jerusalem and of Cyprus on one of his younger sons. These allegations, which were used to explain Philip's action against the Temple, circulated at the Curia, in Aragon, England, Italy, and even in Cyprus. It was also said that Philip wished to make one of his sons master of a new united military order and king of Jerusalem and to secure for him all the possessions of the Temple.[56] Dubois's 'Oppinio

---

[54] Pierre Dubois, *De Recuperatione*, ed. Langlois, p. 129 n. 1.

[55] Id., 'Oppinio cujusdam', pp. 154–62. The 'Pro facto Terre Sancte' was composed during the interregnum between the death of Albert (1 May 1308) and the election of Henry VII of Luxemburg (28 Oct. 1308): id., 'Pro facto Terre Sancte', in 'Notices et extraits des documents inédits relatifs à l'histoire de France sous Philippe le Bel', ed. E. P. Boutaric, *Notices et extraits des manuscrits de la Bibliothéque Nationale*, 20 (1862), 186–9. For the passagium, see below, ch. 7.

[56] *Papsttum*, vol. ii, no. 34, p. 51; no. 75, p. 118; no. 101, p. 183; Adam Murimuth, *Continuatio chronicarum*, ed. E. M. Thompson (RS, London, 1889), 16; Thomas Walsingham, *Historia Anglicana*, ed. H. T. Riley (RS, London, 1863–4), i. 127; Geoffrey le Baker, *Chronicon*, ed. E. M. Thompson (Oxford, 1889), 5; Hill, *Cyprus*, ii. 239.

cujusdam' presented his own most elaborate plan to achieve these same objectives. He proposed that the king of France should be granted the rights to the kingdom of Jerusalem held by Raoul of Brienne, count of Eu and high constable of France. The claims of Charles II of Anjou to this kingdom should be simply annulled by the general council of the Catholic princes assembled to provide for the Holy Land, and this on the ground that the sale of the rights of Maria of Antioch to Charles I of Anjou had not been valid and that the kingdom belonged by right to the king of Cyprus. The latter, whom Dubois apparently confused with Hetoum II of Lesser Armenia, murdered on 13 August 1307, should be made the master-general of a new royal military order to be substituted for the Hospital and other orders founded for the benefit of the Holy Land. On his joining the proposed order he should make over to it all his property, especially his claim, if he had any, to the kingdom of Jerusalem. After him the succession should go to any other Catholic king of Jerusalem belonging to the order. This *rex ordinis*, as Dubois called him, should be required to employ all his forces to aid the new kingdoms of Egypt, Acre, and other Catholic kingdoms against the infidels and schismatics. Thus Dubois conspicuously distinguished between the mastership of the new military order and the crown of Jerusalem. This was perhaps due to his conviction that the son of Philip IV, when made king of 'Acre, Babylon, Egypt, and the Assyrians', could hardly command the forces of the new order. As to the Temple, it should be abolished by the advice of the council and its property devoted to the aid of the new order. The king of Sicily, i.e. Frederick III whose official title was the 'king of Trinacria', should take part in the conquest of the kingdom of Jerusalem. As compensation for losing his rights in that kingdom, he should be granted the kingdom of Tunis. The count of Eu should become king of Cyprus. The brother of the king of Cyprus, Amalric the lord of Tyre, should apparently be granted a good county in the Holy Land or elsewhere. The expedition should be led by the king's sons Louis, Charles, and Philip or the king's brother Louis of Évreux. The king himself, as Dubois suggested already in 'De Recuperatione' should remain in his kingdom, even though he took the Cross, because of the danger of disease and of confusion in the administration of his kingdom. Thus whereas in the 'De Recuperatione' and 'Pro facto Terre Sancte', Dubois suggested that the general crusade should take land-routes, he now referred to opinions pronounced by prudent and experienced men that only by sea was there easy access for an army to Egypt. It would be best, following its blockade, to invade Egypt by sea, after directing a larger army towards Acre, so that Egypt, denuded of warriors, could be quickly overrun. Once conquered, it would be of greater value to the king than his kingdom of France, since all the rural population were slaves and the land was very fertile. The strategy suggested in the 'Oppinio cujusdam' seems prima-facie close to that of Fulk of Villaret. As far as the maritime blockade is concerned, it seems to have been influenced by the papal plan for the 'particular' expedition of 1309. However, its embellishment by flights of fantasy, and above all by Dubois's attempt to assure France of world domination, made it hardly realistic.[57]

Whereas Dubois, a French chauvinist by any standards, reflected what he thought to be the French interest in the crusade, Hayton represented the interest of the unionist party in the kingdom of Lesser Armenia. Focused on Mongol cooperation and the necessity of landing in Armenia, Hayton's project recalls that of Fidenzio of Padua, but is far removed from those of Fulk of Villaret, James of Molay, Marino Sanudo Torsello, and Henry II of Cyprus, all of whom were as closely acquainted with the Orient as Hayton himself. James of Molay, for example, argued that the fate of a 'particular' expedition to be dispatched to Lesser Armenia would be sealed unless it was numerically capable of stemming the tide of the Egyptian forces. This would require some 12,000 to 15,000 knights and 40,000 to 50,000 infantry. He estimated that from 4000 knights to be dispatched to Armenia, no more than 500 would survive. In the case of a general crusade, he objected to a landing in Lesser Armenia on the ground of its climatic conditions, of the differences between western and Armenian tactics, of Armenian suspicions as to the intentions of the Franks, and of the presence of a hostile, native population which included Turks, Kurds, and Bedouins.[58]

[57] Pierre Dubois, 'Oppinio cujusdam', pp. 155–61; id., *Recovery*, trans. Brandt, pp. 177–8; Hillgarth, *Ramon Lull*, p. 60 and n. 38.
[58] James of Molay, 'Concilium', pp. 146–7.

Molay regarded the landing in Lesser Armenia with much the same grave misgivings as Henry II of Cyprus. The latter argued that the kingdom of Armenia was weak and its inhabitants unreliable; the march on Egypt through Gaza would be long and difficult. He therefore preferred the sea-route through Cyprus. A direct descent on Egypt, following the example of St Louis, had numerous advantages: the sea-crossing from Cyprus to the Egyptian coast was short, five to six days only; the landing should present no serious difficulties, especially if the sultan was left in the dark as to the destination of the crusader troops; his contingents in Syria would be unable to leave their posts and join their master in Egypt for fear of a Mongol invasion by land and a Cypriot invasion by sea. Egypt, contrary to Lesser Armenia, had other advantages: it was fertile and could easily provide for an army; once the invasion had been crowned with success, it would be easy to conquer Syria. The voyage by sea took only four or five days and, deprived of the hope of rescue from Egypt, Syria would offer no resistance. On the whole the plan of the king of Cyprus attempted not only to assure his realm of a dominant part in the crusade, but also in the conquest of Egypt. It was actually the latter consideration that conditioned his attitude to the route via Armenia. Thus, like Marino Sanudo Torsello but for different reasons, Henry's project proclaimed Egypt as the key to the Holy Land. Therein lay the importance of the report. Though not immediately successful, the policy outlined by King Henry became the crusader policy of the kings of Cyprus during the rest of the fourteenth century. It was precisely this policy which was followed by King Peter of Cyprus in his crusade of 1365 against Alexandria.[59]

Conversely, Hayton's argument stressed that if the 'particular' expedition succeeded in establishing a fortified foothold in Syria, the general crusade should depart from Cyprus directly to the Holy Land; or, alternatively, after a period of repose in Cyprus until Michaelmas Day (29 September) in order to avoid the excessive heat of the Armenian summer, to Tarsus in Lesser Armenia. From Tarsus, accompanied by a Mongol force of 10,000 men, the army should proceed to Antioch which could

---

[59] Henry II of Lusignan, 'Consilium', pp. 122–4.

be easily taken, and then on by the coast of Syria. Relying upon Mongol co-operation, Hayton's plan was based on an ephemeral situation which ceased to exist shortly after he presented his 'Flos historiarum Terre Orientis' to Clement V (August 1307). The coup of the Mongol chieftain Bilargou, a Moslem fanatic, which ended in the assassination of Hetoum II and Leo IV (17 November 1307), put an end to the friendly relations between Lesser Armenia and the Mongols of Persia which existed under Hetoum II. Ghazan's brother and successor Oljaitu (1304–16), called by the westerners Carbenda or Carbanda, despite his Moslem piety, during the first years of his reign followed the policy of his father Arghun and sought alliance with the West. His embassy to Europe, led by Tuman, that is Tommaso Ugi of Siena, the khan's swordbearer, arrived in Venice in 1306 and in England in the autumn of the next year. From 26 June until 4 August 1307 Mongol envoys stayed at the papal court in Poitiers and it follows from Pope Clement's letter of 1 March 1308 to Oljaitu that the latter offered to support any crusade by putting into the field 100,000 horsemen and to supply the crusaders with 200,000 horses and 200,000 loads of grain which would be in Lesser Armenia when the crusader army arrived there. Encouraged by these promises and probably also by Hayton's insistence upon the zeal of the Mongols for the project of conquering Syria, Clement V (as well as Charles II of Anjou) contemplated at that time the possibility that 'the Mongols may conquer the Holy Land and offer it to Christendom before the launching of a crusade'.[60] Clement's correspondence reveals his energetic efforts to win the il-khan's consent for the missionary activities of the Dominicans and Franciscans in the Mongol realm, as well as to assure Mongol help for a crusade. Writing on 1 March 1308 to the khan, the pope referred to the promise of aid against the Saracens. Obviously he also exhorted the il-khan to accept the Christian faith.[61]

[60] Hayton, 'Flos', pp. 359–63; Marino Sanudo Torsello, 'Liber', pp. 242–3; *Papsttum*, vol. ii, no. 25, p. 38 and n. 3; Clement V, *Regestum*, no. 2269; *Vitae Paparum*, vol. iii, no. 158, p. 133; A. Mostaert and F. W. Cleaves, 'Trois documents mongols des archives secrètes vaticanes', *Harvard Journal of Asiatic Studies*, 15 (1952), 5–6; L. Petech, 'Marchands italiens dans l'Empire mongol', *Journal Asiatique*, 250 (1962), 566–7; J. A. Boyle, 'Il-Khans of Persia and the Prince of Europe', *Central Asiatic Journal*, 20 (1976), 38–9.

[61] Clement V, *Regestum*, nos. 2216–21, 2300–1, 3549, 3582, 7480–2; Thier, *Kreuzzugsbemühungen*, p. 64 n. 3.

Though the message of the Mongol embassy to the West in 1307 seemed promising, it was the last appeal of the Mongol il-khans of Persia to Europe. Oljaitu was also to launch the last attack on the Mamluks. In the winter of 1312–13, encouraged by certain dissident Syrian emirs, he invaded Rahbatal-Shām (Meyadin). However, fierce resistance forced the Mongols to withdraw across the Euphrates never to return again. From then on, hostilities between Mongols and Mamluks became less frequent and peace was concluded in 1322. Commercial relations and missions continued, but the military and political relations between the West and the Mongols of Persia came to a standstill. But of course in 1307 the disappearance of the Mongols of Persia from the European scene could not have been foreseen. The belief that they would co-operate in a crusade seems to have been still strong; it was still assumed by William of Nogaret and Henry II of Cyprus, as well as writers influenced by Hayton's 'Flos'—Marino Sanudo Torsello, William Adam (c.1317), the anonymous Dominican author of the 'Directorium ad faciendum Passagium Transmarinum' (1332), and the anonymous authors of 'Via ad Terram Sanctam' and 'Memoria Terrae Sanctae'.[62]

The attempts to win over the Mongols to co-operate appeared in crusade-planning as part of a general search for allies outside Europe. Thus Marino Sanudo Torsello aimed at an alliance with Christian Nubia (Ethiopia), a kingdom which, as he argued, was ready to invade Egypt from the south during the second expedition. Hayton claimed that the Ethiopians, converted by St Thomas, could invade the territory of the sultan through the desert and thus inflict damage on Egypt and occupy some of its forces. The Georgians should be appealed to as they had many troops of high quality and were close to the kingdom of Lesser Armenia.[63] This search for allies was possibly also linked with the almost generally accepted view that

---

[62] Boyle, 'Il-Khans', pp. 38–9.
[63] William of Nogaret, 'Quae sunt advertenda pro passagio ultramarino et que sunt petenda a papa pro persecutione negocii', in 'Notices et extraits des documents inédits relatifs à l'histoire de France sous Philippe le Bel', p. 123; Henry II of Lusignan, 'Consilium', pp. 124–5; C. Kohler, 'Deux projets de croisade en Terre Sainte', *ROL* 10 (1903/4), 406–57; William Adam, 'De modo Sarracenos extirpandi', *RHC Hist. arm.* ii. 530–5; Heidelberger, *Kreuzzugsversuche*, p. 74 n. 46; Thier, *Kreuzzugsbemühungen*, p. 68 n. 31; J. Richard, 'The Mongols and the Franks', *Journal of Asian History*, 3 (1969), 56–7;

some successful and meaningful achievements should precede the launching of the general crusade. There was almost unanimous agreement as to the necessity of weakening Egypt through a maritime blockade; in addition there was need to establish bridgeheads in the East to assist the disembarkation of the general crusade. This reasoning seems to have been the result of accurate knowledge and realistic evaluation of the Mamluk military forces.

The Mamluk army was basically a force of mounted archers. As such it had all the virtues and shortcomings of a military class of noble horsemen. Prominent among its weaknesses was an outspoken disregard of naval power. The Egyptian government did not keep a fleet on a permanent footing, and such naval resources as it possessed were only an unimportant branch or appendix of an army which, at the height of its power, was the strongest in Islam and one of the most powerful in the world. Western writers on the theme of *de recuperatione Terrae Sanctae* were therefore right in pointing out that Europe's naval forces were far superior to those of the sultanate. The Christian West was probably inferior to the sultanate as far as land-armies were concerned. It has to be kept in mind that whereas the Mamluk Empire could, in time of extreme necessity, such as that created by a Christian invasion, mobilize forces from all of its vast territories, the West was too politically divided to do so. Only if effectively united in a major crusade could its armies achieve numerical equality with or superiority to those of the Egyptians. Europe had therefore to make the most of its naval preponderance. There was thus much knowledge and no less wisdom in the advice given by the crusade-experts on the effective use of armed shipping. But was the Christian naval superiority enough to bring down the Mamluk Empire? It is impossible to be certain. If there is any indication at all, then it should be remembered that in the Christian–Moslem confrontation in the second half of the thirteenth century the naval might of Europe was not sufficient to turn the scales.[64]

Petech, 'Marchands italiens', pp. 567–70. For the authorship of the *Directorium* see C. Kohler, 'Quel est l'auteur du *Directorium ad Passagium Faciendum?*' *ROL* 12 (1911), 104–11; C. R. Beazley, 'Directorium ad Faciendum Passagium Transmarinus', *AHR* 12 (1906/7), 810–13.

[64] Marino Sanudo Torsello, 'Liber', p. 36; Hayton, 'Flos', p. 358.

As already stated, the search for allies was prompted not only by the necessity to augment and strengthen the Christian hosts, but also by another part of the envisaged programme of military and territorial gains which would precede and usher in the general expedition. Such considerations indicated the conquest of Constantinople. This was propagated not only by Philip IV and his advisers William of Nogaret and Pierre Dubois, but also by Ramon Lull and William Durant. James II of Aragon and his spokesmen advocated on their part the conquest of Granada and North Africa. Whatever the proposed chronology of events, and whatever ideas there were about alliances in Europe and allies outside it, there was an almost unanimous consensus regarding the reform of the military orders. Ramon Lull, Pierre Dubois, and William of Nogaret all emphasized the necessity of merging them. Lull in his 'Liber de fine' combined the idea of the warrior–king with the fusion of the existing orders, precisely as Charles II of Anjou did in his 'Conseil' of c.1292–4: the warrior–king of a royal house was to command a new united order created by the merging of those already existing. Later, however, in the 'Liber de acquisitione', Lull proposed instead of his former warrior–king of royal blood, a master-general who should be a member of the order, strictly obedient to and dependent on the pope, to head the new united order. 'When one king is dead' he argued, 'his son may well not have the same devotion to the Holy Land that his father had. Furthermore some [kings] wish to acquire the Holy Land *for their sons*'. This was, as said already, precisely, the aim of Pierre Dubois.[65] It was actually in Dubois's plans that the insistence upon the various objectives to be achieved as the preamble to a crusade became almost obsessive. With him the crusade became a part of a general reform in all branches of society as well as its vehicle.[66] This came to characterize crusade-planning, and the debates of the Council of Vienne showed that it was not

[65] D. Ayalon, 'Mamluks and Naval Power', in id., *Studies on the Mamluks of Egypt 1250–1517* (London, 1977), 1–12; Atiya, *Crusade*, pp. 482–3.

[66] Ramon Lull, 'Liber de fine', pp. 270–3; id., 'Liber de acquisitione', p. 114; Pierre Dubois, 'Remonstrance du peuple de France', in 'Notices et extraits des documents inédits relatifs à l'histoire de France sous Philippe le Bel', pp. 175–9; id., 'De facto Templariorum', in 'Notices et extraits des documents inédits relatifs à l'histoire de France sous Philippe le Bel', pp. 180–1; William of Nogaret, 'Quae sunt advertenda', pp. 117, 120.

limited to crusade-theoreticians. As already stated, it was during the pontificate of Nicholas IV that the main lines of crusading policy were sketched, but it was only during the first years of Clement V (1305–9) that those lines were more clearly and firmly drawn. The *de recuperatione Terrae Sanctae* treatises composed after 1309 added very little or nothing to the ideas already expressed. Characteristically, Marino Sanudo Torsello despite the various versions of his own treatise, did not alter his basic ideas until his death (*c.*1343). His plan was originally completed two generations earlier in 1309.

7

1309

THE PAPAL–HOSPITALLER *PASSAGIUM PARTICULARE* AND THE 'CRUSADE OF THE POOR'

Marino Sanudo Torsello submitted his 'Conditiones Terrae Sanctae' to Clement V in the course of 1309.[1] By that time the pope was already engaged in what Marino Sanudo and most of the crusade theoreticians had been recommending since the loss of the Holy Land as the first stage of a successful crusade.

Clement V decided in favour of a plan put forward by the Hospitallers. This follows from a recently published manuscript, the BN lat. 7470, of a crusader memoir of Hospitaller origin, entitled the 'Tractatus dudum habitus ultra mare per magistrum et conventum Hospitalis et per alios probos viros qui diu steterunt ultra mare: Qualiter Terra Sancta possit per Christianos recuperari'. It is undated but it appears that the plan was formulated between September 1306 and the summer of 1307 in the East, possibly by Fulk of Villaret himself. The memoir in question not only supplements Fulk of Villaret's plan of *c.*1305 as far as crusade-strategy is concerned, but provides the plan of action of the papal–Hospitaller *passagium particulare*. Proclaimed on 12 August 1308, this *passagium* was seen by Clement V as intended to prepare, over a period of five years, the way for a general crusade by defending Cyprus and Lesser Armenia and by preventing illegal commerce with Moslems. Fulk of Villaret was even more optimistic about its prospects and saw the expedition as one which could also be used to complete the conquest of Rhodes, launch an attack on Byzantium, and even recapture Antioch or Jerusalem within five years.[2]

[1] See above, ch. 6 n. 46.
[2] BN MS lat. 7470, fos. 172^r–8^v. See the plan in B. Z. Kedar and S. Schein, 'Un projet de passage particulier proposé par l'ordre de l'Hôpital 1306–1307', *BEC* 137 (1979), 211–26, esp. pp. 220–6. The plan is mentioned by J. Delaville le Roulx, *France en Orient au SIV^e siècle: Expeditions du maréchal Boucicaut* (Paris, 1885–6), i. 80 n., who dated it

The Papal–Hospitaller Crusade

Clement's decision in favour of the Hospitaller plan was taken after a meeting at Poitiers in May 1308 with Philip IV, who was accompanied by his sons Philip, count of Poitou and Burgundy and Louis, king of Navarre.[3] Philip IV did not appear in Poitiers only to discuss plans for a crusade. His main aim was to force the pope to capitulate over the case of the Templars. With a pope as keen on a crusade as Clement V, bargaining power was in favour of the king. Not only was the crusade in the balance, but he threatened Clement V to proceed with the trial against Boniface VIII and even to accuse that pope of heresy. William of Plaisians brought before the papal consistory on 29 May 1308 the confessions which the agents of the king and the Inquisition had extracted by torture from the Templars in France. Addressing the pope on the same occasion, William linked the loss of the Holy Land in 1291 with the Templars: 'per defectum ipsorum Terra Sancta dicitur perdita et pactiones secretas cum Soldano cepius dicuntur fecisse'.[4] This allegation was picked up by French chroniclers, thus linking in the minds of contemporaries, especially the French, the fate of the Templars, the Holy Land, and the crusade.[5] The affair of the Templars, on which was focused the attention of

---

wrongly as having been composed between the years 1323 and 1328. The prolonged sojourns of Fulk of Villaret at the Curia and in France in the course of 1307–8 can be explained as intended to win French and papal support for this plan. See above, ch. 5 n. 35, and A. Luttrell, 'Hospitallers in Rhodes 1306–1421', in *Setton's History*, iii. 285 n. 6; see also *AA*, vol. iii, no. 97, pp. 207–11. For Fulk of Villaret's statements see *AA*, vol. ii, no. 91, p. 199; *Papsttum und Untergang des Templerordens*, ed. H. Finke (Münster, 1907), vol. ii, no. 126, p. 243.

[3] *Papsttum*, vol. ii, no. 86, p. 134. See also L. Levillain, 'A propos d'un texte inédit relatif au séjour du pape Clement V à Poitiers en 1307', *Moyen Âge*, 1 (1897), 73–86.

[4] *Papsttum*, vol. ii, no. 87, p. 139; no. 88, p. 145; *Dossier de l'affaire des Templiers*, ed. G. Lizerand (Paris, 1964), 122–3.

[5] See e.g. *Annales Regis Edwardi Primi*, ed. H. T. Riley (RS, London, 1865), 495; *Vitae Paparum Avenionensium*, ed. G. Mollat (Paris, 1914–27), i. 95; Rostanh Berenguier, 'Opera', ed. P. Meyer, in Meyer, 'Derniers troubadours de Provence', *BEC* 30 (1869), 497–8; J. Michelet, *Procès des Templiers* (Paris, 1841), ii. 44–5, 644–5; E. Müller, *Konzil von Vienne* (Münster, 1934), 690. This is, however, hardly to say that public opinion was entirely hostile to the Templars: see C. R. Cheney, 'Downfall of the Templars and a Letter in their Defence', in id., *Texts and Studies* (Oxford, 1973), 320–7; J. N. Hillgarth, *Ramon Lull and Lullism in Fourteenth-Century France* (Oxford, 1971), 92, 103–4; M. Barber, *Trial of the Templars* (Cambridge, 1978), 229–30 and *passim*.

European public opinion in the years 1307–12, became thus a powerful factor in crusade-politics. The implications were clear. The sins of the Templars had led to the loss of the Holy Land in 1291. If speedy action were not taken, the wrath of God would equally fall on the kingdom of France which sheltered the perpetrators of the 'unnatural acts'. The Church and the people had to eradicate this danger. Such was the theme disseminated by French royal propaganda, a theme easily understood and accepted. The original and often-repeated plan for the union of the military orders was superseded by a radical plan for the suppression of the Order of the Temple. The pope was now bullied (5 July 1308) into granting the diocesan bishops permission to proceed against the Templars and to revoke his suspension of the Inquisition in France.[6]

Having forced the pope to agree to proceed against the Templars, Philip was ready to accept and support Clement V's plan in the East. Both the affair of the Templars and the problem of the crusade were to be put before a general Church council. On 12 August 1308, by the bull *Regnans in excelsis*, Clement convoked a general council to Vienne for 1 October 1310 to deal with the following subjects: the fate of the Order of the Temple and its property, the *recuperatio et subsidium Terrae Sanctae*, and the reform of the Church.[7] As the planned general crusade to be led by the king of France could not be launched immediately, and as the kingdoms of Cyprus and Lesser Armenia were under the imminent danger of destruction, immediate action was needed. So Clement argued in the bull *Exsurgat Deus* of 11 August 1308. Therefore, he continued, it had been decided with the advice of the grand masters of the military orders, familiar with the state of the Holy Land and the eastern kingdoms concerned, and in agreement with the king of France, to take immediate action. This would take the form of dispatching a Hospitaller expendition for the defence and

---

[6] M. Barber, 'Propaganda in the Middle Ages: The Charges against the Templars', *Nottingham Medieval Studies*, 17 (1973), 56–7; id., 'James of Molay, the Last Grand Master of the Temple', *Studia Monastica*, 14 (1972), 111.

[7] *Conciliorum Oecumenicorum Decreta*, ed. J. Alberigo *et al.* (Basel, 1962), 312–19; Müller, *Konzil von Vienne*, pp. 196–200. On 4 Apr. 1310 a new bull postponed the opening of the council for a year to 1 Oct. 1311: Clement V, *Regestum Clementis Pape V*, ed. monks of the Order of St Benedict (Rome, 1885–92), no. 6293.

safe-keeping of the menaced kingdoms: they were to prepare suitable obstacles for perfidious Christians who carried victuals and prohibited wares to the Saracens, and to fight those Saracens. Their main aim, however, would be: 'pontes et vias ad idem generale passagium preparando'. This expedition, which was to last five years, was to be mainly financed not from the tithes and their like, but directly from the treasury of the Curia and a subsidy of the king of France.[8] The pope was to give 300,000 gold florins and Philip another 100,000 to cover the entire cost of 400,000 gold florins.[9] The prelates were nevertheless exhorted to encourage the faithful to contribute money to the Order of St John. All believers, clerics and laymen alike, who helped the expedition in that way were to be granted large indulgences: those who contributed the sum needed to pay for a passage to the Holy Land, or even half that sum, would be granted a plenary indulgence; those who gave 24 *livres tournois* would be granted twenty-four years' indulgence and so on, according to a kind of sliding scale. This money, to be kept in special collection chests placed for this purpose in parochial churches, was to be supervised by the bishops and transferred before the coming Feast of the Resurrection to the papal Camera. Like his predecessors, Clement ordered prayers 'contra paganorum perfidiam' (namely the *Omnipotens sempiterne Deus*, *Sacrificium Domine*, and the *Protector*) to be said during services. The granting of indulgences was proclaimed in churches and the mendicants carried the tidings all over the dioceses. The *Esurgat Deus*, like the other bulls which announced the *passagium particulare*, explicitly abstained from inducing believers to take the Cross and accompany the expedition. This was to be carried out exclusively by the Hospitallers.[10]

The papal bulls also abstained from mentioning that at that time the Hospitallers were engaged in the conquest of Byzantine Rhodes, a conquest already advocated by Fidenzio of

---

[8] Clement V, *Regestum*, nos. 2987–92; *Cartulaire général de l'ordre des Hospitaliers de St Jean de Jérusalem 1100–1310*, ed. J. Delvaille le Roulx (Paris, 1894–1906), no. 4807, and cf. the Hospitaller plan in Kedar and Schein, 'Un projet', pp. 221–6.

[9] *Papsstum*, vol. ii, no. 92, pp. 157–8.

[10] *Cartulaire*, no. 4807; Clement V, *Regestum*, nos. 2987, 2996–7; G. Lizerand, *Clément V et Philippe IV le Bel* (Paris, 1910), 278–80; J. A. Brundage, *Medieval Canon Law and the Crusader* (London, 1969), 131–8.

Padua and in 1305 by Ramon Lull.[11] The acquisition of Rhodes, which was destined to become the Hospitallers' headquarters for the next two hundred years (till 1522), began in rather obscure circumstances. It began with an ignoble alliance between the notorious Genoese pirate Vignolo de Vignoli and Fulk of Villaret. The two met secretly in May 1306 off the coast near Limassol. Vignolo landed there in disregard of Cypriot prohibitions that pirates should be given shelter in that neighbourhood. Vignolo, who had obtained from Emperor Andronicus II Palaeologus the lease of the islands of Cos and Leros, suggested to the master-general that they should join in conquering the whole Dodecanese. An agreement was drawn up on 27 May stipulating that one-third of the captured islands would go to Vignolo and two-thirds to the Hospital, whereas Rhodes, Cos, and Leros would belong entirely to the Hospital, except for two 'casalia' on Rhodes, one already in Vignolo's possession. On 23 June 1306 a Hospitaller fleet with over 500 men, supported by two Genoese galleys, proceeded from Limassol, landed in Rhodes, and slowly began the reduction of the island. On 20 September 1306 the castle of Pheraclos fell to the invaders. In November the great castle of Philermos was captured. The year of the fall of the city of Rhodes which concluded the conquest is less clear and opinions vary between 15 August 1306, 1308, 1309, or 1310.[12] The available evidence seems to suggest, however, that the conquest of Rhodes dragged on till 1310. Clement's bull of 5 September 1307, which confirmed the order's possession of Rhodes, was therefore prospective rather than retrospective. Fulk of Villaret's protracted visit to the Curia (August 1307) and his sojourn in Europe until autumn 1309, argue in the same direction. Fulk of Villaret was occupied with the organization of the *passagium particulare* he was to lead and which he intended to use to consolidate the conquest of Rhodes. And indeed in 1310 the

---

[11] Fidenzio of Padua, 'Liber recuperationis Terrae Sanctae', in Golubovich, i. 49; Ramon Lull, 'Liber de fine', ed. A. Madre, *CCCM* xxxv (1981), 86. As pointed out by Luttrell, the scheme emerged in a climate of opinion in which the Latins were planning a new series of crusades designed to conquer the Greek islands. In 1305 an abortive expedition to conquer Rhodes was launched by Frederick III of Sicily and Sancho of Aragon. See A. Luttrell, 'Hospitallers in Cyprus after 1291', *Acts of the First International Congress of Cypriot Studies* (Nicosia, 1972), 165.

[12] Luttrell, 'Hospitallers in Rhodes', p. 284 and n. 5.

passagium sailed to this island. Once in Rhodes, it completed its subjugation. Only when Rhodes was conquered in its entirety and the adjoining islands secured did the Hospitallers proceed to the main aim of the *passagium*, namely the naval blockade of Egypt.[13]

This main aim of the 'particular expedition', the naval blockage of the sultanate, though well known to the many authors who had drawn up written plans for a crusade, was still a new idea to Europeans as a whole. This explains why they misunderstood and therefore misinterpreted the aims of the papal–Hospitaller *passagium*. Some, like Bernard Gui, presented the expedition as aimed exclusively at the conquest of Rhodes because this island 'had many ports and was situated on the route to the Holy Land'. This was in itself a rather crude exaggeration.[14] Another interpretation of the aims of the expedition originated at the Curia and implied that it was intended for the defence of Cyprus. It was rumoured that the pope had received news from Hayton and Amalric brother of the king of Cyprus, alleging that the sultan was equipping eighty galleys for use against Cyprus.[15] Amalric, it seems, was trying through exaggerated description of Cyprus's situation to gain papal approval for his acts and confirm him for life in the office of governor of Cyprus.[16]

Intensive efforts to organize the expedition began in September 1308. Galleys were ordered in Marseilles, Catalonia, Narbonne, Pisa, Venice, and Genoa; the Teutonic Knights were requested to provide the Hospitallers with troops and galleys for a five-year-long expedition; Genoa on her part was requested to provide arms and food as well as troops; the kings of Europe were requested to allow the Hospitallers to take out

---

[13] Clement V, *Regestum*, no. 2148; *Cartulaire*, no. 4751; Luttrell, 'Hospitallers in Rhodes', pp. 284–7; id., 'Hospitallers in Cyprus', pp. 164–7. The Hospitaller treatise of c.1306–7 explicitly links the conquest of Rhodes with the other aims of the preparatory *passagium*.

[14] Bernard Gui, 'Flores Chronicorum', *RHGF* xxi. 716, 719–20; William of Nangis, *Chronique latine de Guillaume de Nangis de 1113 à 1300 avec les continuations de cette chronique de 1300 à 1368*, ed. H. Géraud, i (Paris, 1843), 359, 376; *Grandes Chroniques de France*, ed. J. Viard, viii (Paris, 1834), 271; 'Chronica XXIV Generalium Ordinis Minorum', *Analecta Franciscana*, 3 (1897), 457.

[15] *Papsttum*, vol. ii., no. 92, p. 158; no. 101, p. 184.

[16] *Vitae Paparum*, vol. iii, p. 86.

from their estates whatever was needed.[17] The Franciscan bishop of Rodez, Peter of Pleine Chassagne, later nominal patriarch of Jerusalem (1314–18), was appointed papal legate to be associated in the leadership of the expedition with Fulk of Villaret.[18] On 12 October 1308 Clement V published a bull proclaiming an embargo on the export of arms, horses, iron, timber, and food to the lands of the sultan. No one would be absolved from excommunication unless he contributed to the cause of the Holy Land an amount equal to the value of the goods exported. Moreover, no absolution would be given but by a special mandate of the Apostolic See. The prelates of Ancona, Genoa, Pisa, and Venice (that is, the main maritime cities) were ordered to make the bull public. The legislation of Clement was not more radical than that of his predecessors. It did not satisfy crusade-planners like Marino Sanudo Torsello who advocated that the 'mauvais crestiens' should be simply proclaimed heretics. One way or another the new proclamation was as ineffective as those which had preceded it. The papal–Hospitaller expedition depended on the maritime powers for the supply of galleys and therefore such absolutions had to be granted.[19]

The expedition—1,000 knights, 4,000 foot-soldiers, and forty galleys strong—was scheduled to depart on 24 June 1309.[20] However, almost as soon as the crusade was announced, difficulties multiplied. European public opinion was not altogether in favour of the papal–Hospitaller enterprise. Since for over a decade it had been revised to see the papacy seriously undertaking a crusade, Clement's intentions were not entirely believed. In Genoa, for example, people suspected its itinerary. As Christian Spinola, the Genoese merchant who served for a quarter of a century as informant of James II of Aragon,

[17] Clement V, *Regestum*, nos. 2986, 2996, 3218–19, 3753, 4392, 4459, 4494–517; *Cartulaire*, nos. 4820, 4821, 4828, 4830, 4835, 4841, 4844, 4860, 4862, 4866, 4882; *AA*, vol. iii, no. 88, p. 91; E. Baratier, *Histoire du commerce de Marseille* (Paris, 1951), 214–15.

[18] On Peter of Pleine Chassagne see M. L. de Mas-Latrie, 'Patriarches latins de Jérusalem', *ROL* 1 (1893), 28; G. Golubovich, 'Fr. Pietro da Pleine Chassagne, O. F. M.', *Archivum Franciscanum Historicum*, 9 (1916), 51–90.

[19] Clement V, *Regestum*, nos. 2294–5, 3088, 6438, 7118–19. For absolutions see ibid., nos. 223, 3978–9, 5090. See also Marino Sanudo Torsello, 'Liber Secretorum Fidelium Crucis, in *Gesta Dei per Francos*, ed. J. Bongars (Hannau, 1611; repr. Jerusalem, 1973), ii. 28, 30–1; W. Heyd, *Histoire du commerce de Levant au Moyen Âge*, trans. F. Raynaud (Leipzig, 1936), ii. 42.   [20] *Cartulaire*, no. 4807; *AA*, vol. iii, no. 88, p. 191.

reported to the latter on 8 January 1309, 'gentes vero de terra nostra suspicionem habent quod accedere non debeat armamentum. Et a quibusdam dicitur pro regno Sicilie invadendo'.[21] Moreover, there was criticism of the whole crusading movement voiced in the form of a forged letter, allegedly dispatched by a sultan named 'Warradach' to Clement V upon hearing that 'Philippus rex Francie et rex Anglie multique principes et barones expedicionem in Terram Sanctam post aliquot annos contra Sarracenos indierunt.' The sultan's letter, forged by a German, accused the pope of being more of a stepfather than a father to Christendom, because he compelled his sons to set out for a certain death: 'Qualis pater, qui filios non naturali morte, sed subita perdere non formidas? . . . Filios tuos ad nos ire compellis ut sagittas nostras in se recondant et vivaces animas in formas mortales'. The last phrase in the alleged sultan's letter sounds familiar. It was actually one of the explanations current after the failure of the Second Crusade and was finally formulated in a famous saying by Humbert of Romans in the second half of the thirteenth century. Now, however, not God but the pope was accused of filling Heaven with martyrs and of caring more for the reconquest of Jerusalem, Acre, and Tripoli than the fate of his sons. The sultan's advice to the Christians is therefore to remain at home 'ut non martires sed confessores morientur'.[22] Yet this sort of opinion should be seen in its proper proportions. As a matter of fact, on the eve of its departure, the expedition encountered difficulties of a precisely opposite nature, namely an excess of popular zeal, which found its expression in the so-called 'Crusade of the Poor'. Secondly, the papal–Hospitaller enterprise met with the opposition of two secular heads of Europe, Philip IV of France and James II of Aragon. This opposition almost caused its total failure.

Philip's co-operation was particularly important. Not only because, as Philip stated, the major part of the Hospitallers would come from France, but also because he had undertaken

---

[21] *AA*, vol. ii, no. 88, pp. 191–2.
[22] W. Wattenbach, 'Fausse correspondance du sultan avec Clément V', *AOL* 2 (1884, 297–303; 'Martini Continuatio Coloniensis', ed. G. Waitz, in *Chronica Regia Coloniensis, MGH SRG* xi (Hanover, 1880), 364–6; J. Prawer, *Histoire du royaume latin de Jérusalem* (Paris, 1969), i. 388, ii. 384.

to finance the crusade to the tune of 100,000 gold florins. Clement, relying on the king's co-operation, stressed his initiative time and again. It was also to please Philip that Clement nominated the bishop of Rodez as his legate for the *passagium*. The king, who basically favoured a general passage rather than a particular one, gave his consent to the latter, assuming his own leadership to be self-evident. Realizing his mistake, he could not tolerate the independent papal–Hospitaller enterprise which deprived him of complete control of the movement. He refused to transfer to the master general of the Hospital the sum of money he had promised. This money, it seems, never reached its promised goal.[23] Instead, he dispatched (October 1308) to Avignon a certain Robert, an English knight, who offered a contingent of 500 English knights, ready to depart at their own expense with the papal crusade. Philip suggested that these should replace the Hospitallers in the enterprise. Clement rejected the offer as there was no proof of its feasibility. As this failed, Philip made a direct attack on the Hospital's master and leader of the planned crusade, Fulk of Villaret. In the beginning of 1309 he complained to the pope that he had not been sufficiently informed by Fulk of Villaret, about the preparations and that the French Hospitallers had not been given their due importance. This, he wrote to Peter of Peredo, his nuncio at the Curia, was an insult to the status and honour of the French kingdom (*regnum Francie*) and even the French nation (*natio Francorum*). He demanded that the master-general apologize and that: 'tot et tales de fratribus regni Francie ad hoc passagium assumantur, quod rex possit confidere de cisdem'. Peter of Peredo was also instructed to emphasize that the honour of the French nation was involved in the crusade. Philip, it seems, regarded the attempt to launch a papal–Hospitaller crusade as an insult to his dignity. It was in line with that feeling that he insisted (24 August 1312) 'quod per sedem apostolicam sic dictorum Hospitaliorum ordo regularetur et reformaretur tam in capite quam in membris'.[24]

---

[23] Clement V, *Regestum*, nos. 2986, 3218, 5384; *Cartulaire*, nos. 4807, 4820, 4831, 4835, 4863, 4884; *Vitae Paparum*, vol. iii, pp. 105–8, 113–15, 120–1, 129; *Papsttum*, vol. ii, no. 75, pp. 118–19; Hillgarth, *Ramon Lull*, pp. 76–82.

[24] *Vitae Paparum*, vol. iii, pp. 89–90; *Cartulaire*, nos. 4831, 4841; *Dossier*, p. 200; S. Schein, 'Philip IV the Fair and the Crusade: A Reconsideration', in P. W. Edbury (ed.), *Crusade and Settlement* (Cardiff, 1985), 123.

While King Philip IV tried to secure control of the expedition and to turn it into a French enterprise, James II of Aragon aimed at its deviation to the Iberian peninsula. James's unbounded enthusiasm for a general crusade to the Holy Land in 1305 gave way two years later to a crusade-plan against Granada.[25] In March 1309 James, who rejected the papal–Hospitaller plan, produced a project of his own. In his opinion: 'Desa se diu, quels Espitalers donen entendre al senyor papa, que dins V annys li auran per guerra o per plet Jherusalem o Antiochia, de la qual cosa non creega regu, que no es cosa possible, que per aquesta armada ne per aquesta gent re pusha fer'. He also argued that the crusade should cost the pope as little as possible and that it could be launched with just twenty galleys and 300 to 400 knights (instead of forty galleys and 1,000 knights), so that it would not arouse the suspicion of the sultan and thus endanger Cyprus and Lesser Armenia. As he rightly pointed out, the conquest of Rhodes as a base for Christian activities against the sultanate was a mistake as it was too far from Egypt to be of much use and nobody sailed through it to Alexandria. Therefore James, as his brother Frederick III of Sicily had already argued in 1304, recommended that the islands of Romania should be conquered. As in his opinion the Hospitallers were too involved in Rhodes, the leadership of the expedition should be given to the lords of the Dodecanese.[26]

Since the announcement of the expedition, the prevailing view in Aragon was that the Saracens of Spain and the king of Morocco, each as powerful as the sultan of Egypt, should be defeated before an attack on the Nile. Aragonese public opinion was thus as violently opposed as its monarch to the Hospitallers' expedition.[27] In December 1308 the kings of Aragon and Castile agreed on a joint attack against Granada. James argued that both Granada and Ceuta, a permanent menace to Aragon and Castile, could be conquered by a force of 1,000 knights and five galleys. This was not feasible, however, without the aid of the papacy. Therefore from the beginning of 1309 James's nuncios at Avignon lobbied the pope and cardinals to grant to

---

[25] *Papsttum*, vol. ii, no. 125, p. 234; no. 134, p. 266; no. 142, pp. 289–91; A. J. Forey, *Templars in the Corona de Aragón* (Oxford, 1973), 140–1, 359ff.

[26] *AA*, vol. iii, no. 91, pp. 198–9; no. 126, p. 243; Kedar and Schein, 'Un projet', p. 218.

[27] *AA*, vol. ii, no. 477, pp. 764–5.

the Aragonese expedition the same status as that of the Hospitallers, namely the status of a crusade. Accordingly he also demanded that a five-year ecclesiastical tithe be granted in all his dominions and that those who had already undertaken to participate in person or money in the Hospitaller expedition should be absolved from their oath and so be free to participate in the expedition against Granada. The pope had also to be persuaded to appoint James as its 'capitaneus, dominus atque caput'. James even managed to get Christian Spinola of Genoa and Ramon Lull to act on behalf of his plan in the Curia. Ramon Lull actually intended in September 1309 to go to the Curia to urge aid for the Spanish crusade.[28] On 21 March 1309 Clement gave in to some of James's demands. He was granted the tithe of Aragon for one year. On 24 April 1309 clergymen willing to participate in the expedition against Granada were granted a subsidy to the amount of two years' revenue from their benefices. On the same date Clement also officially recognized the expedition as a crusade and granted to its participants the same privileges as those granted for the Holy Land. The tithe granted to James was then extended to three years in all his dominions.[29] The pope's generosity, however, did not satisfy the king of Aragon. In the spring of 1309, he tried to use for his own purposes the Hospitallers who were gathering in Spain for their expedition. He forbade their departure from Aragon and imposed a prohibition on the export of horses and transfer of subsidies from his kingdom. In the face of Clement's protests, James gave in (June 1309). A year later the attack on Granada proved disastrous.[30]

Thus by the spring of 1309 it seemed that the whole project of the papal–Hospitaller expedition was on the brink of collapse. As has been seen already, it was torn between France and Aragon and it could not depart on the fixed date (24 June 1309) because of lack of ships and funds. The building of galleys was

---

[28] *AA*, vol. ii, no. 478, pp. 766–7; vol. iii, no. 90, pp. 195–7; vol. iii, no. 91, pp. 198–9; *Cerdeña y la expansión mediterránea de la Corona de Aragón*, ed. V. Salavert y Roca (Madrid, 1956), no. 323, pp. 403–4, no. 332, pp. 413–14; Hillgarth, *Ramon Lull*, pp. 67–8; N. J. Housley, 'Pope Clement V and the Crusades of 1309–10', *Journal of Medieval History*, 8 (1982), 32–3.
[29] Clement V, *Regestum*, nos. 3819, 3988–91.
[30] *Cartulaire*, no. 4866; *AA*, vol. iii, no. 91, p. 200; L. Thier, *Kreuzzugsbemühungen unter Papst Clemens V 1305–1314* (Düsseldorf, 1973), 89–93; Housley, 'Clement V', pp. 34–5.

interrupted because some lords in France confiscated the timber destined for their construction. As to the funds, Philip did not keep his part of the agreement and unexpected difficulties occurred in the collection of the believers' contributions. Clement himself grudgingly lent 50,000 gold florins (October 1309), but this was not enough.[31] But Fulk of Villaret doggedly continued with the preparations. Finally in October 1309 Clement was able to inform Philip that the master-general of the Hospital had departed. In September a small fleet carrying the master-general, the papal legate Peter of Pleine Chassagne, and the papal nuncio Raymond of Pins had put to sea. The ships ran into bad weather and had to sail into Brindisi and winter there. Thus further operations had to be postponed until the following year. At this juncture, the expedition was said to consist of some twenty-six galleys, 200–300 knights and 3,000 foot-soldiers—that is, less than a third of the knights envisaged by the Hospitaller plan and the papal bull, and only three-quarters of the foot-soldiers. In the spring of 1310 the force sailed for the East; by May 1310 promises of goodwill sent by Fulk of Villaret from somewhere in Greek waters had arrived in Venice, which had been apprehensive lest the crusading force turn against its Aegean possessions. Having arrived in Rhodes, the expedition assisted the local Hospitallers in completing the conquest of Rhodes and some of the adjacent islands. Rhodes, where the Hospitallers moved their headquarters from Cyprus (1309), now became the main base of their operations.[32]

The significance of the conquest of Rhodes was that it offered the Hospital a prospect of independence it totally lacked in Cyprus as well as an effective base for warfare against the Moslems, both the Turks and the Mamluks. Located northeast of Crete and north-west of Cyprus, Rhodes was not on the

---

[31] Clement V, *Regestum*, nos. 3825, 4771, 4923; William Greenfield, *Register*, ed. W. Brown and A. Hamilton Thompson (Surtees Society, 145, 152, 153; 1931–40), vol. iv, no. 2351a, pp. 363–4; *Vitae Paparum*, vol. iii, pp. 105–8; *Cartulaire*, nos. 4856, 4884. Housley's argument that the *passagium* was also hindered by the papal crusade against Venice (June–Sept. 1309) does not seem convincing: see Housley, 'Clement V', pp. 36–7. Not even the papal Camera was affected: see *Cartulaire*, no. 4884.

[32] *Cartulaire*, nos. 4807, 4883–4; Ptolemy of Lucca, 'Vita Clementis V', in *Vitae Paparum*, vol. i, p. 35; Bernard Gui, 'Vita Clementis V', in *Vitae Paparum*, vol. i, pp. 67–9; Luttrell, 'Hospitallers in Rhodes', p. 285; Kedar and Schein, 'Un projet', pp. 218–19. For a different account of these events see Housley, 'Clement V', pp. 37–8.

most direct European trade routes to Constantinople or Alexandria but its fine harbour added to its considerable strategic importance. Moreover the conquest strengthened the Latin domination in the Aegean. At that time the Frankish states in Greece were the principality of Morea and the duchies of Athens and of the Archipelago ruled by various European dynasties and especially by the Villehardouins and the Angevins. Venice held Negroponte, the 'Black Bridge', between classical Euboea and the mainland, its main base in the Aegean. There it held also various islands, namely Naxos, Skyros, Andros, Cos (Kos), and Crete; the latter bordered both the south-western and the south-eastern entrances to the Aegean and lay on the direct route from the Ionian Sea to Egypt or Syria. In the Ionian Sea the Venetians held Modon and the nearby Coron at the Southern tip of the Morea. Genoa, like Venice, had established colonies in the area during the thirteenth century. The most important were Pera in the harbour of Constantinople and Kaffa on the northern shore of the Black Sea; other Genoese commercial centres were the island of Chios in the Aegean and Focea near Smyrna (Ismir).[33]

The conquest of Rhodes and its neighbouring islands was in itself a blow to the republic of Venice, as it seriously weakened its position in the Aegean. In 1309 it was rumoured that Fulk of Villaret would take forty galleys and a large force to Rhodes, Lesbos, Crete, or Cyprus. The Venetians, having already sent fifty mercenaries to resist the Hospitallers at Cos, now took elaborate measures to protect their Aegean colonies. Attempts by the Hospitallers to stop the trade of the 'mauvais crestiens' with the sultanate were bound to cause even greater friction between the order and the maritime powers, whose prosperity depended to a considerable extent on the Levantine trade. The question was raised in Venice in May 1310; in 1313 the doge refused the order the right to sell in the city indulgences for a new crusade; in 1314 Venice sent an embassy to Rhodes to protest against the seizure of a Venetian galley. The situation was exacerbated when the Hospitallers seized from the Venetian Andrew Carnaro, the island of Carpathos and other

---

[33] J. Longman, 'Frankish States in Greece 1204–1311', in *Setton's History*, ii. 235–74; F. C. Lane, *Venice: A Maritime Republic* (Baltimore and London, 1973), 42–3.

islands between Rhodes and Crete. In 1312 and 1314 the Venetian government seized Hospitaller funds in transit at Venice. Both sides submitted in 1314 to papal arbitration.[34]

The relations with Genoa, an ally of Byzantium, were even worse. In the winter of 1310–11 the Hospitallers captured near Cotrone a Genoese galley going from Alexandria to Brindisi. In 1311 a Genoese embassy, headed by Antonio Spinola, arrived in Rhodes to negotiate its release, having on its way incidentally captured Vignolo. The Hospitallers refused to negotiate. Thereupon the enraged Spinola passed to Menteshe in Asia Minor and convinced its Turkoman emir Orkham to put into prison some 250 men from Rhodes who were then in the place. The emir was even offered 50,000 gold florins to invade and conquer Rhodes from the order. Still not satisfied, Spinola and his companions ordered two Genoese galleys to attack a Hospitaller fleet near Rhodes. One of the Hospitaller ships, which carried a large number of knights, twenty-five horses, and other cargo, was captured, carried to Menteshe, and there sold to the Saracens.[35] A Hospitaller embassy dispatched in the summer to Genoa was refused audience. The government of Genoa obviously supported the policy of its ambassador Spinola. The embassy's visit to the Curia met with more success. Clement ordered the archbishop of Genoa and the bishop of Nola to collect proofs of Genoese contraband. The archbishops of Naples and of Brindisi, charged (17 July 1311) with a similar inquest, had to intervene with Robert of Naples to return to the Hospitallers the merchandise of the captured Genoese galley kept in Brindisi. In November 1311 Clement addressed to Genoa a severe admonition demanding the immediate release, without ransom, of captured Hospitallers. Those involved had to be punished. Nevertheless, Fulk of Villaret had, as an act of reprisal, captured a Genoese fleet (25 May 1312) on its way from Cyprus to Genoa, confiscated its cargo, and imprisoned its crew. In the same year the Hospitaller fleet was able also to pursue twenty-three Turkish ships to Amorgos in the Cyclades;

---

[34] Luttrell, 'Hospitallers in Rhodes', pp. 285–7.
[35] Heyd, *Histoire du commerce*, ii. 36; G. Caro, *Genua und die Mächte am Mittelmeer 1237–1311: Ein Beitrag zur Geschichte des 13. Jahrhunderts* (Halle, 1895–9), ii. 383–4; See also Henry II of Lusignan, 'Consilium', ed. M. L. de Mas Latrie in Mas Latrie, *Histoire de l'île de Chypre sous la règne des princes de la maison de Lusignan* (Paris, 1852–61), ii. 119–20.

when the Turks landed, the Hospitallers burned their ships and destroyed or captured almost the entire force, losing themselves some fifty or more brethren and 300 foot-soldiers.[36] The tensions and conflicts between the Hospitallers and the republics of Genoa and Venice explain one of Marino Sanudo Torsello's recommendations, namely, that the fleet of any future expedition should be provided and equipped by the maritime powers. Only in this way could conflict between those powers and the Church be prevented.[37] Marino was obviously right to argue that an economic blockade would not be successful without the co-operation and support of the maritime communes. The argument was sound, but it may be doubted if such co-operation and support could ever be forthcoming. The trade and the prosperity of Genoa and Venice were at stake and they could hardly support such policy. From the beginning this strategy was doomed to fail, like the papal bulls against commerce with the enemy. This, however, does not necessarily mean that the expedition of 1309-10 failed to fulfil the hopes which the Hospitallers had placed on their project for a *passagium particulare*. The expedition did not, of course, achieve its explicitly stated aims, since its effect on the military and economic standing of the sultanate was negligible. None the less the Hospitallers obtained their immediate objective which, as James II of Aragon only too rightly suspected (17 March 1309), was to 'conqueire la partida que tenen los Turchs en Rodes e les illes de Romania'.[38]

The 'Crusade of the Poor'

Another difficulty which was encountered at the stage of organization and initial operation was, as said above, caused indirectly by papal propaganda. The preparations for the

---

[36] Clement V, *Regestum*, nos. 7118-19, 7631-2; Francesco Amadi, *Chronique, Chroniques d'Amadi et de Strambaldi*, ed. M. L. de Mas-Latrie (Paris, 1891), 395; J. Delaville le Roulx, *Hospitaliers à Rhodes 1310-1421* (Paris, 1913), 10-11; Caro, *Genua*, ii. 385 ff.; Luttrell, 'Hospitallers in Rhodes', pp. 286-7. On the relations between the order and the republics see also Kedar and Schein, 'Un projet', p. 219.
[37] Marino Sanudo Torsello, 'Liber', p. 31.
[38] *AA*, vol. iii, no. 91, p. 199; Kedar and Schein, 'Un projet', pp. 219-20.

expedition now under discussion did not include certain traditional features of a crusade, such as the taking of the Cross and the participation of believers. Such omissions were easily misconstrued, and one of the results of such misinterpretation was the 'Crusade of the Poor'. With the accession of Clement V in 1305 and the death of Edward I in 1307, France inherited the leading place which England had for some time occupied in papal crusading policy.[39] This was well reflected by the chroniclers; after 1305 the French replaced the English as the best informed on crusading affairs and the most involved in them. This change was assisted by the fact that, with the pope in Avignon, the king of France involved in planning a crusade, and the papal requests for financial aid to the Hospital aiming particularly at France, the crusade almost became a local French affair. It was these changed conditions which explain why it was mainly French chroniclers who recorded the crusader indulgences granted by the pope on 11 August 1308. Like the papal bulls, the chroniclers stressed the indulgences accorded for financial aid to the *passagium*. From among the non-French chronicles, the papal crusader propaganda was recorded only by the English author of 'Annales Paulini', by Ptolemy of Lucca, and by Giovanni Villani.[40]

Yet what apparently was aimed at France found most unexpected repercussions in Germany. This was where the papal indulgences, described in the 'Annales Paulini' as 'a saeculo non erant audita', caused the deepest stir. Ecclesiastical chroniclers, mainly German, recorded for 1309 a mass movement which can be best described as a 'crusade of the poor'. The outburst of this popular zeal is recorded all over Germany, in Lübeck, Westphalia, Brunswick, Tiel, Bavaria, Austria, Flanders, Brabant, and even in Pomerania and Silesia in Poland. According to the ecclesiastical annals of St Blasius in Bruns-

[39] This was much in tune with other activities of Clement V, which were all centred in France; see R. Gaignard, 'Gouvernement pontifical au travail: L'Exemple des dernières années du règne de Clément V: 1er août 1311–20 avril 1314' *Annales du Midi*, 72 (1960), 169–214.

[40] John of Paris, 'Memoriale historiarum', *RHGF* xxi. 653; Girard of Frachet, 'Chronicon cum anonyma eiusdem operis continuatione', *RHGF* xxi. 32; *Grandes Chroniques de France*, viii. 265–6; 'Anonymum S. Martialis Chronicon', *RHGF* xxi. 813; 'Annales Paulini', in *Chronicles of the Reigns of Edward I and Edward II*, ed. W. Stubbs (RS, London, 1882–3), i. 266; Ptolemy of Lucca, 'Vita Clementis V', p. 34; Giovanni Villani, 'Historia Universalis', *RIS* xiii. 438.

wick: 'fuit magnus cursus de viris et mulieribus volentibus ire transmare temporibus Clementis pope V. Isti appelabantur *Crutzebrodere*', that is *Kreuxbrueder* or 'Brothers of the Cross'.[41] The German chroniclers, and also Ptolemy of Lucca and Bernard Gui, described them as 'populus sine capite'; other definitions were no more flattering: 'tales in mores muscarum multiplicati super numerum'; 'quasi pauperes', 'multitudo communis plebis', 'parassiti, sectatores otii, post dilapidatas substantias . . . agricultores, sutores, pelliperarii et alii manuales operarii, omnes miseri et mendaci et invalidi, instignate se ut creditur iniquo et deceptore spiritu, cruce signaverunt se ipsos'. Though the ecclesiastical chroniclers emphasized the low social standing of the participants (bankrupt landowners, nobles who had squandered their wealth, peasants and agricultural labourers, poor craftsmen and artisans like tailors and leather makers or furriers) it was still evident that the composition was much more heterogenous and included quite respectable knights and burghers. Though predominantly German, it was almost a pan-European movement. It was seen as such by the Franciscan annals of Ghent: 'innumerabiles vulgares de terra Angliae, Picardiae, Flandriae, Brabantiae, Alemaniae'. Obviously those who joined regarded it as a crusader movement with clear crusader aims of the conquest of the Holy Land, and the participants actually took the Cross (*crucesignati*) and wished to join the Hospitaller expedition, either 'pro recuperatione terre sancte Iherusalem' or, for the salvation of their souls.[42]

[41] 'Annales Sancti Blasii Brunsvicenses', *MGH SS* xxiv. 825; 'Annales Tielenses', *MGH SS* xxiv. 26; 'Continuatio Florianensis', *MGH SS* ix. 752–3; 'Annales Gandenses', *MGH SS* xvi. 596; 'Annales Lubicenses', *MGH SS* xvi. 421; John of Victring, 'Chronicon Carinthiae', in *Fontes rerum Germanicarum* ed. J. F. Böhmer (Stuttgart, 1854–68), i. 884; 'Continuatio Canonicorum Sancti Rudberti Salisburgensis', *MGH SS* ix. 819; *Chronique liégeoise de 1402*, ed. E. Bacha (Brussels, 1900), 257–8; John of Hocsem, in *Gesta pontificium Leodensium*, ed. J. Chapeauville (Liège, 1613), ii. 128; Levold of Northof, 'Chronik de Grafen von der Mark', ed. F. Zschaeck, *MGH SRG* vi (Berlin, 1929), 64; 'Annales Cisterciensium in Heinrichow', *MGH SS* xix. 545; 'Chronicon Elwacense', *MGH SS* x. 39; 'Annales Colbazienses', *MGH SS* xix. 717; 'Annales Terrae Prussicae', *MGH SS* xix. 692.

[42] See n. 41 above. See also William of Egmont, 'Chronicon', in *Veteris aevi Analecta seu Vetera Monumenta*, ed. A. Matthaeus (Halle, 1738), 577; Ptolemy of Lucca, 'Vita Clementis V', p. 34; Clement V, *Regestum*, no. 4400; J. Stengers, *Juifs dans les Pays-Bas au Moyen Âge* (Brussels, 1950), 16 and n. 71. It is known that burghers departed from Münster. See H. Lahrkamp 'Nord west deutsche Jerusalemwallfahrten im Spiegel der Pilgerberichte', *Oriens Christianus*, 40 (1956), 117–18.

The movement of 1309 originated, like all similar movements, in a rather complex set of circumstances. Like *Pastoraux* of 1251 and of 1320, as well as the popular ferment that the capture of Smyrna (1344) aroused in Italy, it reacted to events connected with the Holy Land. In 1309 the papal proclamation of the Hospitaller expedition, accompanied by the grant of indulgences, was probably interpreted as a proclamation of a general crusade. According to the monastic 'Chronicon Elwacense' the aim of this popular crusade was 'to sail to the Holy Land and to remove it by force from the rule of the Mongols'. Such a statement connected the movement with the events of 1300. The excitement was compounded by an outbreak of serious famines in Picardy, the Low Countries, and the Lower Rhine. It was therefore by no chance coincidence that it was precisely in these areas that the movement is better known and resulted in a typical 'crusade of the poor', a popular movement under the banner of *iter Terrae Sanctae*.[43]

In the spring of 1309 the bands of 'crusaders' started for Avignon to join the Hospitallers. On their route, like many similar movements before them, they pillaged or begged their way. They sought their victims in the Jewish quarters of the cities, but they also stormed the castles in which the nobility sheltered those valuable sources of their revenue. One of the bands, numbering if the chroniclers may be believed 12,000 men, attacked Genappe, a fortress of John, duke of Brabant. In this incident 200 of the 'crusaders' were killed. In June 1309 a formidable multitude, said to be 30,000 strong, drawn from all part of Europe but mainly from Germany and England, finally gathered in Avignon.[44]

The accounts of events which took place in Avignon reflected the ecclesiastical disapproval of this and similar movements. A chronicler of the monastery of St Florian in Austria went so far

[43] 'Continuatio Canonicorum Sancti Rudberti Salisburgensis', p. 819; 'Continuatio Florianensis', p. 752; 'Chronicon Elwacense', p. 39. See also N. Cohn, *Pursuit of the Millenium: Revolutionary Millenarians and Mystical Anarchists of the Middle Ages* (London, 1970), 98–102, 133, 346 n. 132; P. Alphandéry and A. Dupront, *Chrétienté et l'idée de croisade* (Paris, 1954–9), ii. 257–72; H. C. Lea, *History of the Inquisition of the Middle Ages* (New York, 1956), i. 268–70; H. H. W. F. Curschmann, *Hungersnöte im Mittelalter* (Leipzig, 1900), 82–5.

[44] 'Annales Colbazienses', p. 717; *Chronique liégeoise*, pp. 257–8; William of Egmont, 'Chronicon', p. 577; 'Continuatio Florianensis', pp. 752–3.

as to describe the 'crusaders' as stirred up by the Devil. The chroniclers looked with disfavour on those who took the Cross without the authorization of the Church.[45] Biased, the chroniclers did not even mention that on 25 June 1309, the pope granted indulgences to all those who gathered for the crusade. This was done on behalf of the 'plures Theutonici plebei' through the intervention of John of Chalon-Arlay, himself a 'crucesignatus'.[46]

The fate of the 'Crusade of the Poor' of 1309 was no more fortunate than that of its predecessors or of its successors. Some, content with the indulgences which they had received, left Avignon and regained their homes. As Ptolemy of Lucca puts it: 'recesserunt ad propria cum scandalo multo'. Others, more determined, who even equipped themselves with crafts for the crossing of the Danube, departed for the Holy Sepulchre, but on their way vanished from history.[47] As for the *res gestae* of the papal–Hospitaller *passagium particulare*, it proves that Europe at the beginning of the fourteenth century was incapable of launching this sort of crusade. A military expedition, organized by the papacy and the Hospitallers, it depended on the cooperation of the secular heads of Europe and the enthusiastic response of the faithful to be tangibly expressed in financial support. This was clearly a miscalculation. Most of the monarchs were reluctant to support an enterprise which did not contribute to their own interests or, at least, to their glory. Despite being a papal–Hospitaller expedition, it was still basically a pan-Christian enterprise. But precisely on this level it confronted the new national, and by definition, particularistic interests. The crusade had to serve such interests, and the only way to assure this, was to dominate it. It was in this light that one should view the reaction of Philip IV of France and of

---

[45] See e.g. 'Annales Lubicenses', p. 421; John of Victring, 'Chronicon Carinthiae', p. 889; 'Annales Gandenses', p. 596; Ptolemy of Lucca, 'Vita Clementis V', p. 34.

[46] Clement V, *Regestum*, no. 4400. In the bull Clement referred to the multitude of believers from Germany who took the Cross and vowed to set out to aid the Holy Land, and who could not fulfil their vows because of the lack of ships. He granted them an indulgence of 100 years. See also *Acta Imperii Angliae et Franciae 1267–1313*, ed. F. Kern (Tübingen, 1911), no. 186, p. 125.

[47] Ptolemy of Lucca, 'Vita Clementis V', p. 34; 'Annales Londonienses', p. 156; *Chronique liégeoise*, p. 258; Levold of Northof, 'Chronike', p. 64; 'Annales Cisterciensium in Heinrichow', p. 545; 'Annales Gandenses', p. 596; 'Chronicon Elwacense', p. 39; 'Annales Tielenses', p. 26; Lahrkamp, 'Orientreisen', p. 118.

James II of Aragon. On the other hand, public reaction showed that the new crusade-strategy, so popular in the circles of crusade-planners, was alien to the masses. Papal crusade-propaganda found no response in the public mind. The aims of the new type of crusade, economic warfare rather than the traditional direct confrontation with the infidels, were not sufficiently explained and certainly had less appeal. In the traditional type of crusade there was at least some glamour. A general, large-scale crusade of the old kind, however, was condemned by the theorists and experts as impracticable in the conditions of the early fourteenth century, unless certain preparatory states could be successfully accomplished. This was to become the prevailing opinion at the fifteenth ecumenical Church council held at Vienne.

8
1310–1314
THE CRUSADE AT THE COUNCIL OF VIENNE

A prominent Church historian has remarked that the Council of Vienne barely justifies its rank as a general council of the Church as in reality it had to deal with one point only, that is, the Templars and their property.[1] This statement is in many ways a crude exaggeration. Three topics of discussion were listed in the bull of convocation *Regnans in excelsis* (12 August 1308): the affair of the Order of the Templars, the recovery of the Holy Land, and the reform of the Church.[2] The papal ordering of the subjects already pointed to the place each was to have at the council. The crusade took second place after the Templars but, as will be shown, hardly a negligible one. Contrary to custom not all the bishops were summoned to the council. The full *Regnans in excelsis* introduced a distinction between the prelates invited by name (*personaliter nominati*) and those who were to be only represented. The latter accorded *plenaria potestas* to those who represented them *per publica documenta*, or they were to be represented by *procuratores idoneos cum potestate simili*. This procedure was endorsed to ensure the exercise of pastoral duties throughout the duration of the council. The selection of prelates to receive direct invitations had previously been discussed with the king of France. The definitive list contained some 253 names; ultimately 4 patriarchs, 75 archbishops, 151 bishops, 13 abbots, 7 masters general of orders, and 3 priors, accepted the invitation. The French and the Italians constituted the largest groups at the council.[3]

The bull *Regnans in excelsis* was addressed not only to the different ecclesiastical corps of Christendom, but also to the

---
[1] W. Ullman, *Short History of the Papacy in the Middle Ages* (London, 1974), 281.
[2] See above, ch. 7 n. 7.
[3] *Papsttum und Untergang des Templerordens*, ed. H. Finke (Münster, 1907, vol. ii, no. 184, pp. 303–6; E. Müller, *Konzil von Vienne* (Münster, 1934), 24, 663–70.

monarchs of Europe: Philip IV the Fair, Edward II of England, Charles II of Anjou, James II of Aragon, Frederick III of Sicily, Henry II of Cyprus, James II of Majorca, Ferdinand of Castile, and others. They were invited to attend the council, but of all the heads of state, only Philip IV attended in person. Henry VII of Luxembourg the emperor-elect (since 26 July 1309), was absent and this was crucial to the fate of the council. Clement tried hard by his conspicuous benevolence to Henry VII to set him against Philip the Fair, whose ever-increasing protectiveness he found oppressive. Moreover, the emperor-elect, as much as Philip, promised to take the Cross. The embassy dispatched by Henry to Clement early in June 1309 took the oath in the king's name (26 July 1309). Henry promised that, once crowned, he would immediately lead a crusade to liberate the Holy Land. Peter of Zittau, author of the 'Annales Aulae Regiae', states that almost from the time of his royal election Henry was convinced that it was his mission to recover the Holy Land. However, during the council, Henry was busy with Italy. Crowned emperor on 22 June 1312 in St John Lateran in Rome, he died on 24 August 1312 still fighting in Italy.[4]

The Suppression of the Order of the Knights Templar

Clement V, following in the footsteps of Innocent III and Gregory X, ordered in *Regnans in excelsis* that written memoranda be submitted before the council met. It would be the duty of the council, wrote the pontiff (again imitating Gregory X), to find means not only for the recovery of the Holy Land, but also for its future preservation. Nevertheless, in comparison with the encyclicals of Innocent III and Gregory X, little was said of the crusade in *Regnans in excelsis*. Its major part dealt with the trial of the Templars, summing up all its phases until August 1308. But the link between the Templars and the fate of the Holy Land was emphasized. It was for the sake of the Holy

---

[4] Clement V, *Regestum Clementis Papae V*, ed. Monks of the Order of St Benedict (Rome, 1885–92), nos. 3626–7; Peter of Zittau, 'Annales Aulae Regiae', in *Fontes rerum Austriacarum Scriptores*, ed. H. Pez (Vienna, 1855–75), viii. 338–9; W. M. Bowsky, *Henry VIII in Italy: The Conflict of Empire and City-State 1310–1313* (Nebr., 1960), 22, 45, 184, 257 n. 44; Müller, *Konzil von Vienne*, p. 158.

Land, the pontiff recalled, that the Order of the Temple was
founded and enriched by lavish liberties and privileges. Unfortunately apostasy, idolatry, sodomy, and heresy destroyed
it. Clement also repeated the statement made by William of
Plaisians in May 1308 at the Curia: the king of France was
acting against the Temple 'non tipo avaricie cum de bonis
Templariorum nichil sibi vendicare vel appropriare intendat,
sed fidei orthodoxe fervore'.[5] This statement was an attempt to
refute contemporary opinions regarding the motivations of
Philip's attack against the Temple. Some contemporaries
praised him as 'zelo fidei ductus' but some argued that the king
of France was acting out of jealousy and avarice. Thus Christian Spinola of Genoa writing to James II of Aragon (2
November 1307): 'the pope and the king [of France] are doing
this for the sake of money, and because they wish to make of the
Hospital, the Temple, and all the other military orders, one
united order; the king wishes and intends to make one of his
sons master-general [of the order]. The Temple, however,
stood out strongly against these [proposals] and would not
consent to them'. Basically the same explanation was advanced
by an unnamed cardinal in an interview with the representatives of James II at Avignon in March 1309: 'the king of France
was attempting to secure all the goods of the Temple, wherever
they are, for his son, who should be king of Jerusalem'. The
same view was expressed in another independent source,
namely in Adam Murimuth's *Continuatio chronicarum* followed
by later English chroniclers.[6]

These allegations were only partially correct. From the very
beginning Clement staunchly opposed the plans of Philip
regarding the property of the Temple. Since Philip could not
persuade either the pope or the council to agree to his plan for a
united military order under his own command, he was forced to
agree at Vienne to Clement's original plan to transfer the goods

---

[5] William Greenfield, *Register*, ed. W. Brown and A. Hamilton Thompson (Surtees Society, 145, 152, 153; 1931–40), iv. 311–18. For William of Plaisians's speech, see *Papsttum*, vol. ii, no. 88, pp. 142–3.

[6] *Papsttum*, vol. ii, no. 34, p. 51; no. 101, p. 183; Adam Murimuth, *Continuatio chronicarum*, ed. E. M. Thompson (RS, London, 1889), 16. See also Thomas de la Moore, 'Vita et Mors Edwardi Secundi', in *Chronicles of the Reigns of Edward I and Edward II*, ed. W. Stubbs (RS, London, 1882–3), ii. 298–9; J. N. Hillgarth, *Ramon Lull and Lullism in Fourteenth-Century France* (Oxford, 1971), 93–4.

of the Temple to the Hospital, although he managed to retain much of them for himself.[7] Philip's attack on the Temple may be explained by a complex set of considerations. Obviously Philip needed money, but there were as well the motives suggested by Spinola, avarice combined with an inordinate ambition for aggrandizement and no doubt a craving for crusading glory. This interpretation, also advanced by a notary in the royal chancery, seems to fit with Philip's policies. The events of 1305–12 prove, as rightly emphasized by H. Finke and F. Heidelberger, the very close link between the destruction of the Templars and Philip's crusade-project. In his view the successful future of a crusade depended on that move as much as on the conquest of Constantinople by a French prince.[8] As for, Clement, the pope was little to blame and his approval was inevitable under the circumstances. He could not take any decision about the crusade, which would benefit from French support, without first coming to agreement with Philip over the issue of the Temple. Clement feared that a far greater scandal would be provoked if Philip IV realized his threat to revive the proceedings against Boniface VIII unless he did not agree with Philip over the issue of the Templars. He also hoped that by a quick suppression of the order the way would be cleared for a new crusade: one led by the king of France. On 3 April 1312, the day when the suppression of the order was published, it was also announced that a decision had been taken regarding the crusade. The bull *Vox in excelso* (promulgated on 22 March 1312), suppressing the order, was immediately followed by letters patent bearing the seal of France, in which Philip IV promised to take the Cross during the next year and to go on a crusade.[9] At the same session of the council (3 April 1312), the general crusade was solemnly proclaimed and scheduled to depart six years, from 1 March 1313. The projected general crusade was to be led by Philip and his son Louis, king of Navarre. The council approved a six years' ecclesiastical tithe,

[7] *Dossier de l'affaire des Templiers*, ed. G. Lizerand (Paris, 1964), 196–203; below, n. 28.

[8] *Papsttum*, vol. i, p. 365; F. Heidelberger, *Kreuzzugsversuche um die Wende des 13. Jahrhunderts* (Leipzig and Berlin, 1911), 54; Hillgarth, *Ramon Lull*, pp. 86–95; below, n. 16.

[9] *Papsttum*, vol. ii, no. 140, p. 288; no. 144, p. 193; G. Lizerand, *Clément V et Philippe le IV le Bel* (Paris, 1910), 284–8.

the first year's revenue assigned to Philip. Some time later on 10 June 1312 Philip was granted the revenues of three additional years. By the same *auctoritas apostolica et approbatio concilii*, the properties of the dissolved Order of the Temple, excepting the possessions situated in Castile, Aragon, Portugal, and Majorca, were given to the Hospitallers. As far as the crusade was concerned, this was the culmination of years of vacillations, efforts, and propaganda.[10]

The suppression of the Order of the Knights Templar can be thus explained mainly in terms of several factors, all external to the order, namely, the financial needs and the political ambitions of King Philip IV, and Pope Clement's wish to stop the proceedings against the memory of Pope Boniface VIII as well as to realize his plan for a crusade to be led by the king of France.[11] Additionally it can be explained by the impact of the loss of the Holy Land upon the attitude of Christendom to the military orders of the Latin Kingdom. After 1291 both the two major orders of the kingdom, Hospitallers as well as Templars, were made scapegoats to account for the fall of Acre. It was only after their arrest and during their trial (1307–11) that the Templars alone were made responsible for the fate of the Holy Land. Emphasized by the papal full *Regnans in excelsis* (12 August 1308), the link between the Templars, their heresy, and the fate of the Holy Land was picked up by public opinion as well as its makers, like William of Nogaret. The knights were consequently accused not only of the fall of Acre but also of co-operating in the past with Moslem rulers including Saladin, as well as of the loss of the island of Ruad off Tortosa in 1302.[12] There were, however, many who had doubts with regard to this as well as to other of the accusations heaped on the Templars. Some even came, during their trial, to their defence. In his

---

[10] *Conciliorum Oecumenicorum Decreta*, ed. J. Alberigo *et al.* (Basel, 1962), 320, 327; Clement V, *Regestum*, nos. 7885–6, 7952, 8783–4, 8973, 9983–4; *Papsttum*, vol. ii, nos. 144–5, pp. 292–8; Müller, *Konzil von Vienne*, pp. 170–5, 655–78; J. R. Strayer and C. H. Taylor, *Studies in Early French Taxation* (Cambridge, Mass., 1939), 95–7; G. Mollat, *Popes at Avignon 1305–1378*, trans. J. Löve (London, 1963), 250.

[11] M. Barber, *Trial of the Templars* (Cambridge, 1978), 221–47; above, n. 8.

[12] See above, ch. 3, pp. 130–1; S. Schein, 'Image of the Crusader Kingdom of Jerusalem in the Thirteenth Century', *Revue belge de philologie et d'histoire*, 64 (1986), 704–7; Barber *Trial*, pp. 128–9; S. Menache, 'Contemporary Attitudes concerning the Templars' Affair: Propaganda's Fiasco?', *Journal of Medieval History*, 8 (1982), 136–45.

defence of the Templars, for example, Jacques de Plany, a knight who was present at the fall of Acre, testified in Cyprus (May 1310) that he had seen many Templars spill their blood on behalf of the Christian faith and die for its sake, unlike many other knights who fled the battlefield.[13]

Another controversial issue was the fate of the Temple's properties. Already during their trial some of the Knights Templar had expressed their concern regarding the property of their order, arguing that it should not be dissipated but applied solely to the aid of the Holy Land as it had been originally given to the order for this purpose. In Vienne most of the Fathers of the council as well as Philip IV the Fair favoured the creation of a new order which was to be granted the Templars' property. Pope Clement himself preferred that the property be transferred to the Hospital, a solution which would avoid the creation of a new order and which would prevent the property being annexed to any order which had a more specifically national or regional character. The pope wanted thus to ensure that the property would be used exclusively for the sake of the Holy Land.[14] After persuading the king of France that the transfer to the Hospital could be subjected to certain conditions favourable to the French crown, the pope, with the assent of the council, transferred the property of the Templars to the Hospital.[15] However, though solemnly granted to the Order of St John, a large proportion of the Templars' riches never reached the Hospital and some not until the 1320s. It is true that already by the time of the Council of Vienne the Hospital was richer and more powerful than before but this was due less to the Templars' property than to that of Rhodes and the Templars' former estates in Cyprus. Actually in the 1310s the Hospitallers experienced difficulties in gaining control over Templars' lands in France, Germany, and Italy. Though in England the Hospital did not manage to take over completely the Templar lands up to 1338, and in France they were obliged to pay, over the years, large sums to the French royal treasury for them, within ten years of the Council of Vienne the Hospital succeeded in gaining the greater part of the Templar property.

[13] Barber, *Trial*, pp. 218–19.
[14] See ibid. 230–1.
[15] See above, n. 10; Barber, *Trial*, p. 231.

Thus Giovanni Villani's statement that 'the Hospital was poorer in its property after it was granted that of the Templars than before' is purely exaggeration.[16]

Though the acquisition of the extensive Templar estates had ultimately enriched the Hospital, on the whole the suppression of the Order of the Temple had a negative influence on the Hospital as well as the cause of the Holy Land. During the years 1291–1312 the Order of St John was the most active and reliable single factor in Christendom's struggle against Islam, perhaps more so than the Temple. This was partly caused by the sometimes harmful but often most productive competition between the two orders. With the Order of the Temple gone the Hospital became more powerful both economically and politically, but in the long run the lack of competition led to its internal decline as well as to stagnation. On the whole, in the years following the Council of Vienne, due to internal problems, the Hospital was to become a less important factor in crusade-planning and -making than before. Since it was based in Rhodes it had very often its own objectives which sometimes were opposed to those of the Holy Land. In 1343 Clement VI declared that it was 'the virtually unanimous and popular opinion of the clergy and laity' that the order was doing almost nothing for the defence of the faith; at that time suggestions were made to the pope that part of the order's wealth should be confiscated and used to create a new order whose zeal and rivalry would stir the Hospitallers into action.[17]

The suppression of the Temple had also a negative influence on crusade-making. First of all the military orders had always formed an important part of the crusading armies. The Templars with their rich estates all over western Europe should have been able by 1307, the time of their arrest in France, to recruit 1,000 cavalry and 4,000 infantry from their priories, as Hospitallers were to do for their crusade of 1309,[18] and hire some more. Thus with the Templars gone the regular core of the potential crusading army was meaningfully reduced and

---

[16] Barber, *Trial*, pp. 230–8.

[17] N. J. Housley, *The Avignon Papacy and the Crusades 1305–1378* (Oxford, 1986), 267 and n. 19.

[18] B. Z. Kedar and S. Schein, 'Un projet de passage particulier proposé par l'ordre de l'Hospital 1306–1307', *BEC* 137 (1979), 216, 224.

weakened. But the suppression of the Order of the Temple harmed the case of the crusade in yet another way. The arrest and trial of its members and finally the suppression of the order, who were for almost two centuries supported and protected by the papacy as those who were particularly dedicated to the Holy Land, created an atmosphere of suspicions regarding crusade-politics. The causes of both the Templars and the Holy Land were too closely connected in the European public opinion not to be now intermingled. Indeed this is reflected in the European reaction to the decisions taken in Vienne: both the pope's intentions along with those of Philip IV were suspected. The verdict passed on the Templars and the council's approval of six years' ecclesiastical tithe for the financing of the *passagium generale* scheduled to depart in 1319 were described as motivated by avarice and as pretexts to extort money from the Church.[19] The suppression of the Templars seems thus to have considerable weakened the belief in the sincerity of papal plan for a crusade and consequently it affected negatively the crusade planned for 1319.

For a *passagium generale*

Once the Order of the Temple disappeared, Clement could proceed with obtaining the approval of the council for the organization of the second stage of his plan for a crusade, namely the general expedition. Thus with the closing of the third session of the council, a major crusade, the first since the loss of the Holy Land, was on the way to realization. Above and beyond these moves and counter-moves of *Machtpolitik*, and yet linked to them, another decision was taken during the fourth and last plenary session (6 May 1312). The council ordered (canon 11), that 'Catholic men' were to teach Hebrew, Arabic Greek, and 'Chaldean' (i.e. Syriac) at the universities in Paris, Oxford, Bologna, Salamanca, and at the Roman Curia. They were charged with translating works from these languages into Latin, and their pupils were expected to become missionaries. The king of France was to finance the chairs at Paris; the

[19] See below, nn. 37, 38.

prelates and monasteries of England those at Oxford; the bishops and monasteries of Italy those at Bologna, and the Church of Spain the chairs at Salamanca. The Holy See itself was to be responsible for the chairs to be established at the Curia. The practical missionary aims of this decision clearly reveal Ramon Lull's intervention. The close correspondence between the opinions of Lull and of Philip leads to the conjecture that Philip might have been instrumental in the passing of this decision, which revealed the council's recognition of the link between mission and crusade, and of the importance of the mission *per se* in the battle against Islam.[20]

The decisions of Vienne regarding the crusade, taken in the shadow of the all-powerful Philip,[21] raise the question how far they were representative of the opinions of the Fathers of the council or of currents of European public opinion. The answer should be sought in opinions expressed in the course of the years 1311–12 either at Vienne or immediately after the end of the council. These opinions, either treatises prepared in writing for the council or views expressed at the council itself, are of particular interest even in a wider context. As had happened in the Church councils of 1291–2, the opinions expressed evolved from a wider circle than that of the supporters of the crusade only. On both occasions they came mainly, if not entirely, from the upper stratum of churchmen. The comparison between these two sets of opinions, one generation apart, is invaluable in attempting to appreciate European attitudes to the crusades after the loss of the Holy Land. Compared with those of 1291–2, let alone those expressed at the Second Council of Lyons, the opinions of Vienne were certainly more in tune with the new thinking about the strategy of the crusade. Various aspects of that strategy were in fact the main subjects of discussions. There is scarcely a trace of evidence that ideology or moral

---

[20] *Conciliorum Decreta*, pp. 355–6; J. Lecler, *Vienne* (Paris, 1964), 193–4. For Ramon Lull's participation in the council see his 'Vita Coaetanea', ed. H. Harada, *CCCM* xxxiv (1980), 303; Hillgarth, *Ramon Lull*, pp. 126–7. For the decree see R. Weiss, 'England and the Decree of the Council of Vienne on the Teaching of Greek, Arabic, Hebrew and Syrian', *Bibliothèque d'Humanisme et de Renaissance*, 14 (1952), 1–3; E. Bellone, 'Cultura e studi nei progetti di riforma presentati al concilio di Vienne 1311–1312', *Annuarium Historiae Conciliorum*, 9 (1977), 67–111.

[21] This was seen clearly by Philip's contemporaries: see *Papsttum*, vol. ii, no. 135, p. 270; no. 137, p. 274.

issues were deliberated at Vienne as they had been at the Second Council of Lyons. These minute and careful examinations of strategy prove that, after two decades of failures and disasters, the handling of the problem had become more cautious. This caution, apart from the obvious psychological reasons, was to a large extent determined by the flourishing literary genre, *de recuperatione Terrae Sanctae*, which envisaged the future crusade as a minutely planned military expedition, and moreover as one evolving from a complexity of politico-economic considerations. The crusade was treated in Vienne as an extremely difficult problem which needed technical solutions to be worked out in the course of long-term preparations, but which had very little ideological content. Conspicuous was the total absence of the pan-Christian declarations which were so prominent in the Church councils of 1292. Neither the insistence on proclaiming an emperor as leader of the crusade, nor the call to establish general peace, were voiced. Conversely, all pointed to the French monarch, the most Christian king, as a new species of leader and promoter of the crusade. Obviously William of Nogaret and Pierre Dubois had their own variants, substituting French hegemony for the Roman Empire. But this they shared with many others, including Ramon Lull, who by 1314 was claiming that 'imperium est propter hoc ut teneat iustitiam, et cum gladio defendat Romanam Ecclesiam ... contra infideles qui possident Terram Sanctam'.[22]

The treatises prepared in writing for the council were of two types: those, already well known, on the theme of *de recuperatione Terrae Sanctae*, and those of a more general character, which dealt with the various subjects proposed for the deliberations of the council, among them the problem of the crusade. Three *de recuperatione Terrae Sanctae* treatises are extant: the plans of Henry II of Cyprus, Ramon Lull, and William of Nogaret which have already been discussed. It has been stated also that Lull's ideas on the crusade did not differ at that time from those

---

[22] William of Nogaret, 'Quae sunt advertenda pro passagio ultramarino et que sunt petenda a papa pro persecutione negocii', in 'Notices et extraits des documents inédits relatifs à l'histoire de France sous Philippe le Bel', ed. E. P. Boutaric, *Notices et extraits des manuscrits de la Bibliothèque Nationale*, 20 (1862), 199–205; F. J. Peques, *Lawyers of the Last Capetians* (Princeton, NJ, 1962), 39–40, 98–99. For Ramon Lull, see Hillgarth, *Ramon Lull*, p. 59 n. 34.

at the court of France.[23] It follows thus that thinking in French royal circles was represented in Vienne by two of the three *de recuperatione* treatises.

The most important aspect of Nogaret's plan was perhaps its author's conclusion that prospects for the crusade were extremely difficult. The preparations, he calculated, would take ten to twenty years.[24] Bringing the difficulties into the open in this way was typical not only of the *de recuperatione* memoirs but also of the whole Council of Vienne. Long-term preparations were an idea shared by many prelates and especially those from France, including William Durant the Younger. The author of the famous *Tractatus de modo generalis concilii celebrandi* (*c.*1310) became known for his involvement with the affairs of the Holy Land in the 1320s, when he was made responsible by Pope John XXII (1323), together with Amauri II, the viscount of Narbonne, for the preparations of a crusade. Yet already before January 1313[25] he had composed an 'Informatio brevis super hiis que viderentur ex nunc fore providenda quantum ad passagium, divina favente gracia, faciendum'. Written shortly after Vienne, it no doubt recorded opinions expressed by Durant during the council. Though his *Weltanschauung* hardly corresponded to that of Nogaret, they shared complementary opinions regarding the crusade. Durant indicated as necessary preliminaries: general peace; assistance of the maritime cities; careful preparations of troops, arms, horses, ships, and provisions. He also supported the French plan for the conquest of Constantinople before that of the Holy Land: the crusade should go by sea to Asia Minor and then, through Constantinople, by land to Syria. Simultaneously a separate force of Hospitallers should disembark in the Holy Land to establish a bridgehead for the expected forces from Constantinople. There

---

[23] Ramon Lull, 'Petitio Raymundi in concilio generali ad adquirendam Terram Sanctam (1311)', ed. H. Wieruszowski, in Wieruszowski, 'Ramon Lull et l'idée de la Cité de Dieu', *Estudis Franciscans*, 47 (1935), 103–9; Hillgarth, *Ramon Lull*, p. 127 n. 342.

[24] Above, n. 13.

[25] William Durant (the Younger), 'Informatio brevis super hiis que viderentur ex nunc fore providenda quantum ad passagium, divina favente gracia, faciendum', ed. G. Düzzholder, in Düzzholder, *Kreuzzugspolitik unter Papst Johann XXII 1316–1334* (Strasburg, 1913), 104–10. For the date of William Durant's 'Informatio' see C. J. Tyerman, 'Philip V of France, the Assemblies of 1319–1320 and the Crusade', *BIHR* 57 (1984), 27–8.

was, however, a basic difference between Durant and Nogaret. Because of his ecclesiastical outlook, Durant aimed at the protection of the rights and property of the Church. Whereas Nogaret's plan was based on the co-operation between his *rex et ecclesia*, Durant's referred to the pope and the Church as the most important factor in the preparations as well as in the expedition itself. The churchmen, cardinals, bishops, abbots, and priors should take part in the expedition. Moreover, they should stay one year in Outremer. The task of the Church appeared to Durant as one of capital importance. The priests and friars should preach, exhort, and counsel in time of need, and offer a laudable example to the armed forces. As a preliminary to the expedition, both secular and regular clergy should undertake missionary work among the infidels, the schismatics, and especially the Greeks.[26]

Another French participant in the council was William Le Maire, bishop of Angers (1291–1317). Le Maire presented himself as a typical European prelate: 'in tam arduis negociis maxime inexperto'.[27] The subject of the *subsidium Terre Sancte* actually belonged 'ad sapientissimos viros in mundanis et expertissimos in rebus bellicis'. Yet, out of obedience to the request of the pope, he presented his view. What he had to say was rather pessimistic. Almost echoing the old Joachimite statement, Le Maire argued that the time for the liberation of Jerusalem had not yet come. Despite this, because of the devotion of the people, and above all that of Pope Clement V, a general crusade might be launched in ten to twelve years' time, that is to say, approximately as advocated by Nogaret. In the meantime the crusade should be preached all over Christendom. Liberal indulgences, like those granted to the Hospitallers at their expedition, should be offered. Le Maire and Durant, both high prelates, were much concerned with the rights of the Church, mainly with regard to property. The goods and possessions of the Templars should be administered by prelates. He

---

[26] William Durant (the Younger), 'Informatio', pp. 104–10.
[27] William Le Maire, *Liber*, ed. C. Port (Paris, 1877), 474. Port, (*Liber*, p. 474) and consequently Heidelberger (*Kreuzzugsversuche*, p. 76) were mistaken when they asserted that Le Maire was kept by illness from attending the council: see Lecler, *Vienne*, p. 58 n. 86. Le Maire dealt with all three subjects discussed at the council in his memorandum presented before the session of 3 Apr. 1312: see id., *Liber*, p. 472 and n. 1.

also resisted new taxation of the Church in the interests of a crusade. Le Maire's recommendation regarding the property of the Templars as well as crusade finances was thus aimed in the opposite direction from that of Nogaret. Like Durant, he wanted to establish ecclesiastical control of the crusade's funds and to place on secular society the main burden of providing finance.[28]

Much the same differences appeared at the council itself. The opinion of the participants are known to us almost exclusively through the reports of the envoys of James II of Aragon at Vienne.[29] The impression given by these reports is that the council focused on the dilemma of crusade to the Holy Land versus crusade to Granada rather than on the problems of organizing a crusade for the recovery of the Holy Land. This impression, only partially correct, is the result of the particular interest of James II in getting a decision in favour of a crusade against Granada. He advocated at the time of the council a plan which combined the crusade against Granada with a crusade to the Holy Land. The presence of the Saracens in Spain, James II argued, provided an obstacle to a general crusade, since they immobilized the potential help from the peninsula. The expedition should be preceded by a naval blockade of Granada by twenty galleys held in the sea-district of Gibraltar. It would need only a year to capture the famine-stricken area. Following the conquest of Granada, the crusading troops should conquer Morocco, and then, by way of Majorca, Minorca, Sardinia, and Sicily (where they could be reinforced and provisioned) they would continue to the Holy Land.[30] Such a plan, pointing clearly to the Aragonese–Catalan interest, could have hardly appealed to Philip IV. Clement V rejected the project on the ground that such a plan would be rejected by the council and especially by the French and English prelates.[31] Although the pope repeatedly rejected the project during the audiences of 16 October and of 4 November 1311, James II was not discouraged. This was possibly due to the fact that when Clement

---

[28] Ibid. 474–6; Müller, *Konzil von Vienne*, pp. 148–50.
[29] See *Papsttum*, vol. ii, pp. 230–306.
[30] Ibid., vol. ii, no. 125, pp. 234–7; *AA*, vol. iii, no. 100, p. 219; Their, *Kreuzzugsbemühungen*, pp. 28–39.
[31] *Papsttum*, vol. ii, no. 131, p. 256.

brought up the question of the crusade at the first session of the council (16 October 1311) he did not specify its course. According to the Aragonese report: 'Enapres parla del fet del passatge; recomtam com seria necessaria cosa e profitosa enantar a impugnatio dels enamichs de la fe e daquesta rahio di[x] molt en paraules generals, no specifican en altra manera'.[32] Besides, not all the French prelates at Vienne were hostile to James's plan. Amaneu VII d'Albert, a relative of the pope, the vice-chancellor of the Church Cardinal Arnald Novelli, Cardinal Arnald de Pellagrua, Cardinal Bernard de Garvo, nephew of Clement V, all of them French, actually supported the plan for a crusade against Granada. Amaneu VII d'Albert even promised his intervention with the pope in its favour. Arnald Novelli thought that a general crusade to the Holy Land by sea was not practicable. He recommended a crusade to the Holy Land, which would go through Granada and Barbery or else by land through Greece and Lesser Armenia. The last part should be easy as 'gens Cathalans et Aragoneses, qui son ja en Romania, qui han subjugades moltes terres, los quals serien ab vos en totes coses ab ço que han subjugat, et los quals han axi esgleyats los Grechs, que, segons que dix lo dit vice-canceller, ni Fra[n]seses ni altres gens del mon no temen, sino les vostres'.[33] Arnald de Pellagrua held the opinion that the pope should order a two-pronged expedition, one of which should be directed against Granada. Bernard de Garvo, who wished to be dispatched by the pope on a mission to Spain, supported the Aragonese project, and opposed that for the conquest of Constantinople. The views of the cardinals show clearly that their views differed as to the route to be taken by a crusade. Moreover, some did not believe at all in the possibility of such an undertaking. Cardinal Berenger Fredol, for instance, thought that 'en fet de negun passatge se faes res en aquest Concili'. He was most sceptical as to the interest of the monarchs of France and England and whether their prelates would give their consent to an ecclesiastical tithe for a crusade. He also had doubts about Fulk of Villaret's promise to bring in five years the keys of Jerusalem. Such scepticism, it is worth

---

[32] *Papsttum*, vol. ii, no. 126, p. 239; Müller, *Konzil von Vienne*, pp. 164–5.
[33] *Papsttum*, vol. ii, no. 130, pp. 253–4.

mentioning, also characterized in 1313 the opinion of the powerful chamberlain of Philip IV, Enguerrand de Marigny.[34]

The opinions of the cardinals who supported the Granada crusade were, however, not only unrepresentative of the prelates at large, but probably described their own position during the first months of the council, October–December 1311. The declarations of the various *naciones*, when summoned by the pope in mid-January 1312, show their recommendations to be far more strongly coloured by their 'national' interest, or at least, and this was particularly true of the French, by the interest of their monarch. The *naciones* appeared in the following order: Germany, England, Arles and Provence, Spain, Italy, and France. The prelates of Germany, England, and the kingdom of Arles declared themselves in favour of a crusade to the Holy Land and gave their agreement to a six years' tithe. The Italians also gave their consent, but made it conditional on the expedition proceeding by way of Constantinople. Following the Spaniards, they also recommended that a crusade against Granada be launched during the preparations for a general crusade. The Spanish prelates agreed to a tithe; regarding the crusade, they declared: 'que en aquests affers sabien mes los princeps, per que conseylaven, quel senyor papa volgues haver conseyl sobre aco dels princeps Despanya e senyaladament de vos, senyor, qui sabiets mes en aquests affers, ey erets pus exercitat, que negu dels altres'. None the less they insisted on the urgency of an expedition against Granada. The Spanish aid for the Holy Land could not be assured, they said, as long as the Moorish danger in Spain persisted. The prelates of France were the last to be consulted. They, however, remained silent, and, as the ambassadors of James II sardonically remarked: 'Lo concili ... es en aquest estament quant al espeegament, quel senyor papa esperava lo rey de Franca: "Sine quo factum est nichil"'.[35] Later the French declared explicitly that, without the approval of their sovereign, they could not give their consent to the tithe. And indeed, it was only after the arrival in Vienne of Philip IV (20 March 1312) that decisions were taken.[36]

[34] Ibid., no. 126, p. 243; no. 130, pp. 254–5; Hillgarth, *Ramon Lull*, p. 83.
[35] *Papsttum*, vol. ii, no. 135, p. 270.
[36] Ibid., no. 140, p. 286; no. 144, p. 293.

The Aftermath of Vienne

European reaction to the crusade decisions taken at Vienne was not always positive. Criticism was mainly directed against papal financial policies. As such, this was typical of the Avignonese papacy as a whole. The 'state of the Holy Land', as discussed at Vienne, was presented by the chroniclers, in accordance with the papal bulls, as a secondary item. The decisions, if compared with the accounts of the verdict passed on the Temple, were discussed *en passant* only. The English chroniclers were the most outspoken in their criticism of the papacy and its exactions. This opposition, a marked feature of the last years and especially the very last year (1307) of the reign of Edward I (though quite prominent already with Matthew Paris) culminated in the accounts of the council. The pope's intentions, along with those of Philip IV, were suspected. The verdict passed on the Templars was described, by Thomas de la Moore, as 'propter importunam pecuniae exactionem'. It was said that the tithe was decided without the consent of the council. The convocation of the council, like the proclamation of the crusade, was but another pretext to exact money from the Church and especially from the Church in England.[37] Similar opinions, we know, circulated in Germany and in Italy. Another source for suspicion was the alleged endowment of the Hospital with the goods of the Temple. This was little believed, especially in England and in Italy. It was rumoured that the properties were being given to Philip IV or at least divided between him and the Hospitallers.[38]

---

[37] For the diversion of the proceeds of the Vienna tenth to crusades against European lay rulers see N. J. Housley, *Italian Crusades: The Papal–Angevin Alliance and the Crusades against Christian Lay Powers 1254–1343* (Oxford, 1982), 104–5.

[38] Thomas de la Moore, 'Vita et Mors Edwardi Secundi', pp. 298–9; William of Nangis, *Chronique latine de Guillaume de Nangis de 1113 à 1300 avec les continuations de cette chronique de 1300 à 1368*, ed. H. Géraud, i (Paris, 1843), 389–92; Bernard Gui, 'Vita Clementis V', in *Vitae Paparum Avenionensium*, ed. G. Mollat (Paris, 1914–27), vol. i, pp. 71–2; Geoffroy of Paris, *Chronique metrique attribuée à Geoffroy de Paris*, ed. A. Diverrès (Strasbourg, 1956), 175–7; Girard of Frachet, 'Chronicon cum anonyma eiusdem operis continuatione', *RHGF* xxi. 37; John of Paris, 'Vita Clementis V', in *Vitae Paparum*, vol. i, p. 19; *Grandes Chroniques de France*, ed. J. Viard, viii (Paris, 1834), 285–6; Francisco Pipino, 'Chronicon', *RIS* ix. 748. See also Paulinus of Venice, 'Vita Clementis V', in *Vitae Paparum*, vol. i, p. 82; above, n. 6; Walter of Guisborough, *Chronicle*, ed. H. Rothwell (London, 1957), 396; 'Martini Continuatio Brabantina', *MGH SS* xxiv. 262;

The epilogue to Vienne took place on 5 June 1313, when Philip IV of France received the Cross from the hands of the papal legate at a ceremony in the Île-de-la-Cité. On the same occasion the Cross was taken by the king's three sons Louis of Navarre, Philip, and Charles, and by the king's two brothers Charles of Valois and Louis of Evreux. It was also taken by his son-in-law Edward II of England, and a large party of nobles from France and England. John of Paris reported that the Cross was taken by many barons, newly dubbed knights, and moreover, by their wives. All this took place amidst superb feasts which lasted eight days and impressed the burghers of Paris. Geoffroy of Paris, 'the echo of the public in Paris', described the event in the following verses: 'La se croisa le roi de France | Et ses trois enfans en presence | Et du royaume la valor | Et d'Angleterre aussi la flor | Avecques le roy d'Angleterre | Por fere contre paiens guerre | . . . Ceste croiseree fu fait | . . . A Paris on pré que l'en clame | De touz l'isle de Notre Dame'.[39]

It was as if a kind of crusading frenzy had struck the French nobles. Moreover, these powerful sentiments were not confined to the nobility only. As anonymous source, probably written in the first years of Philip V (1316–22), reveals that

'et fut faicte si grant croiserie par toutes les autre citez et villes de France, que c'estoit une grant merveille de la dévocion et de la bonne voulente que tout le peuple auoit de visiter le saint sepulchre de Nostre seigneur et les autres lieux sainctz où Jhesu Crist conversa en terre, et avoient grant dueil au cuer de ce que la Terre Saincte estoit en

---

Thomas Walsingham, *Historia Anglicana*, ed. H. T. Riley (RS, London, 1863–4), i. 127; Geoffrey le Baker, *Chronicon*, ed. E. M. Thompson (Oxford, 1889), 5; Adam Murimuth, *Continuatio*, p. 16; William Ventura, 'Memoriale', ed. C. Combetti, in *Monumenta Historiae Patriae*, ed. C. Albert (Turin, 1836–1955), iii. 737; 'Corpus Chronicorum Bononiensum', *RIS* NS xviii. 323; *AA*, vol. i, no. 216, p. 316; A. J. Forey, 'Military Orders in the Crusading Proposals of the Late Thirteenth and Early Fourteenth Centuries', *Traditio*, 36 (1980), 332–3.

[39] Geoffroy of Paris, *Chronique*, pp. 183–4; *Grandes Chroniques*, viii. 287–90; Bernard Gui, 'Vita Clementis V', p. 75; William of Nangis, *Chronique latine*, i. 396; Girard of Frachet, 'Chronicon', p. 38; John of Paris, 'Vita Clementis V', pp. 656–7; *'Merton' Flores* in *Flores Historiarum*, ed. H. R. Luard (RS, London, 1890), 337–8. See also *AA*, vol. i, no. 308, pp. 259–60; Hillgarth, *Ramon Lull*, p. 82 n. 133. For the taking of the Cross by members of the French nobility see also *Registres du trésor des chartes*, i. *Règne de Philippe le Bel*, ed. M. R. Fawtier et al. (Paris, 1958), nos. 1913, 1944.

la main des ennemis a Nostre Sauveur Jhesu Crist, et estoient appareilliez d'espandre leur sang pour l'amour de Jhesu Cristi et pour conquere son heritage, aussy comme Nostre Seigneur espandi le sien sanc pour nous conquerre paradis'.[40]

This testimony, marked by an anti-papal and anticlerical attitude, reveals the persistence of a popular pro-crusade sentiment. This seems to have been endemic in France throughout the twelfth and thirteenth centuries. It was to lead to another of the *Pastoraux* movements in 1320.

In 1313 the preparations for the crusade were in full progress. The persistent papal support of Philip of Taranto's plan to capture Constantinople, as well as rumours from the papal court, show that the pope was contemplating a plan by which the crusader host would conquer the Byzantine Empire, then go through Asia Minor to Lesser Armenia, and from there to the Holy Land.[41] At the same time (13 June 1313) Clement considered the dispatch of an embassy to the sultan of Egypt, since he had been led to believe that the latter was willing to become a Christian and hand back the Holy Land. Clement did not, it seems, dispatch an embassy himself, but he authorized (August 1313) the dispatch of an Aragonese one.[42] Within a few months, however, the plan for the crusade of 1318 collapsed. The deaths of William of Nogaret (April 1313), Clement V (20 April 1314), and Philip IV (29 November 1314) marked the end of the great crusader upheaval which had begun almost a decade earlier and closed a well-defined period of crusade-planning and crusade-making. The three were dead shortly before or after James of Molay and Geoffrey of Charney were burnt at the stake as relapsed heretics (18 March 1314) on the Île-des-Javiaux in Paris. The coincidence and some after-

---

[40] 'Fragment d'une chronique anonyme', *RHGF* xxi. 150.

[41] *Registres du trésor*, nos. 1976, 2170-1, 2173-7; C. Du Cange, *Histoire de l'Empire de Constantinople sous les empereurs français*, ed. J. A. C. Bouchon (Paris, 1826), ii. 366-8; *Papsttum*, vol. ii, no. 130, pp. 253-4; Heidelberger, *Kreuzzugsversuche*, pp. 57-62.

[42] *AA*, vol. iii, no. 467, pp. 751-2; G. Lizerand, *Clément V et Philippe IV le Bel* (Paris, 1910), 479-82; A. S. Atiya, *Egypt and Aragon: Embassies and Diplomatic Correspondence between 1300 and 1330* AD (Leipzig, 1938), 35-41; Hillgarth, *Ramon Lull*, pp. 81-2; Heidelberger, *Kreuzzugsversuche*, pp. 60-1.

thoughts were at the cradle of the legend which made James of Molay summon them before the tribunal of God.[43]

[43] H. C. Lea, *History of the Inquisition in the Middle Ages* (New York, 1956), iii. 326–7; Barber, *Trial*, p. 242. In the final codicil of his will, drafted on 28 Nov. 1314, Philip IV stated that unless his son Louis set out on a crusade before the end of the six years within which he himself had pledged to leave, 100,000 *livres tournois* would be donated to the Holy Land to be spent by the nearest of his successors who agreed to fulfil his commitment: 'Notices et extraits des documents inédits relatifs a l'histoire de France sous Philippe le Bel', pp. 151–2; E. A. R. Brown, 'Royal Salvation and Needs of State in Late Capetian France', in W. C. Jordan *et al.* (eds.), *Order and Innovation in the Middle Ages: Essays in Honour of J. R. Strayer* (Princeton, NJ, 1976), 372–3 and n. 47; S. Schein, 'Philip IV the fair and the Crusade: A Reconsideration', in P. W. Edbury (ed.), *Crusade and Settlement* (Cardiff, 1985), 126 nn. 31, 33.

CONCLUSION

An attempt can now be made to state some conclusions on the major issues discussed in this study: how did the loss of the Holy Land affect the concept of the crusade and its strategy; what was the place of the crusade in European politics, in the preoccupations and currents of thought of the period; what was the attitude of Europeans at large to the crusade aimed at the recovery of the Holy Land; and finally, why was there no major crusade to the Holy Land following the crusade of St Louis to Tunis (1270)?

Crusade-Strategy

The enlargement of the chronological framework, as well as the analysis of a wider range of sources than have been used by earlier historians has led, first of all, to a conclusion which is basic for the understanding of the whole complex of problems connected with crusading ideology after the end of the 'classical' age of crusades in the twelfth and thirteenth centuries. This conclusion can be summarized as follows: The loss of the Holy Land was regarded by contemporaries as a tragic but only a temporary setback; it was generally expected that the situation would soon be reversed, the Holy Land recovered and a new kingdom created. Being temporary, the loss of the Holy Land did not necessitate, at least not in the first years after the event, any radical transformation in the concept of the crusade. Therefore there remained a pronounced feature of traditionalism in the conceptual realm of the crusade. There was, however, a marked change in ideas about its strategy. These included a blockade of Egypt as a necessary prologue to a major general crusade, the reform of the military orders, and an alliance with the Mongols of Persia. These measures were already discussed at the time of the Second Council of Lyons and were current in the crusade-planning in the period preceding the loss of the Holy Land. It was, however, Nicholas IV who inaugurated a new period by attempting to put them into action. Thus, evidently, it was not the loss of the Holy Land but the generation which followed the last of the great crusades, that of St Louis to Tunis (1270), which marked a new beginning in crusade-thinking and -planning.

The impact of the loss of the Holy Land made itself felt in the attempts to translate theory into action. Most of the actions of Pope Nicholas IV and his advisors dictated the very lines of policy which were to be closely followed during the entire period dealt with in this study, and beyond. This was true of Clement V and also of most of the *de recuperatione Terrae Sanctae* treatises which appeared during his pontificate. The latter added little, if at all, to the contents of the memoirs written for Nicholas IV. The repercussions of the loss of the Holy Land were also felt in rather unexpected quarters, namely in the new hiearchy of importance taken by the different political forces in Europe which contributed or participated in the crusade. Of momentous importance was the reappearance of the king of France in crusading affairs. This was an unforeseen result of Clement V's crusade-policy and one which had not been intended. France and its monarch became the centre of the crusading movement, as they had been during the reign of St Louis, although after 1270 the centre passed to Edward I of England. With the accession of Clement V (1305), France regained its former dominant position.

Another trend which emerged following the loss of the Holy Land and was particularly felt during the pontificate of Clement V, and which was to remain a permanent characteristic of the future, was the growing and later absolute dependence of the crusade on the military orders. The fall of Acre caused a marked increase in the already abundant criticism of the military orders of the Latin East. This criticism, which poured from every quarter of Christendom, culminated in the demand for their reform and very often for their unification. The loss of Outremer, however, made the orders of the Temple and of St John, which had now taken refuge in Cyprus, a more important factor than ever in crusade-planning and -organization. They became the most active and reliable single factor in Christendom's struggle against Islam. Their active efforts since 1292 to enforce the naval blockade against Egypt, their attempts to recover the Holy Land in conjunction with a Mongol ally (1299–1303), the papal attitude as well as that of the authors of *de recuperatione Terrae Sanctae* treatises, prove that no crusade could have been envisaged, let alone realized, without assigning to the much-criticized orders a preponderant and indeed decisive role. Therefore also the suppression of the Order of the Templars (22 March 1312) was a hard blow to the cause of the Holy Land. From now on crusader plans had to rely on the Hospital alone, which won a leading role in Clement V's crusader policy as the only international order and the only one involved solely in the affairs of the Holy Land.

The new dominant role played by the Order of St John in the plans of Clement V mirrored to a large extent the modifications that took place in the strategy of the crusade. Papal crusade-projects, individual memoirs, Church councils, opinions put forward by rulers, other single factors concerned with the Holy Land, and finally the replies of the *naciones* at the Council of Vienne—all reflect the attempt to adjust the traditional crusade-model to the very fact that following 1291 the goal of the crusade to the East was no longer the defence of the crusader kingdom but the recovery of what had been completely lost. They all reflect also the attempt to adjust it to the economic and military exigencies as perceived by crusader experts and particular, 'national' interests. Tradition dictated the conventional pattern of the general crusade of devoted warriors, inspired from Heaven. This was never formally abandoned, not even by those who may be called the most 'radical' of the crusade-theorists. At the same time trends of secular, political, and economic thinking conditioned new ways of action and new strategies. Their most marked result was the change in the function of the general crusade, the *passagium generale*, which was now being reduced to the role of the epilogue in the naval–economic war directed against Egypt, the key to the recovery of the Holy Land. Christendom was taking stock and assessing its naval power and supremacy on the seas. This new strategy, a sober military and economic enterprise with clearly fixed aims, was replacing the traditional notions and pattern of the general crusade. However, realistic in itself, this new crusade-strategy was impractical in the face of political and economic reality. The economic blockade of Egypt was impractical because the chief maritime powers of the West, and especially Genoa and Venice, would not enforce it as their commercial relations with the Mamluks of Egypt were too profitable to be given up. The measures taken by the papacy against these relations in order to mount an economic blockade were never radical or firm enough as the papacy itself was much dependent in its crusade-planning on the maritime power of Genoa and Venice, since no major crusade could be realized without their co-operation.

Another result of the new ways of thinking was the reassessment of the thirteenth-century military tenet, the impossibility of a direct attack on the Holy Land. What was now advocated was the establishment of Mediterranean bases, whether in the lands of Islam or Byzantium, from which a final attack would be launched. This apparently logical demand, conditioned by military exigencies and supported by the historical experience of the foregoing century, brought in its wake quite unexpected results by introducing 'national' interest into the planning of the future crusade. Thus, on the basis of strategic considerations, the conquest of Constantinople, of Rhodes, of Granada, or even of Sicily, could be demanded as a preliminary essential for the recovery of the Holy Land. The needs of the crusade and the necessities of the Holy Land served many a papal enterprise during the thirteenth century and even earlier. What was new in the final quarter of the thirteenth century, and especially during the pontificate of Boniface VIII, was the professional and expert character of the explanations. It was not only the papacy which utilized the crusades for its own particular aims. The royal and princely heads of Europe, the potential crusaders and leaders of a crusade, like Philip IV of France, Edward I of England, and James II of Aragon, used the crusade as a mean of being granted tithes, a real peacetime subsidy for their monarchies. The crusade, moreover, was used as a propaganda weapon against their enemies, who, it could be alleged, were delaying the expedition. Above all the Holy Land and the crusade were turned into a powerful vehicle for the territorial expansion and even the internal aims of these early national monarchies. Thus Philip IV, the 'most Christian King', used the slogan of the recovery of the Holy Land in his domestic affairs, as for example in his war against Flanders, and his attacks on Boniface VIII and the Order of the Temple. This certainly introduced a new note into European politics, and the king's son Philip V (1316–22), would go even further. In the name and interest of the crusade (which he would never accomplish), he requested a series of reforms of the French monetary system, the unification of weights and measures, and the recovery of what he argued was illegally alienated royal domain. This was an extreme case of using the crusade for the particularistic needs of an early 'nation-state'.[1]

The Concept of the Crusade and its Instruments

The new strategic thinking was in the long term not without impact on the concept of the crusade. According to the ideas of crusade-planners, ideas which found a particularly attentive ear at the Curia of Pope Clement V, the crusade became a strictly military expedition, minutely planned and to be executed by a professional class of fighters. Thus this new concept did not include the most traditional features of a crusade, i.e. the taking of the Cross and the participation of all believers without any class or professional distinction; the ordinary believers could still obtain the full crusade-indulgence, but only through a financial contribution to a crusade executed by professional soldiers. The new concept of the crusade meant, as was clearly reflected in the papal–Hospitaller *passagium particulare* of 1309, the absolute exclusion of ordinary believers from participation. The fact that the papacy readily adopted this new concept of a crusade in 1309 shows that papal policy was less conservative than has been assumed.

Nevertheless it cannot be said that the crusades in Italy and Syria were preached and organized in the same manner[2] and in fact papal attitude to the European crusades clearly reflects an amount of conservatism and even traditionalism. It is true that the *crucesignati* received the same privileges and undertook the same obligations; but there were also significant differences.

[1] C. H. Taylor, 'French Assemblies and Subsidy in 1321', *Speculum*, 43 (1968), 217–44, esp. pp. 220–30.
[2] For such conclusions see N. J. Housley, *Italian Crusades: The Papal–Angevin Alliance and the Crusade against Christian Lay Powers 1254–1343* (Oxford, 1982), 256–7.

First of all, papal propaganda was far more readily inclined to use the crusade to the Holy Land, as a spiritual weapon, than crusades within Italy. Pope Boniface VIII, for whom the recovery of the papal fief of Sicily became an essential precondition for that of the Holy Land, never rebuked Philip IV, during their conflict of 1302–3, for damaging the cause of the crusade against the Sicilians, although he did rebuke him time and again for injuring the cause of carrying aid to the Holy Land. In the famous bull *Asculta fili* of 5 December 1301, Pope Boniface reproached King Philip for his neglect of the business of the Holy Land, for which the Mongols, pagans that they were, cared more than the king of France.[3] It emerges, too, from N. J. Housley's book that the popes of the period 1291–1344 justified their Italian crusades not so much with reference to the needs of defending the faith as to those of recovering Jerusalem.[4]

There were also fluctuations in papal administration of the instruments of the crusades. For example, from the time of Pope Nicholas IV to the close of the pontificate of Boniface VIII there was a tendency to make the attainment of crusader vows more difficult. In 1297 Boniface ordered dispensations to be granted only to the old and the disabled, and then only for the considerable price of 300 *livres tournois*.[5] As to commutations of vows, they were easily obtained up to the pontificate of Pope Gregory X, who attempted to reverse the direction of his predecessors' policy of commuting vows away from the Holy Land. There are no references to the commutation of crusading vows to the Holy Land between the pontificate of Gregory X and the loss of the Holy Land in 1291,[6] and indeed up to the end of the pontificate of Boniface VIII, who even in 1302 forbade the commutation of vows made to go on crusade to the Holy Land in order to join the crusade against Sicily.[7] The

[3] Boniface VIII, *Registres de Boniface VIII*, ed. G. Digard et al. (Paris, 1884–1939), nos. 4424, 5342; S. Schein, '*Gesta Dei per Mongolos 1300*: The Genesis of a Non-Event', *EHR* 95 (1979), 816; above, ch. 5.

[4] Cf. Housley, *Italian Crusades*, pp. 35–70, esp. pp. 69–70.

[5] See above, ch. 5.

[6] Above, ch. 1, pp. 42, 153; M. Purcell, *Papal Crusading Policy: The Chief Instruments of Papal Policy and Crusade to the Holy Land from the Final Loss of Jerusalem to the Fall of Acre 1244–1291* (Studies in the History of Christian Thought, 11; Leiden, 1975), 114; J. A. Brundage, *Medieval Canon Law and the Crusader* (London, 1969), 113.

[7] Boniface VIII, *Registres*, no. 4625; Housley, *Italian Crusades*, p. 99 and n. 112.

obtaining of both redemptions and commutations of crusader vows became increasingly easy during the pontificates of Benedict XI and Clement V (1303–14). The first allowed vows to be commuted from the Holy Land to Byzantium.[8] The latter encouraged the redemption of vows on the eve of the crusade of 1309 and freely allowed commutation from the Holy Land to Byzantium (1306) or the Italian crusades.[9] On the whole, from the pontificate of Martin IV onwards, crusade-indulgences were more easily obtained than before. For example, whereas Innocent IV granted a maximum of forty days for attending crusade-sermons, Martin IV granted an indulgence of a year and forty days, and John XXII raised it to a year and a hundred days.[10] As to the plenary indulgence, it was Clement V who granted it for smaller payments than even before, and who was the first to allow it to beneficiaries without requiring them to take the Cross. In 1308 he reduced the contribution which earned a partial indulgence to 1 *denier tournois* for each year of remission granted.[11] In this aspect the policy of the popes of Avignon was begun during the pontificate of Clement V.

The *Fideles Crucis*

The vicissitudes and metamorphosis of crusade-strategy as well as ideology were not the only aspects of the overall impact of the loss of the Holy Land. This found its complement outside the realm of theoretical planning and practical attempts, in the emotional stir that immediately followed the fall of Acre. Its scope reflected the deep, one might say passionate, involvement of almost all sections of European society in the affairs of the Holy Land. The reaction to the events of 1299–1300 in Syria; the 'Crusade of the Poor' of 1309, which was not, as ecclesiastical chroniclers tried to present it, a popular movement of

---

[8] Benedict XI, *Registres de Benoît XI*, ed. C. Grandjean (Paris, 1905), nos. 1007–8; above, ch. 5, p. 177.

[9] Clement V, *Regestum Clementis Papae V*, ed. the monks of the Order of St Benedict (Rome, 1885–92), no. 243; above, ch. 6.

[10] Purcell, *Papal Crusading Policy*, pp. 62–6; Housley, *Italian Crusades*, p. 125.

[11] Clement V, *Regestum*, no. 2989; W. E. Lunt, *Papal Revenues in the Middle Ages* (Columbia University Records of Civilization, 19; New York, 1934), i. 114–18.

the lowest classes, but included burghers, clergy, and nobility; and finally, the massive taking of the Cross in France in 1313—all reveal a sustained and vivid interest in the fate of the Holy Land. Generally speaking there was much uniformity in the reaction, though its intensity was not the same in all regions. In Italy the fall of Acre gave rise to an antagonistic attitude to the crusade to the East, which was to characterize the peninsula for many decades, and although the immediate German, and above all, English, reactions were particularly strong, we witness after 1291 a weakening of German and English involvements. At the same time, we observe a growing interest from France, especially from 1305, when Philip IV became the pivot of crusader efforts. This reached its apogee in 1313 when the king, his family, and the high nobility took the Cross.

The three major outbreaks of popular interest in the Holy Land that took place during the period under consideration (1274–1314), were, generally speaking, marked by a rise and intensification of eschatological expectations (in 1291–92, 1300, and 1309). Immediately after the fall of Acre the loss of the Holy Land can be found neatly inserted into an apocalyptic setting and thus presented as a basically temporary loss, part of the events of the Last Days. In the years 1300–1, when the news of a Mongol conquest of the Holy Land reached Europe, and even in 1309 during the preparation of a papal–Hospitaller expedition to the East, there was a marked growth in millenarian eschatological expectations, which found its expression in visions of a peaceful and idyllic Christian return to the Holy Land. It is probably not a coincidence but part of the involvement of the Europeans with the crusade, that the beginning of the fourteenth century witnessed reactions ranging from the apocalyptical and mystical to the sober and materialistic striving for the recovery of the Holy Land.

At the conclusion of this study it is appropriate to ask, what was the place of the crusade in the preoccupations and currents of thought of the period? In the years which immediately preceded and followed the loss of the Holy Land in 1291 the crusade, whether as an idea or a plan for action, no longer held the same central place as it had after Jerusalem was lost in 1187. Yet this hardly warrants J. Delaville le Roulx's verdict that it was a period of 'sterile projects and abortive attempts',[12] of that of A. S. Atiya, who saw it as characterized only by 'active propaganda'.[13] The fact that no large-scale crusade was launched does not prove that such an expedition was not feasible, nor that those who wrote or spoke out in its favour were either hypocrites or utopians. The crusade continued to be genuinely desired among large sections of the population in different regions of Europe; although it was no longer a mass movement, neither aristocratic nor popular enthusiasm for the crusade had faded. Moreover, it still was a major concern of the popes of the period and can surely be classified as one of the permanent factors with which, apart from any personal inclination, the rulers of Europe had to reckon.[14] This was particularly true of Philip IV. To him the crusade was no longer an independent venture, with its own traditional justifications, as it had been for his father and grandfather. He saw it wholly within the context of what J. R. Strayer has called 'the religion of the monarchy'.[15] The importance of the ideology embodied in that religion as a source of inspiration to Philip the Fair cannot be doubted and leadership of the crusade was one of the obligations which it imposed on the king. The question of the sincerity of Philip's interest in launching a crusade is still doubted by those to whom this king's personality remains an enigma. It is however, obvious that Philip being sincere in his pursuit of the 'ideology of monarchy' had to be sincere about crusade. As one of the most important tasks of the *christianissimus rex* and *defensor fidei* was that of aiding the Holy Land, Philip could not possibly ignore this task and indeed in the years

---

[12] J. Delaville le Roulx, *France en Orient au XIVe siècle: Expeditions du maréchal Boucicaut* (Paris, 1885–6), 110.

[13] A. S. Atiya, *Crusade in the Late Middle Ages* (London, 1938), 10, 127: 'the distinctive feature of the crusade during the first three or four decades of the fourteenth century is the abundance of propagandists who exhorted all good Catholics to uphold the cause of the holy war ... The labours and enthusiasm of the theorists of this era were at last rewarded when the West embarked on a number of expeditions to the East in the remaining part of the century.'

[14] For much the same opinion though referring to a later period (1316–43) see C. J. Tyerman, 'Marino Sanudo Torsello and the Lost Crusade: Lobbying in the Fourteenth Century', *TRHS* 32 (1982), 57–73; id., 'Philip VI and the Recovery of the Holy Land', *EHR* 143 (1985), 25–52.

[15] J. R. Strayer, *Reign of Philip the Fair* (Princeton, NJ, 1980), 32–5.

1301–14 he did not ignore it.[16] Nevertheless, like King Philip IV, and also other rulers of the West, e.g. Edward I of England, James II of Aragon, Rudolph and Albert I of Habsburg, Charles II of Anjou, and Charles of Valois, while considering the reconquest of the Holy Land part of their duty, were never, unlike St Louis, willing to make it their chief priority. There were vested interests and attempts were often made to draw from an expedition to the Orient particular and not always marginal profits. The specific interests of monarchies like those of France, England, and Aragon often clashed with those of the crusade. The latter, instigated and guided by the papacy, was an undertaking traditionally launched by an appeal to the whole Christian West. Its success depended on the enthusiastic response from the faithful, on their military and financial support, and on co-operation between the secular rulers who took part. On either side of 1300, however, the ideal of common action by western Christians grew weaker than it had been before. It was now confronted by the narrow concerns of royal governments and their subjects. Most of the European secular rulers, now almost *imperatores in regnis suis*, were reluctant to support an enterprise which did not directly contribute to the welfare and glory of their kingdoms.[17] This was apparent in the attempts to organize a crusade in the time of Gregory X and again in 1308–9, in the reactions to the events in the Holy Land in 1299–1300, and most of all in the proceedings at the Council of Vienne.

All this explains why there was no major crusade to the Holy Land in the years since the Crusade of St Louis to Tunis (1270), and especially in the years that immediately preceded and followed the loss of the Holy Land in 1291. As said, at this time the crusade no longer held the same central place as it had after the fall of Jerusalem (1187). This is especially true of the papacy, the main pillar and the *spiritus vivens* of the crusading movement. Between the years 1282–1302 it was not the Holy Land crusade but that for the recovery of the papal fief of Sicily which was the focus of papal attention, efforts, and resources.

---

[16] S. Schein, 'Philip IV the Fair and the Crusade: A Reconsideration', in P. W. Edbury (ed.), *Crusade and Settlement* (Cardiff, 1985), 121–6.

[17] For this topic see ead., 'The Future *Regnum Hierusalem*: A Chapter in Medieval State Planning', *Journal of Medieval History*, 10 (1984), 95–105.

After 1302, when the Sicilian problem was finally solved, the Holy Land crusade became again the centre of papal activities. By then, however, the traditional crusade, the *passagium generale*, came to be regarded as the epilogue to a naval–economic war. Clement V, a great supporter of this strategy, attempted, up to the Council of Vienne, where a general crusade was proclaimed, to realize it and like his predecessors he failed. Still whether as an ultimate aim or as an instrument for achieving other goals, the crusade remained a live issue in European politics. Moreover, because it remained directly and permanently linked to the papacy, and was one of the chief papal instruments in maintaining its own universal authority,[18] the crusade was integrated into the plans of the secular rulers, who faced the papacy on an international level. Though it seldom became a major consideration with them, and more often than not remained marginal, there is no ground to doubt the sincerity of their belief in the need for a crusade. Though the ultimate aim was unanimous—the recovery of the Holy Land—this neither meant nor postulated common, let alone identical, interests. This did not augur well for their future co-operation. The common agreement was channelled into too many riverbeds, each with its own outlet. Divergent interests were more evident and more openly pronounced than ever before, all claiming the hallowed name of the crusade.

[18] According to J. R. Strayer ('Political Crusades of the Thirteenth Century', in *Setton's History*, ii. 102–3), since 1285 the crusade had ceased to be 'a regular and reliable instrument of papal policy ... deprived of the steady support of the French king, the pope was in a poor position to combat the rising tide of secularism and indifference'. This opinion, quoted by K. M. Setton, *The Papacy and the Levant 1204–1571*, i (Phil., 1976), 146, is not justified.

APPENDIX 1

List of *de recuperatione Terrae Sanctae* Treatises 1274–1314

1274	Humbert of Romans, 'Opus tripartitum'
	William of Tripoli, 'Tractatus de Statu Saracenorum'
	Gilbert of Tournai, 'Collectio de Scandalis Ecclesiae'
1289–91	'La Devise des chemins de Babiloine'
1290	Fidenzio of Padua, 'Liber recuperationis Terrae Sanctae'
1291–5	Galvano of Levanto, 'Liber sancti passagii Christicolarum contra Saracenos pro recuperatione Terrae Sanctae'
1292	Ramon Lull, 'Epistola Summo Pontifici pro recuperatione Terrae Sanctae'
	Ramon Lull, 'Tractatus de modo convertendi infideles seu Lo Passatge'
1292–4	Charles II of Anjou, 'Conseil'
1294	Ramon Lull, 'Petitio Raymundi pro conversione infidelium ad Celestinum V'
1295	Ramon Lull, 'Petitio Raymundi pro conversione infidelium ad Bonifacium VIII'
1300	Pierre Dubois, 'Summaria brevis'
1305	Ramon Lull, 'Liber de fine' (= 'De expugnatione Terrae Sanctae')
	Fulk of Villaret, 'Hec est informatio et instructio nostri magistri Hospitalis super faciendo generali passagio pro recuperatione Terre Sancte'
1305–6	James of Molay, 'Concilium super negotio Terre Sancte'
1306–7	'Tractatus dudum habitus ultramare per magistrum et conventum Hospitalis et per alios probos viros qui diu steterunt ultramare qualiter Terra Sancta possit per Christianos recuperari'
	Pierre Dubois, 'De Recuperatione Terre Sancte'
1307	Hayton, 'Flos historiarum Terre Orientis'
1308	Pierre Dubois, 'Oppinio cujusdam suadentis regi Franciae ut regnum Jerosolimitanum et Cipri acquireret'
	Pierre Dubois, 'Pro facto Terre Sancte'
1309	Ramon Lull, 'Liber de acquisitione Terrae Sanctae'
	Marino Sanudo Torsello, 'Conditiones Terrae Sanctae'
1310–11	Henry II of Lusignan (Cyprus), 'Consilium'

	William of Nogaret, 'Quae sunt advertenda pro passagio ultramarino et que sunt petenda a papa pro persecutione negocii'
1311	Ramon Lull, 'Petitio Raymundi in concilio generali ad adquirendam Terram Sanctam'
1313	William Durant (the Younger), 'Informatio brevis super hiis que viderentur ex nunc fore providenda quantum ad passagium, divina favente gracia, faciendum'

APPENDIX 2

The Monetary System and Prices

Nearly all monetary systems in Western medieval Europe were based upon the following units of account: £1 (= pound, *libra*, lira, *livre*, mark) = 20s. (= shillings, solidi, soldi, sols) = 240d. (= pence, penny, *denarii*, *denari*). During the period which concerns us, the coin most commonly used in the West was the *denier tournois*, a silver coinage. It was issued by the abbey of St Martin from the tenth century up to 1204. Between the years 1204–1649 it was issued by the kings of France. At the time of Philip I Augustus (1180–1223) its weight was 0.78–1.01 grams.

The most important gold coin used in Western Europe was the gold florin (*fiorino d'oro*), issued in 1252 by the Florentine government; in 1271 the government decreed that it should be exchanged for £1. 9s. or 348d. The gold florin weighed c.3.55 grams of fine gold; when it was made of 3.66 grams of gold, the ratio between 1 florin and 1 ounce of gold was 1:5. The ratio between the gold florin and the silver *livre tournois* was c.1:10 and the ratio between 1 ounce of gold and 1 *livre tournois* was 1:5. In the Eastern Mediterranean the important gold coins were the Byzantine gold coin, the besant, made of c.4.55 grams of gold and the Moslem dinar made of c.4.55 grams of gold.[1]

The following examples should clarify the costs of the crusades and the relative amount of crusade taxation. In 1274 Pope Gregory X required a sum equivalent to 1 *denier tournois* a year from every Christian for six years.[2] Around 1297 the price of redemption of crusading vow was 300 *livres tournois*.[3] In 1308 Pope Clement V estimated the costs of the Hospitaller *passagium particulare* at 400,000

---

[1] This account is based on the following studies: C. M. Cipolla, *Money, Prices and Civilization in the Mediterranean World: Fifth to Seventh Century* (Princeton, NJ, 1956), 13–37; H. A. Miskimin, *Money, Prices and Foreign Exchange in Fourteenth-Century France* (New Haven, Conn., and London, 1963), 1–52; P. Spufford, 'Coinage and Currency', in *Cambridge Economic History*, ed. M. M. Postan *et al.* (Cambridge, 1952; 1987 edn.), ii. 788–873. According to the Florentine Francesco di Pegolotti's *Pratica della mercatura* (c.1310–40), £5. 3s. 6d. petty *tournois* were then equivalent to 1 gold florin. See Francesco di Pegolotti, *Pratica della mercatura*, ed. A. Evans (Cambridge, Mass., 1936), xx, 258.
[2] See above, ch. 1 n. 70.
[3] See above, ch. 5 n. 36.

gold florins, i.e. 2,000,000 *livres tournois* or 480,000,000 *deniers tournois*.[4] At that time the pay of a knight ranged from 7 *sols* 6 *deniers tournois* to 10 *sols tournois* a day.[5]

[4] See above, ch. 7 n. 9.
[5] N. J. Housley, *The Avignon Papacy and the Crusades 1305–1378* (Oxford, 1980), 159–60.

BIBLIOGRAPHY

1. MANUSCRIPT SOURCES

BM Cotton. MS Vesp. B XI, fos. 47^v–8^v.
BN MS lat. 7470, fos. 172^r–8^v.
Humbert of Romans, 'De praedicatione crucis', incunabulum of Bibliothèque Mazarine, XV cent. no. 259 (= 12360).

2. PRINTED SOURCES

Collections of Sources

Acta Aragonensia, ed. H. Finke (Leipzig and Berlin, 1908–22).
Acta Imperii Angliae et Franciae 1267–1313, ed. F. Kern (Tübingen 1911).
Acta imperii inedita saeculi XIII, ed. E. Winkelmann (Innsbruck, 1880–5).
Acta Romanorum Pontificum ab Innocentio V ad Benedictum X^3, ed. A. Tantu, vol. v, pt. 2 (Vatican, 1954).
Annales Ecclesiastici ab anno 1198 usque ad annum 1565, ed. O. Raynaldus et al. (Lucca, 1738–59).
Annales minorum seu historia trium ordinum a S. Francisco Institutorum, ed. L. Wadding (Lyons, 1625–48).
Annales monastici, ed. H. R. Luard (RS, London, 1864–9).
Aus Tagen Bonifaz VIII, ed. H. Finke (Münster, 1902).
Biblioteca bio-bibliografica della Terra Santa e dell'Oriente francescano, ed. G. Golubovich (Quaracchi, 1906–27).
Biblioteca Geographica Palestinae, ed. R. Röhricht (Berlin, 1840; repr. Jerusalem, 1963).
Bibliothèque des croisades, ed. J. Michaud (Paris, 1829–30).
Cartulaire général de l'ordre des Hospitaliers de St Jean de Jérusalem 1100–1310, ed. J. Delaville le Roulx (Paris, 1894–1906).
Cerdeña y la expansión mediterránea de la Corona de Aragón, ed. V. Salavert y Roca (Madrid, 1956).
Chronicles of the Reigns of Edward I and Edward II, ed. W. Stubbs (RS, London, 1882–3).
Codice diplomatico del sacro militare ordine gerosolimitano oggi di Malta, ed. S. Pauli (Lucca, 1733–7).
Concilia Magnae Britanniae et Hiberniae, ed. D. Wilkins (London, 1737).

Conciliorum Oecumenicorum Decreta, ed. J. Alberigo et al. (Basel, 1962).
Corpus Chronicorum Flandriae, ed. C. de Smedt (Brussels, 1837–65).
Councils and Synods with other Documents relating to the English Church, ed. F. M. Powicke and C. R. Cheney (Oxford, 1964).
Diplomatari de l'Orient Català 1302–1409, ed. A. Rubio y Lluch (Barcelona, 1947).
Diplomatarium Veneto-Levantinum, ed. G. M. Thomas (Venice, 1880–99).
'Documents chypriotes du début de XIVe siècle', ed. C. Kohler, *ROL* xi (1905/1908).
'Documents inédits concernant l'Orient latin et les croisés (XIIe–XIVe siècles)', ed. C. Kohler, *ROL* 7 (1900).
Documents relatifs aux états generaux et assemblées réunies sous Philippe le Bel, ed. M. G. Picot (Paris, 1901).
Le Dossier de l'affaire des Templiers, ed. G. Lizerand (Paris, 1964).
Durham Annals and Documents of the Thirteenth Century, ed. F. Barlow (Surtees Society, 155; 1945).
Excerpta Cypria: Materials for a History of Cyprus, ed. C. D. Cobham (Cambridge, 1908).
Florilegium testamentorum ab imperatoribus et regibus sive principibus nobilibus conditorum ab anno 1189 usque ad annum electionis Rudolfi illustri regis Romanorum perductum, ed. G. Wolf (Heidelberg, 1956).
Foedera, conventiones, litterae et cuiuscumque generis acta publica inter reges Angliae et alios quosvis imperatores, reges, pontifices, principes vel communitates 1101–1654, ed. T. Rymer et al. (Hague, 1745).
Fontes rerum Austriacarum Scriptores, ed. H. Pez (Vienna, 1855–75).
Fontes rerum Germanicarum, ed. J. F. Böhmer (Stuttgart, 1854–68).
'Formulaires de lettres du XIIe, du XIIIe et du XIVe siècles' ed. C. V. Langlois, in *Notices et extraits des manuscrits de la Bibliothèque Nationale*, 34 (1891).
Gesta Dei per Francos, ed. J. Bongars (Hannau, 1611; repr. Jerusalem, 1973).
Histoire de l'Empire de Constantinople sous les empereurs français, ed. C. Du Cange and J. A. C. Buchon (Paris, 1826).
Histoire du differend d'entre le pape Boniface VIII et Philippe le Bel roy de France, ed. P. Dupuy (Paris, 1655; repr. Paris, 1963).
Histoire littéraire de la France (Paris, 1733–).
Historical Papers and Letters from the Northern Registers, ed. J. Raine (RS, London, 1873).
Itinéraires à Jérusalem et descriptions de la Terre Sainte, ed. H. Michelant and G. Raynaud (Paris, 1882; repr. Osnabrück, 1966).
Lettres de rois, reines, et autres personnages de cours de France et d'Angleterre,

ed. J. J. Champollion Figeac (Collection des documents inédits, 29; Paris, 1839).
'Lettres inédites concernant les croisades 1275–1307', ed. C. Kohler and C. V. Langlois, *BEC* 3 (1891).
I Misti del senato della repubblica veneta 1293–1331, ed. G. Giacomo (Venice, 1887; repr. Amsterdam, 1970).
Monumenta Historiae Patriae, ed. C. Albert (Turin, 1836–1955).
Monumenta Germaniae Historica: Legum Sectio IV, Constitutiones, ed. J. Schwalm (Hanover, 1904–6), iii, iv.
Monumenta Germaniae Historica: Scriptores, ed. G. H. Pertz *et al.* (Hanover, Weimar, Berlin, Stuttgart, and Cologne, 1826–1934).
'Notices et extraits des documents inédits relatifs à l'histoire de France sous Philippe le Bel', ed. E. P. Boutaric, *Notices et extraits des manuscrits de la Bibliothèque Nationale*, 20 (1862).
'Notices et extraits des manuscrits', ed. C. Kohler, *ROL* 5 (1897).
Nova Alammaniae, ed. E. E. Stengel (Berlin, 1921).
Ordonnances des roys de France de la troisième race, ed. M. de Laurière, 22 vols, (Paris, 1723–1849).
Papsttum und Untergang des Templerordens, ed. H. Finke (Münster, 1907).
Political Songs, ed. T. Wright (Camden Society, London, 1839).
Quinti Belli Sacri Scriptores Minores, ed. R. Röhricht (Geneva, 1879).
Recueil des historiens des croisades: Historiens arméniens (Paris, 1869–1906).
Recueil des historiens des croisades: Historiens occidentaux (Paris, 1844–95).
Recueil des historiens des Gaules et de la France, ed. M. Bouquet *et al.* (Paris, 1737–1904).
Regesta Pontificum Romanorum inde ab anno post Christum natum 1198 ad annum 1304, ed. A. Potthast (Berlin, 1874–5).
Regesta regni Hierosolymitani 1097–1291, ed. R. Röhricht (Innsbruck, 1893), *Additamentum* (1904).
Registres du trésor des chartes, I. Règne de Philippe le Bel, ed. M. R. Fawtier *et al.* (Paris, 1958).
I Registri della cancelleria angioina ricostruiti, ed. R. Filangieri di Candida *et al.*, i–xxvii (Naples, 1950–80).
Rerum Italicarum Scriptores, ed. L. A. Muratori (Milan, 1723–38); NS ed. G. Carducci *et al.* (Città di Castello and Bologna, 1900).
(Rolls Series). *Rerum Britannicarum Medii Aevi Scriptores* (London, 1858–96).
Sacrorum conciliorum nova et amplissima collectio, ed. G. D. Mansi (Florence and Venice, 1759–98).
Sacrosancta Concilia, ed. P. Labbé and G. Cossart (Paris, 1671–2).
Scriptores rerum Austriacarum veteres, ed. H. Pez (Leipzig, 1721–45).
Scriptores rerum Prussicarum, ed. T. Hirsh *et al.* (Leipzig, 1861–74).

Scriptores rerum Suecicarum medii aevi, ed. E. M. Fant (Uppsala, 1818–71).
Spicilegium sive collectio veterum aliquot scriptorum, ed. L. d'Achery (Paris, 1723).
Tabulae Ordinis Theutonici, ed. E. Strehlke (Berlin, 1869; repr. Jerusalem, 1975).
Thesaurus novus anecdotorum, ed. E. Martène and U. Durand (Paris, 1717).
Urkunden zur Ältern Handels- und Staatsgeschichte der Republik Venedig, ed. G. L. Tafel and G. M. Thomas (Vienna, 1856–7; repr. Amsterdam, 1964).
Veteris aevi Analecta seu Vetera Monumenta, ed. A. Matthaeus (Halle, 1738).
Veterum Scriptorum et Monumentorum Amplissima Collectio, ed. E. Martène and U. Durand (Paris, 1724–33).
Vitae Paparum Avenionensium, ed. E. Baluze (Paris 1693); edn ed. G. Mollat (Paris, 1914–27).

Individual Sources

'Actes passés à Famagouste de 1299 à 1301 par le notaire génois Lamberto di Sambuceto', ed. C. Desimoni, *AOL* 2 (1884).
ADAM MURIMUTH, *Continuatio chronicarum*, ed. E. M. Thompson (RS, London, 1889).
AEGIDIUS LI MUISIS (GILLES LE MUISIT), 'Chronica', in *Corpus Chronicorum Flandriae*, ii.
AMBROISE, *L'Estoire de la guerre sainte*, ed. G. Paris (Paris, 1897).
ANDREA DANDULO, 'Chronica Brevis', *RIS* NS xii.
ANDREA NAUGER, 'Historia Veneta', *RIS* xxiii.
Annales Angliae et Scotiae, ed. H. T. Riley (RS, London, 1865).
'Annales Basilenses', *MGH SS* xvii.
'Annales Blandinienses', *MGH SS* v.
'Annales Caesarienses', ed. G. Leidinger as 'Annales Caesarienses seu Kaisheimer Jahrbucher', *Sitzungsberichte der Königlich Bayerischen Akademie der Wissenschaften*, 7 (1910).
'Annales Caesenates', *RIS* xiv.
'Annales Cisterciensium in Heinrichow', *MGH SS* xix.
'Annales Colbazienses', *MGH SS* xix.
'Annales Colmarienses Maiores', *MGH SS* xvii.
'Annales de Terre Sainte', ed. R. Röhricht, *AOL* 2 (1884).
Annales Dunelmenses: Durham Annals and Documents of the Thirteenth Century, ed. F. Barlow (Surtees Society, 155; 1945).
'Annales Forolivienses', *RIS* xxii.
'Annales Frisacenses', *MGH SS* xxiv.

'Annales Gandenses', *MGH SS* xvi.
'Annales Hibernie', in *Chartularies of St Mary's Abbey, Dublin*, ed. J. T. Gilbert, ii (RS, London, 1884).
'Annales Londonienses', in *Chronicles of the Reigns of Edward I and Edward II*, i.
'Annales Lubicenses', *MGH SS* xvi.
'Annales Mellicenses', *MGH SS* ix.
'Annales monasterii de Osneia', in *Annales monastici*, iv.
'Annales monasterii de Waverleia', in *Annales monastici*, ii.
'Annales Neresheimenses', *MGH SS* x.
'Annales Paulini', in *Chronicles of the Reigns of Edward I and Edward II*, i.
'Annales prioratus de Dunstaplia', in *Annales monastici*, ii.
'Annales prioratus de Wigornia', in *Annales monastici*, iv.
'Annales Placentini Gibellini', *MGH SS* xviii.
Annales Regis Edwardi Primi, ed. H. T. Riley (RS, London, 1865),
'Annales Sancti Blasii Brunsvicenses', *MGH SS* xxiv.
'Annales Terrae Prussicae', *MGH SS* xix.
'Annales Tielenses', *MGH SS* xxiv.
'Annales Zwetlenses', *MGH SS*, ix.
'Anonymym S. Martialis Chronicon', *RHGF* xxi.
An Anonymous Short Metrical Chronicle, ed. E. Tettle, *Early English Text Society* (London, 1935).
ANTONY BEK, *The Records of Antony Bek, Bishop and Patriarch 1283–1311*, ed. C. M. Fraser (Surtees Society, 162; 1953).
BALDUIN OF NINOVE, 'Chronicon', *MGH SS* xxv.
BARTHOLOMEW COTTON, *Historia Anglicana: necnon ejusdem Liber de Archiepiscopis et episcopis Angliae* ed. H. R. Luard (RS, London 1859).
BARTHOLOMEW OF NEOCASTRO, 'Historia Sicula', *RIS* NS xiii.
BENEDICT XI, *Les Registres de Benoît XI*, ed. C. Grandjean (Paris, 1883–1905).
BENEVENUTO OF IMOLA, 'Excerpta Historica Ex Commentaris in Coemediam Dantis', *RIS* i.
'Ein Bericht über die Schlacht bei Hems am 23 Dec. 1299', ed. W. Wattenbach, *Neues Archiv*, 4 (1879).
BERNARD GUI, 'Flores Chronicorum', *RHGF* xxi.
—— 'Vita Clementis V', in *Vitae Paparum*, i.
BERNARD OF CLAIRVAUX, *De Laude Novae Miliciae*, *PL* 182.
BOHEMOND OF TRIER, 'Gesta', *MGH SS* xxiv.
BONIFACE VIII, *Les Registres de Boniface VIII*, ed. G. Digard et al. (Paris, 1884–1935).
CAFARO OF CASCHIFELONE, 'De liberatione civitatum Orientis', *RHC Hist. occ.* v.
'Les Casaus de Syr', in *Itinéraires à Jérusalem*.

CHARLES II OF ANJOU, 'Le Conseil du Roi Charles', ed. G. I. Bratianu, *RHSE* 19 (1942).
'Chronica XXIV Generalium Ordinis Minorum', ed. Fathers of the College of St Bonaventure, *Analecta Franciscana*, 3 (1897).
'Chronica Regia Coloniensis', ed. G. Waitz *MGH SRG* xi (Hanover, 1880).
The Chronicle of Furness, ed. R. Howlett (RS, London, 1884).
Chronicon de Lanercost 1201–1346, ed. J. Stevenson (Edinburgh, 1839).
'Chronicon Elwacense', *MGH SS* x.
'Chronicon Estense', *RIS* xv.
'Chronicon Imperatorum et Pontificum Bavaricum', *MGH SS* xxiv.
'Chronicon Lemovicense Brevissimum', *RHGF* xxi.
'Chronicon ordinis equestris Teutonici', in *Veteris aevi annalecta*, v.
'Chronicon Parmense', *MGH SS* xviii.
'Chronicon Salisburgense', in *Scriptores rerum Austriacarum veteres*, i.
'Chronicon Sampetrinum', in *Geschichtsquellen der Provinz Sachsen*, ed. B. Stubel, (Halle, 1870).
'Chronique de royaume de la Petite Arménie', *RHC Hist. arm.* i.
La Chronique liégeoise de 1402, ed. E. Bacha (Brussels, 1900).
Chronographia regum Francorum, ed. H. Moranville (Paris, 1891–3).
CLEMENT V, *Regestum Clementis Papae V*, ed. Monks of the Order of St Benedict (Rome, 1885–92).
'Commendatio Lamentabilis in Transitu Magni Regis Edwardi', in *Chronicles of the Reigns of Edward I and Edward II*, ii.
'Constitutiones pro Zelo Fidei', in Purcell, *Papal Crusading Policy*, app. A, pp. 196–9.
'Continuatio Canonicorum Sancti Rudberti Salisburgensis', *MGH SS* ix.
'Continuatio Florianensis', *MGH SS* ix.
'Continuatio Ratisponensis', *MGH SS* xvii.
The Continuation of the Chronicle of Gervase of Canterbury to 1327, ed. W. Stubbs (RS, London, 1879–80).
'Continuationes Anglicae Fratrum Minorum', *MGH SS* xxiv.
'Corpus Chronicorum Bononiensum', *RIS* NS xviii.
'Cronica Reinhardsbrunnensis', *MGH SS* xxx.
'Cronica S. Petri Erfordensis Moderna', *MGH SS* xxx.
Dante Alighieri, *La Divina Commedia*, ed. C. H. Grandgent and C. S. Singleton, (Cambridge, Mass. 1972).
'La Devise des chemins de Babiloine', in *Itinéraires à Jérusalem*.
DIOMEDES STRAMBALDI, *Chroniques de Chypre d'Amadi et de Strambaldi*, ed. M. L. de Mas Latrie (Paris, 1891–3).
'Directorium ad passagium faciendum', *RHC Hist. arm.* ii.

'The Downfall of the Templars and a Letter in their Defence', ed. C. R. Cheney, in id., *Texts and Studies*.
EBERHARD OF REGENSBURG, 'Annales', *MGH SS* xvii.
ERICH OLAUS, 'Chronica', *Scriptores rerum Suecicarum*, ii.
'L'Estoire d'Eracles empereur et la conqueste de la Terre d'Outrener', *RHC Hist. occ.* ii.
'De Excidio Urbis Acconis Libri II', in *Veterum Scriptorum et Monumentorum Amplissima Collectio*, v.
FERRATUS VINCENTINUS, 'Historia rerum 1250–1318', *RIS* ns ix.
FIDENZIO OF PADUA, 'Liber recuperationis Terrae Sanctae', in Golubovich, i.
Flores Historiarum, ed. H. R. Luard (RS, London, 1890).
FLORIO BUSTRON, *Chronique de l'île de Chypre*, ed. M. L. de Mas-Latrie (Paris, 1886).
'Fragment d'une chronique anonyme', *RHGF* xxi.
FRANCESCO AMADI, *Chronique, Chroniques d'Amadi et de Strambaldi*, ed. M. L. de Mas-Latrie (Paris, 1891).
FRANCESCO DI PEGOLOTTI, *La Pratica della mercatura*, ed. A. Evans (Cambridge, Mass., 1936).
FRANCISCO PIPINO, 'Chronicon', *RIS* ix.
FREIDANK, *Freidanks Bescheidenheit*, ed. Panniez (Leipzig, 1978).
FULK OF VILLARET, 'Hec est informatio et instructio nostri magistri Hospitalis super faciendo generali passagio pro recuperatione Terre Sancte', ed. J. Petit, as 'Mémoire de Foulques de Villaret sur la croisade', *BEC* 9 (1899).
GALVANO OF LEVANTO, 'Liber sancti passagii Christicolarum contra Saracenos pro recuperatione Terrae Sanctae', ed. C. Kohler, as 'Traité du recouvrement de la Terre Sainte adressé, vers l'an 1295, à Philippe le Bel par Galvano de Levanto, médecin génois', *ROL* 6 (1898).
GEOFFREY LE BAKER, *Chronicon*, ed. E. M. Thompson (Oxford, 1889).
GEOFFROY OF PARIS, *La Chronique métrique attribuée à Geoffroy de Paris*, ed. A. Diverrès (Strasburg, 1956).
GEORGIO STELLA, 'Annales Genuenses', *RIS* xvii.
Gesta pontificium Leodensium, ed. J. Chapeauville (Liège, 1613), ii.
GILBERT OF TOURNAI, 'Collectio de Scandalis Ecclesiae', ed. A. Stroick, *Archivum Franciscum Historicum*, xxiv (1931).
GILES OF ROME (AEGIDIUS ROMANUS), *De regimine principium* (Rome, 1607).
GIOVANNI VILLANI, 'Historia Universalis', *RIS* xiii.
GIRARD OF FRACHET, 'Chronicon cum anonyma eiusdem operis continuatione', *RHGF* xxi.
Les Grandes Chroniques de France, ed. J. Viard, viii (Paris, 1834).

GREGORY X, *Les Registres de Grégoire X*, ed. J. Guiraud et al. (Paris, 1892–1906).
HERMAN CORNER, 'Cronica novella', in *Scriptores rerum Suecicarum*, iii.
HAYTON, 'Flos historiarum Terre Orientis', *RHC Hist. arm.* ii.
HENRY OF GHENT, *Aurea Quodlibeta* (Venice, 1616).
HENRY II OF LUSIGNAN (CYPRUS), 'Consilium', ed. M. L. de Mas-Latrie, in Mas-Latrie, *Histoire de l'île de Chypre*, ii.
HONORIUS IV, *Les Registres d'Honorius IV*, ed. M. Prou (Paris, 1886–8).
HUMBERT OF ROMANS, 'Opus tripartitum', ed. E. Brown, in *Fasciculus rerum expetendarum et fugiendarum* (London, 1690).
JACQUES DE VITRY, *Historia Orientalis: Iacobi de Vitriaco Libri Duo* (Douai, 1597).
—— *Letters*, ed. R. B. C. Huygens (Leiden, 1960).
JAMES DORIA, 'Annales Ianuenses', in *Annali genovesi de Caffaro e dei suoi continuatiori*, ed. C. Imperiale de Sant'Angelo, (Rome, 1927).
JAMES I OF ARAGON, *The Chronicle of James I of Aragon*, trans. J. Forster and P. de Gayngos (London, 1883).
JAMES OF MOLAY, 'Concilium super negotio Terre Sancte', in *Vitae Paparum*, iii.
JOHN ELEMOSINA, 'Chronicon', in Golubovich, ii.
—— 'Liber Historiarum', in Golubovich, ii.
JOHN OF HALTON, *Register*, ed. W. N. Thompson and T. F. Tout (Canterbury and York Society, 12–13; 1913).
JOHN OF ROMEYN, *Register* (Surtees Society, 123; 1913–17).
JOHN OF JOINVILLE, 'The Life of Saint Louis', in *Chronicles of the Crusades*, trans. M. R. B. Shaw (London, 1963).
JOHN OF OXENEDES, *Chronica*, ed. H. Ellis (RS, London, 1859).
JOHN OF PARIS, 'Memoriale historiarum', *RHGF* xxi.
—— 'Vita Clementis V', in *Vitae Paparum*, i.
JOHN OF PONTISSARA, *Register*, ed. C. Deeds (Canterbury and York Society, 19, 30; 1915–24).
JOHN OF THILRODE, 'Chronicon', *MGH SS* xxv.
JOHN OF TROKELOWE, *Opus Chronicarum*, ed. H. T. Riley (RS, London, 1866).
JOHN OF VICTRING, 'Chronicon Carinthiae', in *Fontes rerum Germanicarum*, i.
JOHN OF WINTERTHUR, 'Chronica', ed. F. Beathgen, *MGH SRG* iii (Berlin, 1924).
LEVOLD VON NORTHOF, 'Die Chronik de Grafen von der Mark', ed. F. Zschaeck, *MGH SRF* vi (Berlin, 1929).
LUDOLPH OF SUCHEM, *De Itinere Terre Sancte*, ed. F. Deycks (Stuttgart, 1851).

MAKRISI, *Histoire des Sultans Mamluks*, ed. and trans. M. Quatremère (Paris, 1837–45).
MANDEVILLE, *Mandeville's Travels*, ed. M. Seymour (Oxford, 1967).
MARCO POLO, *The Book of Ser Marco Polo*, ed. H. Yule and H. Cordier (London, 1903).
MARINO SANUDO TORSELLO, 'Letters', ed. F. Kunstmann, in 'Studien über Marino Sanudo den Alteren', *Abhandlungen der Bayerischen Akademie der Wissenschaften (Historische Klasse)*, 7 (1853).
—— 'Liber Secretorum Fidelium Crucis', *Gesta Dei per Francos*, ii.
MARINO SANUDO THE YOUNGER, 'Vitae Ducum Venetorum', *RIS* xxii.
MARTIN IV, *Les Registres de Martin IV*, ed. F. Olivier-Martin *et al.* (Paris, 1901–35).
MARTIN OPPAVIENSIS, 'Chronicon', *MGH SS* xxii.
'Martini Continuatio Brabantina', *MGH SS* xxiv.
'Martini Continuatio Coloniensis', in 'Chronica Regia Coloniensis'.
MATTHEW PARIS, *Chronica Majora*, ed. H. R. Luard (RS, London, 1872–83).
'Memoria Terrae Sanctae', ed. C. Kohler, in id., 'Deux projets de croisade en Terre Sainte', *ROL* 10 (1903/4).
MENKO, 'Chronicon et Continuatio', *MGH SS* xxiii.
'Merton Flores Historiarum'; in *Flores Historiarum*.
MOUFAZZAL IBN ABIL FAZAIL, 'Histoire des Sultans Mamluks', ed. E. Blochet, in *Patrologia Orientalis*, 14 (1920); 20 (1929).
NICCOLÒ OF POGGIBONSI, *Libro d'Oltramare 1346–1350*, ed. P. B. Bagatti (Jerusalem, 1945).
NICHOLAS III, *Les Registres de Nicholas III 1277–1280*, ed. J. Gay (Paris, 1898–1938).
NICHOLAS IV, *Les Registres de Nicholas IV*, ed. E. Langlois (Paris, 1886–93).
NICHOLAS OF MARTHONO, 'Liber Peregrinationis ad Loca Sancta', ed. L. le Grand, *ROL* 3 (1895).
NICHOLAS TREVET, *Annales sex regum Angliae*, ed. T. Hog (London, 1845).
Notai genovesi in Oltramare: Atti rogati a Cipro da Lamberto Sambuceto, ed. V. Polonio (Collana storica de fonti e studi dirreta da Geo Pistarino, 31, 32; Geneva, 1982).
OTTOKAR VON STEIERMARK, *Oesterreichische Reimchronik*, ed. J. Seemüller, *MGH SS Deutsche Chroniken*, v (Hanover 1890–3).
PAULINUS OF VENICE, 'Chronicon', in Golubovich, ii.
—— 'Chronologia Magna', in Golubovich, ii.
—— 'Vita Clementis V', in *Vitae Paparum*, i.
PETER OF DUSBERG, 'Cronica Terre Prusie', in *Scriptores rerum Prussicarum*, i.

PETER OF ZITTAU, 'Annales Aulae Regiae', in *Fontes rerum Austriacarum Scriptores*, viii.

PIERRE DUBOIS, 'De facto Templariorum', in 'Notices et extraits des documents inédits relatifs à l'histoire de France sous Philippe le Bel'.

―― *De Recuperatione Terre Sancte*, ed. C. V. Langlois (Paris, 1891); ed. and trans. W. I. Brandt, as *The Recovery of the Holy Land by Pierre Dubois* (New York, 1956).

―― 'Oppinio cujusdam suadentis regi Franciae, ut regnum Jerosolimitanum et Cipri acquireret pro altero filiorum suorum ac de invasione Cipri', in *Vitae Paparum*, iii; trans. W. I. Brandt, in *The Recovery*.

―― 'Pro facto Terre Sancte', in 'Notices et extraits des documents inédits relatifs à l'histoire de France sous Philippe le Bel'.

―― 'Remonstrance du peuple de France', in 'Notices et extraits des documents inédits relatifs à l'histoire de France sous Philippe le Bel'.

―― *Summaria brevis*, ed. H. Kampf (Leipzig and Berlin, 1936).

PIERRE OF LANGTOFT, *Chronicle*, ed. T. Wright (RS, London, 1868).

PIETRO CANTINELLO, 'Chronicon', *RIS* NS xxviii.

PTOLEMY OF LUCCA, 'Die Annalen', ed. B. Schmeidler, *MGH SRG* viii (Berlin, 1955).

―― 'Vita Clementis V', in *Vitae Paparum*, i.

RAMON LULL, *Blanquerna*, ed. M. Aguilar (Madrid, 1944).

―― 'Doctrina pueril', in *Obres de Ramon Lull*, ed. M. Obrador *et al.*, i (Palma de Mallorca, 1906).

―― 'Epistola Summo Pontifici pro recuperatione Terrae Sanctae', ed. J. Rambaud-Buhot, in *Beati Magistri R. Lulli opera latina*, iii (Palma de Mallorca, 1954).

―― 'Letters', in *Thesaurus novus anecdotorum*, i.

―― 'Liber de acquisitione Terrae Sanctae', ed. P. E. Kamar, *Studia Orientalia Christiana Collectanea*, 6 (1961).

―― 'Liber de fine', ed. A. Madre, *CCCM* xxxv (1981).

―― 'Liber natalis pueri parvuli Christi Jesu' (Paris, 1311), ed. H. Harada, *CCCM* xxxii (1975).

―― 'Llibre de contemplació en Déu', in *Obres de Ramon Lull*, iv (Palma de Mallorca, 1910).

―― 'Petitio Raymundi in concilio generali ad adquirendam Terram Sanctam' (1311), ed. H. Wieruszowski, in Wieruszowski, 'Ramon Lull et l'idée de la Cité de Dieu', *Estudis Franciscans*, 47 (1935).

―― 'Petitio Raymundi pro conversione infidelium' (Naples, 1294), in Golubovich, i.

—— 'Petitio Raymundi pro conversione infidelium' (Rome, 1295), ed. Wieruszowski, in *Estudis Franciscans*, 47.
—— 'Tractatus de modo convertendi infideles seu Lo Passatge', ed. J. Rambaud-Buhot, in *Beati Magistri R. Lulli opera latina*, iii.
—— 'Vita Coaetanea', ed. H. Harada, *CCCM* xxxiv (1980).
RASHID-AD-DIN, *Jami-al-tabarikh*, ed. Alizade (Baku, 1957).
—— *Geschichte Gazan-Hari's*, ed. K. Jahn (London, 1940).
RICOLDO OF MONTE CROCE, 'Epistolae V commentatoriae de perditione Acconis 1291', ed. R. Röhricht, *AOL* 2 (1884).
ROBERT WINCHELSEY, *Register*, ed. K. Graham (Canterbury and York Society, 51, 52; 1952–6).
ROGER BACON, 'Opus Tertium', in *Opera Inedita*, ed. J. Brewer (RS, London, 1857).
—— *Opus Majus*, ed. J. H. Bridges, 3 vols. (Oxford, 1897–1900).
ROSTANH BERENGUIER, 'Opera', ed. P. Meyer, in id., 'Les Derniers Troubadours de la Provence', *BEC* 30 (1869).
SALIMBERE, *Chronica*, *MGH SS* xxxii.
SIMON OF GANDAVO, *Register*, ed. C. T. Flower and M. C. B. Dawes (Canterbury and York Society, 40; 1934).
'Le Templier de Tyr', in *Les Gestes de Chiprois*, ed. G. Raynaud (Geneva, 1887).
THADEO OF NAPLES, *Hystoria de desolacione ciritatis Acconensis*, ed. P. Riant (Geneva, 1873).
THOMAS DE LA MOORE, 'Vita et Mors Edwardi Secundi', in *Chronicles of the Reigns of Edward I and Edward II*, ii.
THOMAS TUSCUS, 'Gesta Imperatorum et Pontificum', *MGH SS* xxii.
THOMAS WALSINGHAM, *Historia Anglicana*, ed. H. T. Riley (RS, London, 1863–4).
'Via ad Terram Sanctam', ed. C. Kohler, in id., 'Deux projets de croisades en Terre Sainte', *ROL* 10 (1903/4).
'Visio seu prophetia fratris Johannis', ed. E. Donckel, in *Römische Quartalschrift*, 40 (1932).
'Vita Beati Raymundi Lulli', ed. B. de Gaiffier, *Analecta Bollandiana*, 48 (1930).
Vita Edwardi II, ed. and trans. N. Denholm-Young (London, 1957).
WALTER OF GUISBOROUGH, *Chronicle*, ed. H. Rothwell (London, 1957).
WEICHARD OF POLHAIM, 'Continuatio Weichardi de Polhaim', *MGH SS* xi.
WILLIAM ADAM, 'De modo Sarracenos extirpandi', *RHC Hist. arm.* ii.
WILLIAM DURANT (THE YOUNGER), *Tractatus de modo generalis concilii celebrandi* (Paris, 1691).
—— 'Informatio brevis super hiis que viderentur ex nunc fore providenda quantum ad passagium, divina favente gracia,

faciendum', ed. G. Düzzholder, in Düzzholder, *Die Kreuzzugspolitik unter Papst Johann XXII 1316-1334* (Strasburg, 1913).

WILLIAM GODEL, *Chronicon*, *RHGF* xxi.

WILLIAM GREENFIELD, *Register*, ed. W. Brown and A. Hamilton Thompson (Surtees Society, 145, 152, 153; 1931–40).

WILLIAM LE MAIRE, 'Gesta', in *Spicilegium sive collectio veterum aliquot scriptorum*, ii.

—— *Liber*, ed. C. Port (Paris, 1877).

WILLIAM OF EGMONT, 'Chronicon', in *Veteris aevi Analecta seu Vetera Monumenta*, ii.

WILLIAM OF NANGIS, *Chronique latine de Guillaume de Nangis de 1113 à 1300 avec les continuations de cette chronique de 1300 à 1368*, ed. H. Géraud, i (Paris, 1843).

WILLIAM OF NOGARET, 'Quae sunt advertenda pro passagio ultramarino et que sunt petenda a papa pro persecutione negocii', in 'Notices et extraits des documents inédits relatifs à l'histoire de France sous Philippe le Bel'.

—— 'Allegationes excusatoriae super facto Bonifaciano', in *Histoire du differend d'entre le pape Boniface VIII et Philippe le Bel roy de France*.

WILLIAM OF TRIPOLI, 'Tractatus de Statu Saracenorum', ed. H. Prutz, in Prutz, *Kulturgeschichte der Kreuzzuge*.

WILLIAM RISHANGER, *Chronica et Annales*, ed. H. T. Riley (RS, London, 1865).

WILLIAM VENTURA, 'Memoriale', ed. C. Combetti, in *Monumenta Historiae Patriae*, iii.

3. PRINTED SECONDARY SOURCES

ALEXANDER, P. J., 'Byzantium and the Migration of Literary Works and Motifs: The Legend of the Last Roman Emperor', *Medievalia et Humanistica*, 2 (1961).

—— 'Medieval Apocalypses as Historical Sources', *AHR* 73 (1968).

ALPHANDÉRY, P., and DUPRONT, A., *La Chrétienté et l'idée de croisade* (Paris, 1954–9).

ALTANER, B., 'Glaubenszwang und Glaubensfreiheit in der Missionstheorie des Raymundus Lullus: Ein Beitrag zur Geschichte des Toleranzgedankens', *Historisches Jahrbuch*, 48 (1928).

ANTON, L. G., *Las Uniones aragonesas y las cortes del reino 1283–1301*, (Saragossa, 1975).

ATIYA, A. S., *Egypt and Aragon: Embassies and Diplomatic Correspondence between 1300 and 1330 AD* (Leipzig, 1938).

—— *The Crusade in the Late Middle Ages* (London, 1938).

—— 'The Crusades: Old Ideas and New Conceptions', *Cahiers d'histoire mondiale*, 2 (1954).
AYALON, D., 'The Mamluks and Naval Power', in id., *Studies on the Mamluks of Egypt 1250–1517* (London, 1977).
BARATIER, E., *Histoire du commerce de Marseille* (Paris, 1951).
BARBER, M., 'James of Molay, the Last Grand Master of the Temple', *Studia Monastica*, 14 (1972).
—— 'Propaganda in the Middle Ages: The Charges against the Templars', *Nottingham Medieval Studies*, 17 (1973).
—— *The Trial of the Templars* (Cambridge, 1978).
—— 'The World Picture of Philip the Fair', *Journal of Medieval History*, 8 (1982).
BARKER, E., *The Crusades* (New York, 1923; repr. 1971).
BAUTIER, R. H., 'Diplomatique et histoire politique: Ce que la critique diplomatique nous apprend sur la personnalité de Philippe le Bel', *Revue historique*, 259 (1978).
BEAZLEY, C. R., 'Directorium ad Faciendum Passagium Transmarinum', *AHR* 12 (1905/6); 13 (1907/8).
BEHREND, F., 'Deutsche Pilgerreisen ins Heilige land 1300–1600', in A. Hartmann (ed.), *Festschrift für G. Leidinger* (Munich, 1930).
BEBLONE, E., 'Cultura e studi nei progetti di reforma presentati al concilio di Vienne 1311–1312', *Annuarium Historiae Conciliorum*, 9 (1977).
BOASE, T. S. R., *Boniface VIII* (London, 1933).
—— 'The History of the Kingdom', in id. (ed.), *The Cilician Kingdom of Armenia*.
—— (ed.), *The Cilician Kingdom of Armenia* (Edinburgh and London, 1978).
BOWSKY, W. M., *Henry VII in Italy: The Conflict of Empire and City-State 1310–1313* (Nebr., 1960).
BOYLE, J. A., 'The Il-Khans of Persia and the Princes of Europe', *Central Asiatic Journal*, 20 (1976).
BRANDT, W. I., 'Pierre Dubois: Modern or Medieval', *AHR* 25 (1930).
BRATIANU, G. I., 'Autour du projet de croisade de Nicolas IV: La Guerre ou le commerce avec l'infidèle', *RHSE* 22 (1945).
BRÉHIER, L., *L'Eglise et l'Orient au Moyen Âge: Les Croisades* (Paris, 1907).
BROWN, C., 'Portiuncula', *New Catholic Encyclopedia*, xi (New York, 1967).
BRETT, E. T., *Humbert of Romans: His Life and Views of Thirteenth Century Society* (Toronto, 1984).
BROWN, E. A. R., 'Royal Salvation and Needs of State in Late Capetian France', in W. C. Jordan *et al.* (eds.), *Order and Innovation*

in the Middle Ages: Essays in Honour of J. R. Strayer (Princeton, NJ, 1976).
BRUNDAGE, J. A., *Medieval Canon Law and the Crusader* (London, 1969).
—— 'Holy War and the Medieval Layers', in Murphy (ed.), *The Holy War*.
BULST-THIELE, M. L., *Sacrae Domus Militiae Templi Hierosolymitani Magistri* (Göttingen, 1974).
BURNS, R. I. 'The Catalan Company and the European Powers 1309–1311', *Speculum*, 29 (1954).
—— 'Christian–Islamic Confrontation in the West: The Thirteenth-Century Dream of Conversion', *AHR* 76 (1971).
—— *Islam under the Crusaders: Colonial Survival in Thirteenth-Century Valencia* (Princeton, NJ, 1973).
CAILLET, L., *La Papauté d'Avignon et l'Église de France: La Politique bénéficiale du Pape Jean XXII en France 1316–1334* (Paris, 1975).
CAMPBELL, G. J., 'Clerical Immunites in France during the Reign of Philip III', *Speculum*, 39 (1964).
CARDINI, F., 'Crusade and "Presence of Jerusalem" in Medieval Florence', in B. Z. Kedar et al. (eds.), *Outremer: Studies in the History of the Crusading Kingdom of Jerusalem presented to Joshua Prawer* (Jerusalem, 1982).
CARO, G., *Genusa und die Mächte am Mittelmeer 1237–1311: Ein Beitrag zur Geschichte des 13. Jahrhunderts* (Halle, 1895–9).
CHABOT, J. B., 'Histoire du patriarche Mar Jabalaha et du moine Rabban Cauma, *ROL* 2 (1894).
—— 'Notes sur les relations du roi Argoun avec l'Occident', *ROL* 2 (1894).
Cheney, C. R., 'The Downfall of the Templars and a Letter in their Defence', in id., *Texts and Studies* (Oxford, 1973).
CIPOLLA, C. M., *Money, Prices and Civilization in the Mediterranean World: Fifth to Seventh Century* (Princeton, NJ, 1956).
COHN, N., *The Pursuit of the Millenium: Revolutionary Millenarians and Mystical Anarchists of the Middle Ages* (London, 1970).
CONSTABLE, G., 'The Second Crusade as seen by Contemporaries', *Traditio*, 9 (1953).
CURSCHMANN, H. H. W. F., *Hungersnöte in Mittelalter* (Leipzig, 1900).
DANIEL, E. R., 'Apocalyptic Conversion: The Joachite Alternative to the Crusade', *Traditio*, 25 (1969).
—— *The Franciscan Concept of Mission in the High Middle Ages* (Ky., 1975).
DANIEL, N., *Islam and the West: The Making of an Image* (Edinburgh, 1960).
—— *The Arabs and Medieval Europe* (Edinburgh, 1975).

DELAVILLE LE ROULX, J., *La France en Orient au XIVe siècle: Expéditions du maréchal Boucicaut* (Paris, 1885–6).
—— *Les Hospitaliers en Terre Sainte et à Chypre 1100–1310* (Paris, 1904).
—— *Les Hospitaliers à Rhodes 1310–1421* (Paris, 1913).
DELLE PIANE, M., *Vecchio e nuovo nelle idee politiche di Pietro Dubois* (Florence, 1969).
DER NERSESSIAN, S., 'The Kingdom of Cilician Armenia', in *Setton's History*, ii.
DIGARD, G., *Philippe le Bel et le Saint-Siège* (Paris, 1936).
DONDAINE, A., 'Ricoldo de Monte Croce', *Archivum Fratrum Praedicatorum*, 37 (1967).
DU CANGE, C., *Histoire de l'Empire de Constantinople sous les empereurs français*, ed. J. A. C. Bouchon (Paris, 1826).
EHRLE, F., 'Zur Vorgeschichte des Konzils von Vienne', *Archiv für Litteratur und Kirchengeschichte*, 4 (1888); 5 (1889).
FAWTIER, R., *L'Europe occidentale de 1270 à 1380*, i (*Histoire du Moyen Âge*, vi. *Histoire générale publiée sous la direction de G. Glotz*; Paris, 1940).
FINKE, H., *Konzilienstudien zur Geschichte des 13. Jahrhunderts* (Münster, 1891).
FLAHIFF, G. B., '*Deus non vult*: A Critic of the Third Crusade', *Medieval Studies*, 9 (1947), 162–88.
FLICK, A. C., *The Decline of the Medieval Church* (New York, 1930).
FOLZ, R., *The Concept of Empire in Western Europe from the Fifth to the Fourteenth Century* (New York, 1969).
FOREST, A., et al., *Le Mouvement doctrinal du XIe au XIVe siècle, Histoire de l'Église*, xiii, ed. A. Fliche and A. Martin (Paris, 1951).
FOREY, A. J., 'The Order of Mountjoy', *Speculum*, 46 (1971).
—— *The Templars in the Corona de Aragón* (Oxford, 1973).
—— 'The Military Order of St Thomas', *EHR* 92 (1977).
—— 'The Military Orders in the Crusading Proposals of the Late Thirteenth and Early Fourteenth Centuries', *Traditio*, 36 (1980).
GAIGNARD, R., 'Le Gouvernement pontifical au travail: L'Example des dernières années du règne de Clément V: 1er août 1311–20 avril 1314', *Annales du Midi*, 72 (1960).
GATTO, L., *Il Pontificato di Gregorio X 1272–1276* (Rome, 1956).
—— *La Francia di Filippo il Bello 1284–1314* (Rome, 1973).
GEANAKOPLOS, D., 'Michael VIII Palaeologus and the Union of Lyons 1274', *Harvard Theological Review*, 46 (1953).
—— *Emperor Michael Palaeologus and the West 1258–1282* (Cambridge, Mass., 1959).
—— 'Byzantium and the Crusades 1261–1354', in *Setton's History*, iii.
GIORGI, I., 'Description du *Liber Bellorum Domini*', *AOL* 1 (1881).

GOLUBOVICH, G., 'Fr. Pietro da Pleine Chassaigne, O.F.M.', *Archivum Franciscanum Historicum*, 9 (1916).
GOTTRON, A., *Ramon Lulls Kreuzzugsideen* (Berlin and Keipzig, 1912).
GROUSSET, R., *Histoire des croisades* (Paris, 1924-6).
HAMILTON, B., *The Latin Church in the Crusader States: The Secular Church* (London, 1980).
HAY, D., *Europe: The Emergence of an Idea* (New York, 1966).
HEFELE, C. J., and LECLERQ, H., *Histoire des conciles*, vi (Paris, 1914-15).
HEIDELBERGER, F., *Kreuzzugsversuche um die Wende des 13. Jahrhunderts* (Leipzig and Berlin, 1911).
HENNINGER, J., 'Sur la contribution des missionnaires à la connaissance de l'Islam', *Neue Zeitschrift für Missionswissenschaft*, 9 (1953).
HEYD, W., *Histoire du commerce du Levant au Moyen Âge*, trans. F. Raynaud (Leipzig, 1936).
HILL, G., *A History of Cyprus* (Cambridge, 1940-52).
HILLGARTH, J. N., *Ramon Lull and Lullism in Fourteenth-Century France* (Oxford, 1971).
—— 'The Problem of a Catalan Mediterranean Empire 1229-1327', *EHR* suppl. 8 (London, 1975).
—— *The Spanish Kingdoms 1250-1516*, i (Oxford, 1976).
HOLT, P. M., 'The Sultanate of al-Manṣūr Lāchīn 696-8/1296-9', *Bulletin of the School of Oriental and African Studies, University of London*, 35 (1973).
—— 'Qalawun's treaty with Acre in 1283', *EHR* 91 (1976).
HOUSLEY, N. J., *The Italian Crusades: The Papal-Angevin Alliance and the Crusades against Christian Lay Powers 1254-1343* (Oxford, 1982).
—— 'Politics and Heresy in Italy: Anti-Heretical Crusades, Orders and Confraternities 1200-1500', *Journal of Ecclesiastical History*, 33 (1982).
—— 'Pope Clement V and the Crusades of 1309-10', *Journal of Medieval History*, 8 (1982).
—— 'Charles II of Naples and the Kingdom of Jerusalem', *Byzantium*, 54 (1984).
—— *The Avignon Papacy and the Crusades 1305-1378* (Oxford, 1986).
HUGH, P., *A History of the Church*, iii (London, 1960).
IORGA, N., *Philippe de Mézières (1327-1405) et la croisade au XIVe siècle* (Paris, 1896).
—— *Rhodes sous les Hospitaliers* (Paris and Bucharest, 1931).
JACOBY, D., 'L'Expansion occidentale dans le Levant: Les Vénitiens à Acre dans la seconde moitié de treizième siècle', *Journal of Medieval History*, 3 (1977).
—— 'The Rise of a New Emporium in the Eastern Mediterranean:

Famagusta in the Late Thirteenth Century', in *Meletai kai Hypomnemata*, i (Nicosia, 1984), and in Jacoby, *Studies on the Crusader States and the Venetian Expansion* (Northampton, 1989), no. 8.
JEDIN, H., *Ecumenical Council of the Catholic Church: An Historical Outline* (Freiburg, 1960).
JORDAN, W. C., *Louis IX and the Challenge of the Crusade: A Study in Rulership* (Princeton, NJ, 1979).
KANTOROWICH, E. H., '*Pro patria mori* in Medieval Political Thought', *AHR* 56 (1951).
KEDAR, B. Z., 'The Passenger List of a Crusader Ship 1250: Towards the History of the Popular Element on the Seventh Crusade', *Studi Medievali*, 13 (1972).
—— 'Segurano-Sakrān Salvaygo: un mercante genovese al servizio dei sultani mamlucchi *c.*1303–1322', in *Fatti e idee di storia economica nei secoli XII–XX: Studi dedicati a Franco Borlandi* (Imola, 1976).
—— *Crusade and Mission: The Interplay between Two European Approaches toward the Muslims in Medieval Times* (Princeton, NJ, 1984).
—— and SCHEIN, S., 'Un projet de passage particulier proposé par l'ordre de l'Hôspital 1306–1307', *BEC* 137 (1979).
KING, E. J., *The Knights Hospitallers in the Holy Land* (London, 1931).
KINGSFORD, C. L., 'Sir Otho de Grandison', *TRHS* 3 (1909).
KOHLER, C., 'Deux projets de croisade en Terre Sainte,' *ROL* 10 (1903/4).
—— 'Quel est l'auteur du *Directorium ad Passagium Faciendum*?', *ROL* 12 (1911).
KUNSTMANN, F., 'Studien über Marino Sanudo Torsello den Aelteren', *Abhandlungen der Bayerischen Akademie der Wissenschaften (historische Klasse)*, 7 (1853).
LAHRKAMP, H., 'Nordwestdeutsche Orientreisen und Jerusalemwallfahrten im Spiegel der Pilgerberichte', *Oriens Christianus*, 40 (1956).
LAIOU, A. E., *Constantinople and the Latins: The Foreign Policy of Andronicus II 1282–1328* (Cambridge, Mass., 1972).
LANE, F. C., *Venice: A Maritime Republic* (Baltimore and London, 1973).
LANGLOIS, C. V., 'Documents relatifs à Bertrand de Got (Clément V)', *Revue historique*, 40 (1889).
LAURENT, M. H., *Le Bienheureux Innocent V (Pierre de Tarentaisse) et son temps* (Vat., 1947).
LAURENT, P. V., 'Grégoire X et son projet de ligue anti-turque', *Echos d'Orient*, 37 (1938).
—— 'La Croisade et la question d'Orient sous le pontificat de Grégoire X', *RHSE*, 22 (1945).

LEA, H. C., *A History of the Inquisition of the Middle Ages* (New York, 1956).
LECLER, J., *Vienne* (Paris, 1964).
LE CLERQ, V. M., 'Nicolas de Hanapes, patriarche de Jerusalem', *HL* xx (1842).
—— 'Relation anonyme de la prise d'Acre en 1291', *HL* xx (1842).
—— 'Anonyme continuateur de la chronique de l'abbaye de Werum', *HL* xxi (1847).
LECLERCQ, J., 'Un sermon prononcé pendant la guerre de Flandre sous Philippe le Bel', *Revue du Moyen Âge latin*, 1 (1945).
—— *L'Idée de la royauté du Christ au Moyen Âge* (Paris, 1959).
—— 'Textes et manuscrits cisterciens dans des bibliothèques des États-Unis', *Traditio*, 17 (1961).
—— 'Galvano di Levanto e l'Oriente', in A. Pertusi (ed.), *Venezia e l'Oriente fra tardo medioevo e rinascimento* (Venice, 1966).
LECOY DE LA MARCHE, A., 'La Prédication de la croisade au treiziéme siècle', *Revue des questions historiques*, 48 (1890).
LEMMENS, L., 'Die Heidenmissionen des Spätmittelalters', *Franziskanische Studien*, 5 (1919).
LÉONARD, E. G. *Les Angevins de Naples* (Paris, 1954).
LERNER, R. E. 'An "Angel of Philadelphia" in the Reign of Philip the Fair: The Case of Guiard-Cressonessant', in W. C. Jordan *et al.* (eds.), *Order and Innovation in the Middle Ages: Essays in Honour of J. R. Strayer* (Princeton, NJ, 1976).
—— 'Medieval Prophecy and Religious Dissent', *Past and Present*, 72 (1976).
—— 'Refreshment of the Saints: The Time after Antichrist as a Station for Earthly progress in Medieval Thought', *Traditio*, 32 (1976).
—— *The Powers of Prophecy: The Cedar of Lebanon Vision from the Mongol Onslaught to the Dawn of the Enlightenment* (Berkeley, Calif., 1983).
LEVILLAIN, L., 'A propos d'un texte inédit relatif au séjour du pape Clement V à Poitiers en 1307', *Moyen Âge*, 1 (1897).
LITTLE, D. P., *An Introduction to Mamluk Historiography* (Wiesbaden, 1970).
LIZERAND, G., *Clément V et Philippe IV le Bel* (Paris, 1910).
LOCKHART, L., 'The Relations between Edward I and Edward II of England and the Mongol Il-Khans of Persia', *Iran: Journal of the British Institute of Persian Studies*, 6 (1968).
LONGNON, J., 'The Frankish States in Greece 1204–1311', in *Setton's History*, ii.
LONGPRÉ, E., 'Deux opuscules inédits du B. Raymond Lulle', *France franciscaine*, 18 (1935).

LOPEZ, R. S., *Genova marinara nel ducento: Benedetto Zaccaria ammiraglio e mercante* (Messina and Milan, 1933).
LOURIE, E., 'An Offer of the Suzerainty and Escheat of Cyprus to Alphonso III of Aragon by Hugh de Brienne in 1289', *EHR* 84 (1969).
LÖWE, A., *The Catalan Vengeance* (London, 1972).
LUCAS, S., 'The Low Countries and the Disputed Imperial Election of 1314', *Speculum*, 21 (1946).
LUNT, W. E., 'The Account of a Papal Collector in England in 1304', *EHR* 28 (1913).
—— 'Papal Taxation in England in the Reign of Edward I', *EHR* 30 (1915).
—— 'Collectors' Accounts for the Clerical Tenth levied in England by Order of Nicholas IV', *EHR* 31 (1916).
—— *Papal Revenues in the Middle Ages* (Columbia University Records of Civilization, 19; New York, 1934).
—— *Financial Relations of the Papacy with England to 1327* (Cambridge, Mass., 1939).
LUTTRELL, A., 'Venice and the Knights Hospitallers of Rhodes in the Fourteenth Century', *Papers of the British School of Rome*, 26 (1958).
—— 'The Aragonese Crown and the Knights Hospitallers of Rhodes 1291–1350', *EHR* 75 (1961).
—— 'The Crusade in the Fourteenth Century', in J. R. Hale et al. (eds.), *Europe in the Late Middle Ages* (London, 1965).
—— 'The Hospitallers in Cyprus after 1291', *Acts of the First International Congress of Cypriot Studies* (Nicosia, 1972).
—— 'The Hospitallers in Rhodes 1306–1421', in *Setton's History*, iii.
—— 'The Hospitallers in Cyprus, Rhodes, Greece and the West 1291–1440', in id., *Collected Studies* (London, 1978).
—— 'Emmanuelle Piloti and Criticism of the Knights Hospitallers of Rhodes 1306–1444', in *Collected Studies*.
—— 'The Hospitallers' Intervention in Cilician Armenia 1291–1375', in Boase (ed.), *Cilician Kingdom of Armenia*.
MAGNOCAVALLO, A., *Marin Sanudo il vecchio o il suo progetto di crociata* (Bergamo, 1901).
MANDONNET, P., 'Fra Ricoldo de Monte-Croce, pèlerin en Terre Sainte et missionnaire en Orient', *Revue biblique*, 2 (1893).
MANN, H. K., *The Lives of the Popes in the Middle Ages*, xvii–xviii (London, 1931–2).
MAS-LATRIE, M. L. DE, *Histoire de l'île de Chypre sous la règne des princes de la maison de Lusignan* (Paris, 1852–61).
—— 'Traité des vénitiens avec l'émir d'Acre en 1304', *AOL* 1 (1881).
—— 'Les Patriarches latins de Jerusalem', *ROL* 1 (1893).

MAYER, H. E., *Geschichte der Kreuzzuge* (Stuttgart, 1965); trans. J. Gillingham as *The Crusades* (Oxford, 1972).
MCCREADY, W. D., 'Papalists and Antipapalists: Aspects of the Church/State controversy in the Later Middle Ages', *Viator*, 6 (1975).
MCGINN, B., *Visions of the End: Apocalyptic Traditions in the Middle Ages* (New York, 1979).
MENACHE, S., 'Contemporary Attitudes concerning the Templars' Affair: Propaganda's Fiasco?', *Journal of Medieval History*, 8 (1982).
—— 'Philippe le Bel: Genèse d'une image', *Revue belge de philologie et d'histoire*, 62 (1984).
—— 'Religious Symbols and Royal Propaganda in the Late Middle Ages: The Crusades', in M. Yardeni (ed.), *Idéologie et propagande en France* (Paris, 1987).
MEYVAERT, P. 'An Unknown Letter of Hulagu, Il-Khan of Persia, to King Louis IX of France, *Viator*, 11 (1980).
MICHAUD, J. F., *Histoire des croisades* (Paris, 1824–9).
MICHELET, J., *Le Procès des Templiers* (Paris, 1841).
MILLER, W., 'The Genoese Colonies in Greece, i. The Zaccaria of Phocaea and Chios 1275–1329', in id., *Essays on the Latin Orient* (Cambridge, 1921).
MISKIMIN, H. A., *Money, Prices and Foreign Exchange in Fourteenth-Century France* (New Haven, Conn., and London, 1963).
MOLLAT, G., 'Clement V', *DHGE* xii (1953).
—— *The Popes at Avignon 1305–1378*, trans. J. Löve (London, 1963).
MOLLAT, M., 'Problèmes navals de l'histoire des croisades', *Cahiers de civilisation médiévale*, 10 (1967).
MOLS, R., 'Celestin V', *DHGE* xii (1953).
MONNERET DE VILLARD, M., 'La Vita, le opere e i viaggi di frate Richoldo da Monte Croce OP', *Orientalia christiana periodica*, 10/11 (1944/5).
MORANVILLÉ, H., 'Les Projets de Charles de Valois sur l'empire de Constantinople', *BEC* 51 (1890).
MOSTAERT, A., AND CLEAVES, F. W., 'Trois documents mongols des archives secrètes vaticanes', *Harvard Journal of Asiatic Studies*, 15 (1952).
——,—— *Les Lettres de 1289 et 1305 des il-khans Argun et Oljeitu à Philippe le Bel* (Cambridge, Mass., 1962).
MÜLLER, E., *Das Konzil von Vienne* (Münster, 1934).
NEUMANN, G. A., 'Ludophus de Suchem, *De Itinere Terre Sancte*', *AOL* 2 (1884).
O'CALLAGHAN, J. F., 'The Foundation of the Order of Alcantara 1176–1218', *Catholic Historical Review*, 47 (1962).

—— *A History of Medieval Spain* (New York, 1975).
PALL, F., 'Les Croisades en Orient au Bas Moyen Âge: Observations critiques sur l'ouvrage de M. Atiya', *RHSE* 19 (1942).
PARTEE, C., 'Peter John Olivi: Historical and Doctrinal Study', *Franciscan Studies*, 20 (1960).
PAULUS, N., *Geschichte des Ablasses im Mittelalter von Ursprunge bis zur Mitte des 14. Jahrhunderts* (Paderborn, 1922–3).
PELLIOT, P., 'Les Mongols et la papauté', *Revue de l'Orient chrétien*, 23 (1922–3); 24 (1924); 38 (1931/32).
PEQUES, F. J., *The Lawyers of the Last Capetians* (Princeton, NJ, 1962).
PETECH, L., 'Les Marchands italiens dans l'Empire mongol', *Journal asiatique*, 250 (1962).
PETIT, J., 'Un capitaine du règne de Philippe le Bel, Thibaut de Chepoy', *Moyen Âge*, 2 (1891).
—— *Charles de Valois 1270–1325* (Paris, 1900).
POST, G., 'Two Notes on Nationalism in the Middle Ages', *Traditio*, 9 (1953).
—— *Studies in Medieval Legal Thought: Public Law and the State 1100–1322* (Princeton, NJ, 1964).
POWER, E., 'Pierre de Bois and the Domination of France', in F. J. C. Hearnshaw (ed.), *The Social and Political Ideas of Some Great Medieval Thinkers* (London, 1923).
POWICKE, F. M., 'Pierre Dubois', in T. F. Tout and J. Tait (eds.), *Historical Essays by Members of Owens College, Manchester, published in Commemoration of Its Jubillee 1851–1901* (London, 1902).
—— *The Thirteenth Century* (London, 1953).
PRAWER, J., *Histoire du royaume latin de Jérusalem* (Paris, 1969).
—— *The Latin Kingdom of Jerusalem: European Colonialism in the Middle Ages* (London, 1972).
—— *The World of the Crusaders* (Jerusalem and London, 1972).
PRESTWICH, M., *War, Politics and Finance under Edward I* (London, 1972).
PRUTZ, H., *Kulturgeschichte der Kreuzzuge* (Berlin, 1883).
PURCELL, M., 'Changing Views of the Crusade' *Journal of Religious History*, 7 (1972).
—— *Papal Crusading Policy: The Chief Instruments of Papal Policy and Crusade to the Holy Land from the Final Loss of Jerusalem to the Fall of Acre 1244–1291* (Studies in the History of Christian Thought, 11; Leiden, 1975).
RABANIS, M., *Clement V et Philippe le Bel* (Paris, 1858).
RACHEWILTZ, I. DE, *Papal Envoys to the Great Khans* (Stanford, 1971).
RAYNOUARD, M., *Choix de poësies originales de troubadours*, iv (Paris, 1819).

REEVES, M., 'Joachimist Influences on the Idea of a Last World Emperor', *Traditio*, 17 (1961).
—— *The Influence of Prophecy in the Later Middle Ages: A Study in Joachimism* (Oxford, 1969).
—— *Joachim of Fiore and the Prophetic Future* (London, 1976).
RENAN, E., 'Guillaume de Nogaret, légiste', *HL* xxvii (1877).
—— 'Bertrand de Got: Pape sous le nom de Clément V', in id., *Études sur la politique religieuse du règne de Philippe le Bel* (Paris, 1899).
RENOUARD, Y., *Les Hommes d'affaires italiens au Moyen Âge* (Paris, 1949).
RIANT, P., 'Description du *Liber Bellorum Domini*', *AOL* 1 (1881).
RICHARD, J., 'Le Début des relations entre la papauté et les Mongols de la Perse', *Journal asiatique*, 237 (1949).
—— 'An Account of the Battle of Hattin referring to the Frankish Mercenaries in Oriental Moslem States', *Speculum*, 27 (1952).
—— *Le Royaume latin de Jérusalem* (Paris, 1953).
—— 'Un partage de seigneurie entre Francs et Mamelouks: Les "Casaux de Syr"', *Syria*, 39 (1953).
—— 'La Mission en Europe de Rabban Çauma et l'union des Églises', *Convegno volta*, 12 (Rome, 1957).
—— 'The Mongols and the Franks', *Journal of Asian History*, 3 (1969).
—— 'Isol le Pisan: Un aventurier franc gouverneur d'une province mongole?', *Central Asiatic Journal*, 14 (1970).
—— L'Enseignement des langues orientales en Occident au Moyen Âge', *Revue des études islamiques*, 44 (1976).
—— 'Une ambassade mongole à Paris en 1262', *Journal des savants* (1979).
—— *Saint Louis* (Paris, 1983).
—— 'Le Royaume de Chypre et l'embargo sur le commerce avec l'Égypte (fin XIIIe–debut XIVe siècle), *Académie des Inscriptions et Belles-Lettres: Comtes Rendus des Séances de l'annee 1984* (Paris, 1984).
RILEY-SMITH, J., *The Knights of St John in Jerusalem and Cyprus c.1050–1310* (London, 1967).
—— 'A Note on Confraternities in the Latin Kingdom of Jerusalem', *BIHR*, 44 (1971).
—— *The Feudal Nobility and the Kingdom of Jerusalem 1174–1277* (London, 1973).
—— *What were the Crusades?* (London, 1977).
—— The Templars and the Teutonic Knights in Cilician Armenia', in Boase (ed.), *Cilician Kingdom of Armenia*.
—— and Riley-Smith, L., *The Crusades: Idea and Reality 1095–1274* (London, 1981).
RÖHRICHT, R., 'Die Eroberung Akkas durch die Muslimer im Jahre 1291', *Forschungen zur Deutschen Geschichte*, 20 (1879).

—— 'Études sur les derniers temps du royaume de Jérusalem', *AOL* 1 (1881); 2 (1884).
—— 'Syria Sacra', *ZDPV* 10 (1887).
—— 'Der Untergang des Königreiches Jerusalem', *MIÖG* 15 (1894).
—— *Geschichte des Königreiches Jerusalem* (Innsbruck, 1898).
RUNCIMAN, S., *A History of the Crusades* (Cambridge, 1951–5).
—— *The Sicilian Vespers: A History of the Mediterranean World in the Later Thirteenth Century* (Cambridge, 1961).
—— 'The Decline of the Crusading Ideal', *Sewanee Review*, 79 (1971).
—— 'The Crusader States 1243–1291', in *Setton's History*, ii.
RUSSEL, F. H., *The Just War in the Middle Ages* (Cambridge, 1975).
RYAN, J. D., 'Nicholas IV and the Evolution of the Eastern Missionary Effort', *Archivum historiae pontificiae*, 19 (1981).
SCHEFER, C., 'Étude sur la *Devise des chemins de Babiloine*', *AOL* 2 (1884).
SCHEIN, S., '*Gesta Dei per Mongolos 1300*: The Genesis of a Non-Event', *EHR* 95 (1979).
—— 'The Patriarchs of Jerusalem in the late Thirteenth Century—seigners espiritueles et temporeles?', in B. Z. Kedar et al. (eds.), *Outremer: Studies in the History of the Crusading Kingdom of Jerusalem presented to Joshua Prawer* (Jerusalem, 1982).
—— 'The Future *Regnum Hierusalem*: A Chapter in Medieval State Planning', *Journal of Medieval History*, 10 (1984).
—— 'Philip IV the Fair and the Crusade: A Reconsideration', in P. W. Edbury (ed.), *Crusade and Settlement* (Cardiff, 1985).
—— 'From "Milites Christi" to "Mali Christiani": The Italian Communes in Western Historical Literature', in B. Z. Kedar and G. Airaldi (eds.), *I Comuni italiani nel regno crociato di Gerusaleme* (Genoa, 1986).
—— 'The Image of the Crusader Kingdom of Jerusalem in the Thirteenth Century', *Revue belge de philologie et d'histoire*, 64 (1986).
SCHLUMBERGER, G., *Expédition des 'Almugavares' ou routiers catalans en Orient* (Paris, 1902).
—— 'Prise de Saint Jean d'Acre', in id., *Byzance et croisade* (1927)
SCHOLTZ, R., *Die Publizistik zur Zeit Philippe des Schönen und Bonifaz VIII* (Stuttgart, 1903).
SCHÖPP, N., *Papst Hadrian V (Kardinal Ottobuono Fieschi)* (Heidelberg, 1916).
SCHOTTMÜLLER, K., *Der Untergang des Templerordens* (Berlin, 1887).
SERVOIS, G., 'Emprunts de Saint Louis en Palestine et en Afrique: Appendice', *BEC* 19 (1858).
SETTON, K. M., 'The Catalans in Greece 1311–1380', in *Setton's History*, iii.
—— *The Papacy and the Levant 1204–1571*, i (Phil., 1976).

SETTON, K. M., et al. (eds.), *History of the Crusades* (Madison, 1969–).
SIBERRY, J. E., 'Missionaries and Crusaders: Opponents or Allies 1095–1274', *Studies in Church History*, 20 (1983).
—— *Criticism of Crusading 1095–1274* (Oxford, 1985).
SINOR, D., 'The Mongols and Western Europe', in *Setton's History*, iii.
SIVAN, E., *L'Islam et la croisade: Idéologie et propagande dans les réactions musulmanes aux croisades* (Paris, 1968).
SMAIL, R. C., *Crusading Warfare 1097–1193* (Cambridge, 1956).
SOUTHERN R., *Western Views of Islam in the Middle Ages* (Cambridge, Mass., 1956).
SOMOGYI, J., 'Adh-Dhahabi's Record of the Destruction of Damascus by the Mongols in 699–700/1299–1301', in S. Löwinger and J. Somogyi (eds.), *Ignace Goldziher Memorial Volume*, i (Budapest, 1948).
SPIEGEL, G. M., '"Defense of the Realm": Evolution of a Capetian Propaganda Slogan', *Journal of Medieval History*, 3 (1977).
SPUFFORD, P., 'Coinage and Currency', in *Cambridge Economic History*, ed. M. M. Postan et al. (Cambridge, 1952; 1987 edn.), ii.
STENGERS, J., *Les Juifs dans les Pays-Bas au Moyen-Âge* (Brussels, 1950).
STERNFELD, R., *Der Kardinal Johann Gaetan Orsini (Papst Nikolaus III) 1244–1277* (Berlin, 1905; repr. Vaduz, 1965).
STICKEL, E., *Der Fall von Akkon: Untersuchungen zum Abklingen des Kreuzzugsgedankens am Ende des 13. Jahrhunderts* (Berne and Frankfurt-on-Main, 1975).
STRAYER, J. R., 'The Crusade against Aragon', *Speculum*, 28 (1953).
—— 'France: The Holy Land, the Chosen People and the Most Christian King', in id., *Medieval Statecraft and the Perspectives of History* (Princeton, NJ, 1971).
—— 'The Crusades of Louis IX', in *Setton's History*, ii.
—— 'The Political Crusades of the Thirteenth Century', in *Setton's History*, ii.
—— *The Reign of Philip the Fair* (Princeton, NJ, 1980).
—— and C. H. Taylor, *Studies in Early French Taxation* (Cambridge, Mass., 1939).
SUGRANYES DE FRANCH, R., *Raymond Lulle, docteur des missions*, (Schöneck-Beckenried, 1954).
—— 'Els Projectes de creuada en la doctrina missional de Ramon Llull', *Estudios Lulianos*, 4 (1960).
SUMPTION, J., *Pilgrimage: An Image of Medieval Religion* (London, 1975).
TAYLOR, C. H., 'French Assemblies and Subsidy in 1321', *Speculum*, 43 (1968).

THIER, L., *Kreuzzugsbemühungen unter Papst Clemens V 1305–1314* (Düsseldorf, 1973).
THROOP, P. A., 'Criticism of Papal Crusade Policy in Old French and Provençal', *Speculum*, 13 (1938).
—— *Criticism of the Crusade: A Study of Public Opinion and Crusade Propaganda* (Amsterdam, 1940).
TIERNEY, B., *Foundations of the Conciliar Theory* (Cambridge, 1955).
TOPPING, P., 'The Morea 1311–1364', in *Setton's History*, iii.
—— 'The Catalans in Greece 1311–1380', in *Setton's History*, iii.
TRENCHS ODENA, J., '"De Alexandrinis": El Comercio prohibido con los Musulmanes y el papado de Aviñón durante la primera mitad del siglo XIV', *Anuario del estudios medievales*, 10 (1980).
TYERMAN, C. J., 'Marino Sanudo Torsello and the Lost Crusade: Lobbying in the Fourteenth Century', *TRHS* 32 (1982).
—— 'Sed Nihil Fecit? The Last Capetians and the Recovery of the Holy Land', in J. Gillingham and J. C. Holt (eds.), *War and Government in the Middle Ages* (Bury St Edmunds, 1984).
—— 'Philip V of France, the Assemblies of 1319–1320 and the Crusade', *BIHR* 57 (1984).
—— 'Philip VI and the Recovery of the Holy Land', *EHR* 143 (1985).
—— 'The Holy Land and the Crusades of the Thirteenth and Fourteenth Centuries', in P. W. Edbury (ed.), *Crusade and Settlement* (Cardiff, 1985).
ULLMAN, W., *Medieval Papalism: The Political Theories of the Medieval Canonists* (London, 1949).
—— *A Short History of the Papacy in the Middle Ages* (London, 1974).
VAN DEN GHEYN, S. J., 'Note sur un manuscrit de *l'Excidium Aconis*, en 1291', *ROL* 6 (1898).
VICTOR, J. M., 'Charles de Bovelles and Nicholas de Pax: Two Sixteenth Century Biographies of Ramon Lull', *Traditio*, 32 (1976).
VIOLLET, P., 'Guillaume Durant le Jeune, évêque de Mende,' *HL* xxxv (1921).
WAAS, A., *Geschichte der Kreüzzuge* (Freiburg, 1956).
WALEY, D. P., 'Papal Armies in the Thirteenth Century', *EHR* 72 (1957).
—— *The Papal State in the Thirteenth Century* (London, 1961).
WATTENBACH, W., 'Fausse correspondance du sultan avec Clément V', *AOL* 2 (1884).
WEISS, R., 'England and the Decree of the Council of Vienne on the Teaching of Greek, Arabic, Hebrew and Syrian' *Bibliothèque d'Humanisme et de Renaissance*, 14 (1952).
WENCK, K., *Clemens V und Heinrich VII; Die Anfcänge des französischen Papsttums, Ein Beitrag zur Geschichte des 14. Jahrhunderts* (Halle, 1882).

WENCK, K., *Phillip der Schöne von Frankreich: Seine Persönlichkeit und das Urteil der Zeitgenossen* (Marburg, 1905).
—— 'Aus den Tagen der Zusammenkunft Papst Klemens V und König Philipps des Schönen zu Lyon: November 1305 bis Januar 1306', *Zeitschrift für Kirchengeschichte*, 27 (1906).
WIERUSZOWSKI, H., 'Ramon Lull et l'idée de la Cité de Dieu', *Estudis Franciscans*, 47 (1935).
WILKEN, F., *Geschichte der Kreuzzuge* (Leipzig, 1807–32).
WILKS, M. J., *The Problem of Sovereignty in the Later Middle Ages* (Cambridge, 1963).
ZECK, E., *Der Publizist Pierre Dubois* (Berlin, 1911).

4. UNPUBLISHED PH.D. DISSERTATION

BEEBE, B., 'Edward I and the Crusades' (St. Andrew's, 1970).

INDEX

Abaga, il-khan of Persia 43, 44
Abeldjes, castle of 79
Achaea 30, 58, 86
Acre 1, 3, 10, 12, 17, 20, 25, 40, 41, 44, 50, 59, 61, 67, 68, 69, 70, 71, 72, 73, 80, 83, 98, 99, 106, 109, 122, 132, 155, 164, 165, 167, 211, 212, 226, 244; Dominicans in 125, 126 n.; image of 129–30, 134; Templar Tower 122; war of St Sabas 41
Acre, fall of 73, 74, 77, 83, 84, 89, 90, 92, 94, 102, 113, 115, 116, 119, 121, 123, 125, 126, 127, 128, 130, 134, 193, 201, 204, 243, 259, 265; explanations of 114, 122, 127, 129, 133; reaction to 74–5, 122–34, 265; *see also* Holy Land, loss of
Adam of Murimuth 241; *Continuatio chronicarum* 241
Aden 122 n.
Adolf of Nassau, king of Germany 87, 155
Adriatic Sea 99
Adso of Montien-en-Der 119
Aegean Sea 231
Ain-Jalud, battle of 43
Aix 141
al-Ashraf Khalil, sultan of Egypt 78, 79
Albania 58
Albert I of Austria, king of Germany 87, 186, 187, 188, 210, 267
Aleppo 162
Alexander IV, pope 75
Alexander the Great 119
Alexandria 73, 78, 80, 81, 83, 98, 109, 164, 166, 179, 191, 213, 228
Alfonso III, king of Aragon 57, 72, 84, 86, 190
Alfonso X, king of Castile 46, 53, 55
Alfonso, king of Portugal 53
Amalric, nominal lord of Tyre 163, 164, 194, 195, 224
Amaneu VII, d'Albert 252
Amauri II, viscount of Narbonne 249
Amorgos 232
Anagni 153; congress of 149
Anatolia 99
Ancoma 160, 225
Andalusia 207
Andrew Carnaro 231
Andronius II Palaeologus, Byzantine emperor 87, 179, 184, 185, 209, 223
Andros 231
'Angelic Pope' 141, 142, 146–7, 170
Angevins, dynasty 231
Annales Aulae Regiae 240

Annales Paulini 234
Annales Regis Edwardi Primi 167
Antichrist 119, 120, 123, 134 n., 146, 169, 171
Antioch, principality of 16, 76, 78, 96 n., 99, 125, 214, 219
Antiquorum Relatio, papal bull 169
Antonio Spinola 232
Apostolic Brethren 170
Aquila 141
Arabia 112 n.
Aragon 28, 52, 54, 58, 71, 108, 137, 140, 149, 152, 179, 180, 185, 187, 189, 192, 201, 206, 207, 208, 210, 243, 267; Aragonese at the Council of Vienne 252–3; crusade against 63–7, 84, 85, 135; crusade of against Granada 228–9, 252–3; diplomatic relations with Egypt 190–1; navy of 43, 151
Arghun, il-khan of Persia 87–9, 169, 214
Arles, council of 136
Arles, kingdom of 141, 253
Arnold Nolelli, cardinal 186
Arnold Novelli, cardinal 252
Arnold of Pellagrua, cardinal 252
Arnold of Villanova 123–4, 146–7, 191–2; *De Adventu Antichristi* 123–4, 146–7
Arsenius, Greek monk 132–3
Arsuf 15
Arthur of Brittany 196
Ascuita fili, papal bull 263
Asia Minor 106, 179, 232, 249, 256
Athens 185, 186
Atiya, A. S. 4, 5, 8, 90, 121, 266
Audita tremendi, papal bull 74
Austria 234, 237
Avignon 13, 155, 187, 227, 228, 234, 236, 237, 241, 264
Ayas (Lajazzo) 77, 195

Babylon 113 n., 167; *see also* Cairo; Egypt
Baghdad 125
Baibars, sultan of Egypt 56, 82, 97
Baldwin II, emperor of Constantinople 37, 58
Balkans 99
Barbery 103, 106, 205, 252
Barcelona 27, 40–1, 77, 106
Bartholomew, marshal of the Temple 164
Bartholomew of Neocastro 67, 131
Bavaria 234
Baybars 15

Beaufort 16
Bedouins 212
Beirut 73
Beit-Hânûn, battle of 117
Belfort 15
Bellapaïs 195
Benedetto Zaccazia 165, 166
Benedict XI, pope 8, 123, 140, 143, 160, 175–8, 264; and the crusade 176–8
Berenger Fredol, cardinal 252
Bernard, bishop of Tripoli 84
Bernard of Clairvaux 31, 104
Bernard di Garbo, cardinal 252
Bernard Gui 132, 224, 235
Bertrand de Got, *see* Clement V
Besançon 141
Bethesni, fortress of 79
Béziers, council of 170
Bilargou, Mongol chieftain 214
Bithynia 179
Black Sea 98, 119, 231
Bloemhof (Floridus Hortus) 118
Bohemia 30, 153
Bohemond VII, titular prince of Antioch 60
Bologna 69n., 70, 246
Bonaventura 36
Boniface VIII, pope 8, 16, 112, 123, 140, 143, 144, 146, 147, 163n., 170, 172, 173, 176, 182, 188, 189, 192, 198, 261, 262, 263; and the crusade against the Colonna 160–2, 166–7; and the Holy Land crusade 147–62; and the military orders 76, 153–5; and the Mongol conquest of the Holy Land (1300) 165–8, 173–5; and the papal jubilee of 1300 169; trial of 220, 242–3
Boniface of Calamandrana, Hospitaller grand-commander of Outremer 142, 154, 155, 189
Bordeaux 181, 196
Bosphorus 62
Botrom 69
Bourges 141
Boyle, J. A. 89
Brabant 234
Bratianu, G. I. 110
Brignoles, treaty of 71, 72, 84
Brindisi 99, 184, 230, 232, 233
British Isles 152
Bruno of Olmütz 22, 23
Brunswick 234, 235
Bulgaria 59
Burgundy 52
Buscarel de Ghizolfi 88
Byblos 68, 163
Byzantium 36, 42–3, 44, 46, 52, 53, 58, 59, 60, 61, 72, 87, 103, 105, 107, 138, 157–60, 176, 177, 178, 179, 183, 204, 232, 261; crusade against 46, 53, 58–9, 60–2, 105, 107, 140, 157–60, 176–80, 183–7, 196, 207, 208, 209, 217, 218, 242, 249, 252, 253, 256, 261, 264; *see also* Constantinople

Caesarea 15
Caffa 106
Cairo 72, 163, 165
Calatrava, order of 47
Caltabellota, treaty of 109, 159, 160, 179
Canterbury, council of 136, 138
Carpathos 231
Carthage 101
Cassandreia 185
Castile 28, 53, 54, 55, 56, 58, 201, 207, 228, 243
Catalan Company 166, 177, 179, 180, 184–5
Catalonia 179, 190, 193, 224
Catherine of Courtenay, titular empress of Constantinople 87, 159, 186, 209n.
Catherine of Valois 186
Celestine V, pope 108, 141–4, 182; and the crusade 141–4
Ceuta, emirate of 207, 228
Chabot, J. B. 89
Charlemagne 34, 146, 147, 209, 210
Charles IV, king of France 204, 211, 255
Charles I of Anjou, king of Sicily and Jerusalem 19, 42–3, 45, 46, 53, 55, 58–62, 65, 107, 131, 158, 176, 184, 211
Charles II of Anjou, king of Sicily and Jerusalem 72, 84, 85, 94, 137, 141, 149, 153, 159, 160, 189, 214, 240; and the maritime blockade of Egypt 83, 98, 109; and the military orders 76; claim to the title of king of Jerusalem 211; cross taking 45, 53; plan for crusade 19, 92, 96, 100, 101, 107–11, 217, 267; tithes received 65, 66, 157
Charles of Valois 63, 65, 149, 157–8, 159, 160, 176, 178, 183, 184–6, 187, 193, 196, 198, 208, 209, 267
Chastel Neuf 15
Château Pélerin (Athlit) 73
Chios 166, 204, 231
Chisis, family of 204
Christian Spinola, Genoese merchant 225, 229, 241, 242
Chronicon de Lanercost 68
Chronicon Elwacense 236
Chronicon Sampetrinum 131–2
Church councils, provincial 75, 92, 110, 111, 113; of 1292 110–11, 113, 135–8, 139, 247, 248
Cistercians 47, 69, 117, 118, 167
Clement IV, pope 48, 180
Clement V, pope 5, 8, 12, 16, 17, 76, 173, 176, 180, 181, 182, 188, 190, 192, 193, 195, 196, 200, 201, 202, 204, 207, 208,

210, 218, 234, 236, 237, 250, 251, 256, 259, 260, 262, 264, 268, 271; and the blockade of Egypt 80–1, 232; and the conquest of Constantinople 140, 160, 183–5, 187; and the Holy Land crusade 181–99; and the military orders 197–9, 240–6; and the Mongols of Persia 214–15; and the *passagium particulare* (1309) 219–29; relations with Philip IV, the Fair 181–3, 240–3
Colledimezzo, indulgence of 142–3
Colmar 90
Colonna, family of 140, 160, 161, 162; crusade against 140, 160–2
Commendatio Lamentabilis in Transitu Magni Regis Edwardi 86
commerce, illegal 39–41, 79–83, 103, 151, 190–1, 203; conciliar legislation against 80–1, 178; criticism of 83, 95; fines on 151, 192; of maritime cities 80–3; papal legislation against 178, 203, 221, 225; *see also* trade embargo
Conradin I Hohenstanfen, king of Germany and of Jerusalem 46, 59
Constantine the Great 133
Constantinople 99, 184, 231; crusade for the conquest of 46, 105, 106, 140, 157–60, 176–80, 183, 186, 196, 208–9, 217, 249, 252, 256; recommended as a station on the route of the Holy Land crusade 99, 189, 207, 249, 253; trade of with the Mamluks 205; *see also* Byzantium
Constitutiones pro zelo Fidei 20, 36, 38–42, 47
conversions 103, 104, 106, 119–20, 128, 201; apocalyptic conversion of the Saracens 119–20; of Armenians and Serbs 185; of Christians to Islam 95–6, 97, 103, 125–6, 172; of Greek Orthodox 105, 138; of the Mongols 103, 105, 126, 171, 214; of Moslems 25–8, 101, 103, 105, 106; of schismatics 103
Coron 231
Corsica 65, 152, 159, 192
Cos 223, 231
Cotrone 232
Courtrai, battle of 161
Crete 204, 230, 231, 232
Crimea 204
criticism, of the Franks 95–6, 122, 131; of Christendom's attitude to the Holy Land 96–7, 122; of the crusade 3, 4, 29–31, 35–6, 100, 131, 133, 226, 254; of illegal commerce 83, 95; of the Italian communes 95, 122, 130, 131; of Italian crusaders in Acre 131; of the military orders 76, 97, 130, 131, 153, 220–1, 243, 259; of the papacy 94, 96, 115, 132, 133, 170, 226, 254

cross-taking 45, 46, 53, 58, 69, 70, 72, 107, 171, 176, 196, 235, 255, 262, 265; *see also* crusading vows
Crusade: apologetics of 30–1, 200–2; against Aragon 63–7, 84, 85, 135; against Byzantium 46, 53, 58–9, 60–2, 105, 107, 140, 157–60, 176–80, 183–7, 207, 208, 209, 217, 249, 252, 253, 261, 264; against the Colonna 140, 160–2; criticism of 3, 4, 29–31, 35–6, 100, 131, 133, 226, 254; *excitatoria* 48, 91, 201; fifth 81; financing of 24–5, 32, 39–40, 46–7, 105, 136, 222–7, 250–1, 253; first 18, 73; against Granada 190, 192, 207, 228–9, 251, 252, 253, 261; in Italy 5–7, 62–7, 69, 70, 71, 84–5, 108, 133, 135, 140, 141, 142, 153, 156–7, 160–2, 261, 263; papal-Hospitaller of 1309 18, 193, 204, 212, 219–38, 262, 265, 271; planning of 2, 15–19, 22–49, 91–111, 200–18, 248–51, 259; of the poor of 1309 117, 226, 233–8, 264; of the poor of 1320 117; preaching of 33–5, 48, 79, 86, 250; propaganda for 161–2, 233–4, 238; for the recovery of the Holy Land 37–50, 52–4, 71–91, 96–9, 135–8, 147–62, 181–99, 230, 246–57, 258–68; second 73, 128, 130, 226; against Sicily 62–7, 69, 70–1, 84–5, 108, 133, 135, 140, 141, 142, 153, 156–7, 161, 261, 263; third 18, 40; of 1290 68–71; *see also de recuperatione Terrae sanctae*; general crusade; particular crusade
crusading indulgences 41–2, 48, 58, 65–6, 105, 142–3, 151, 157, 158, 165, 169, 176, 177, 183, 222, 234, 237, 250, 262–4
crusading tithes 46–7, 51, 54–5, 56–7, 141, 142, 152, 157, 161, 183, 242–3; of Council of Vienne 242–3, 253; diversion of from the Holy Land 57–8, 64–5, 142, 157, 161; granted to Charles I of Anjou 45, 61–2, 66, to Charles II of Anjou 65, 141, 157, to Charles of Valois 157, 183, 186; to Edward I 45–6, 54, 71, 142, 152, to James of Majorca 66, to James II of Aragon 157, 192, 229, to Philip III of France 65–6, to Philip IV of France 66, 157, 242, 243; of the Second Council of Lyons 51–2, 54–5, 56–8, 65, 153, 157; *see also* taxation
crusading vows 24, 42, 45, 46, 53, 58, 69, 70, 72, 107, 152–3, 171, 176, 177, 196, 235, 255, 262, 263, 264, 265; commutations of 42, 153, 177, 263; dispensations from 153, 263, 264; redemptions of 24, 42, 152–3, 263, 264; *see also* cross-taking
Cyclades 233

Index

Cyprus, kingdom of 36, 105, 112, 141, 166, 167, 168, 171, 173, 188, 189, 192, 193, 199, 203, 210, 211, 213, 223, 228, 231, 232, 244; aid from the West to 77–8, 98, 194–7, 219, 221, 224; base on the sea route of a general crusade 207; expeditions from to Syria, Palestine, and Egypt 77–8, 163–5; military orders in 154–6, 162–5, 230, 244; navies of 82, 109, 204; point of disembarkation of a general crusade 201; *see also* Henry II

Dalmatia 153
Damascus 101, 112 n., 162, 163 n., 167, 172
Damietta 80
Daniel, N. 97
Dante Alighieri 112, 148
De excidio Urbis Acconis 114–21, 128
Delaville le Roulx, J. 4, 5, 8, 94, 265–6
denier tournois, see monetary system
Denmark 65
de recuperatione Terrae Sanctae treatises 10, 49, 91–111, 114, 135, 138, 153, 173, 181, 200–18, 248–51, 259, 260, 269–70; *see also* crusade, planning; general crusade; particular crusade
La Devise des chemins de Babiloine 50, 92
Deya 27
Digard, G. 137
Directorium ad faciendum Passagium Transmarinum 215
Dirum amaritudinis calicem, papal bull 74–5, 121, 128–9
Dodecanese 223
Dolcino, the leader of the Apostolic Brethren 170
Dominicans 27, 28, 48, 106, 125, 126 n., 127, 132, 134 n., 214
Donation of Constantine 101, 104, 133
Doria, family of 165
Dura nimis, papal bull 74–5, 129
Durazzo 99
Durham 68

Eberhard, archdean of Regensburg 118
Edmund of England, duke of Lancaster 54
Edward I, king of England 19, 37, 44, 45–6, 52, 54, 63, 71, 72, 79, 84, 85, 86, 88, 89, 137, 142, 149, 155, 157, 167, 173, 174, 187, 188, 195, 234, 254, 261, 267; and the Holy Land 44–6, 71–2, 85–6, 187–8, 267; will of 86
Edward II, king of England 240, 255
Egypt 17, 50, 56, 77, 78, 80, 81, 82, 83, 87, 91, 92, 98, 99, 101, 102, 109, 110, 111, 113, 123, 131, 132, 148, 150, 164, 167, 172, 177, 178, 179, 190, 191, 194, 195, 204, 207, 211, 212, 213, 215, 224, 228, 231, 257, 267; army of 96, 216; embassies to from the West 190–1, 256; expeditions against 77–8, 163–5; illegal trade with 79–83, 151, 166, 178, 190–1, 193, 194; maritime blockade of 79–83, 91, 109, 110, 150, 151, 154, 162, 178, 190, 192, 193, 195, 201, 202–3, 204, 205, 206, 216, 219, 224, 228, 231–3, 258, 259, 260–1
Elias, patriarch of Jerusalem 55
England 18, 20, 36, 46, 47, 57, 68, 71, 79, 84, 86, 90, 115, 118, 141, 142, 143, 149, 155, 156, 175, 201, 209, 210, 236, 244, 247, 251, 252, 253, 254, 255, 259, 265, 267
Enguerrand de Marigny 253
Erart of Valeri 18–19, 37, 38
eschatological expectations 113, 115–21, 122–4, 130, 134, 142–3, 169–71, 265; *see also* Last Days
Este 70
Ethiopia 87, 215
Euboea 231
Euphrates, river 215
Exsurgat Deus, papal bull 221

Fakhr-al-Din, emir 191
Famagusta 77, 164, 165
Ferdinand, king of Castile 240
Fidenzio of Padua 22, 49, 75, 76, 82, 83, 89, 92, 93–102, 107, 108 n., 109, 138, 150, 212, 222–3; *Liber recuperationis Terrae Sanctae* 93–102, 107
Finke, H. 5, 242
Ferrara, diocese of 153
Flanders 86, 161, 186, 234, 261
Florence 124, 174, 271; convent of Santa Maria Novella 124, 126 n.
Focea 231
Foucalquier 45
France, kingdom of 18, 37, 56, 58, 71, 79, 84, 86, 90, 115, 118, 132, 144, 149, 155, 156, 175, 176, 180, 186, 188, 199, 201, 202, 206, 208, 212, 222, 226, 227, 230, 239, 241, 249, 259, 265, 267, 271; contingents sent to Acre 19; crusading propaganda in 161, 198, 220–1, 234; crusading taxation of the church 65, 141, 152–3, 177, 183; opposition to crusading taxation 47, 57; prelates from at the Council of Vienne 251, 253; pro-crusade frenzy in 235–7, 255–6; property of the Templars in 244–5; *see also* Louis X, Philip III, Philip IV
Francesco Pipino, Bolognese Dominican and chronicler 57, 132
Franciscans 48, 68, 87, 94, 106, 117, 165, 191, 214

Index 303

Franciscans, Spiritual 142, 143
Frederick I, emperor 18, 209, 210
Frederick II, emperor 81, 131
Frederick III, king of Sicily 149, 156, 159, 160, 170, 177, 178, 187, 189, 191, 192, 211, 223 n., 228, 240
Freidank 131
Frisland 118
Fulk of Villaret, master-general of the Hospital 183, 187, 193 n., 197, 199, 202, 203, 212, 219, 220 n., 223, 227, 230, 231, 232, 252; plan for crusade of 202, 203, 204, 219

Gaeta 66
Galfridus de Semari 167
Gallipoli 180
Galvano of Levanto 92, 144, 145–6; *Liber Sancti passagii Christicolarum contra Saracenos* . . . 145–6; *Thesaurus religiose paupertatis* 146
Gascony 86
Gaza 117, 163, 213
Geanakopolos, D. 159
Genappe 236
general crusade (*passagium generale*) 16–19, 24–5, 37–50, 52, 71–91, 99, 103–4, 106, 107, 109–10, 174, 176, 190, 202, 203, 205, 209, 210, 212, 221, 250, 251, 252, 260, 268; land force of 201, 202, 205, 209, 212; led by warrior king 109–10, 202, 207, 210–11, 217; maritime force of 201, 202, 203, 204, 205; organized by: Boniface VIII 150, Clement V 183, 221, 242–3, 246–7, Gregory X 37–50, Innocent V 52–4, Nicholas IV 71–2, 74–91; route of 99, 106, 109, 138, 209, 210, 211, 213–14, 252; *see also* crusade planning; *de recuperatione Terrae Sanctae* treatises; particular crusade
Genghis Khan 43
Genoa 46, 83, 84, 156, 192, 202, 204, 231, 241; accused of responsibility for the fall of Acre 73; community of in Cyprus and Lesser Armenia 77; community of in Outremer 40–1; crusade of 1300 165–6; crusade of 1309 224–5; crusade against Byzantium 179, 184, 186, 196; crusade planning 99, 105, 106, 210; expeditions against Mamluks 77–8; navy of 43, 77–8, 85, 223, 224; relations with Egypt 72–3, 83; trade with the Moslems 41, 178, 193, 225, 232, 233, 261
Geoffrey of Charney 256
Geoffrey of Paris 255
Geoffrey of Sergines 19
George Metochites, archdeacon of Hagia Sophia 42

Georgia, kingdom of 82, 87, 99, 166, 215
Gerard Picalotti, bishop of Spoleto 84
Gerard of Villiers, preceptor of the Temple in France 165
Gerhoh of Reichsburg 130; *Libellus de Investigatione Antichristi* 130
Germany 16, 18, 36, 47, 57, 134, 149, 153, 175, 201, 209, 234, 235, 236, 237, 244, 253, 254, 265
Ghazan, il-khan of Persia 162, 163, 164, 166, 167, 168, 171–2, 173, 174, 175, 195, 214
Ghent 235
Ghibellines 160
Gibraltar 251
Gilbert of Tournai 16, 19, 22, 23–5, 35, 91; *Collectio de Scandalis Ecclesiae* 23–5, 91
Giles of Rome 145, 146; *De regimine principium* 145
Giles of Sanci 19
Gilles le Muisit (Aegidius di Muisis), abbot of the Benedictine St Martin of Tournai 118
Giordono of Rivalto 113
Giovanni Villani 131, 234, 245
Godfrey of Bouillon 35, 168, 209
Gog and Magog 169
gold florin, *see* monetary system
Gorighos 195
Gotholosa 172
Granada 190, 192, 207, 217, 228, 229, 251, 253, 267; crusade against 190, 192, 207, 217, 228–9, 251, 253, 261
Greece 187, 209, 252
Greek Empire, *see* Byzantium
Greek islands 177, 178, 180
Greek-Orthodox church 30, 46, 158, 250; union with Roman Church 42–3, 46, 105, 107, 158, 159, 208
Gregory I, pope 115, 127; *Morals on Job* 125, 127
Gregory VII, pope 104
Gregory VIII, pope 74
Gregory IX, pope 39
Gregory X, pope 3, 11, 12, 19, 23, 28, 29, 35, 51, 52, 53, 55, 56, 58, 59, 62, 65, 66, 71, 75, 93, 94, 150, 158, 189, 240, 263, 267, 271; and aid for Outremer 21; and the Byzantine empire 42, 43, 87, 158; and commutation of vows 49, 263; and the crusade 20–2, 36–50; and the economic blockade of Egypt 39–40, 82, 50–1; and the Mongols of Persia 43–4; *see also Constitutiones pro zelo Fidei*; Lyons, second council of
Gregory, Armenian patriarch 195
Grimaldi, family of 165
Guelfs 6
Guiscard Bustari 174

Guy of Flanders 85, 153
Guy of Ibelim, lord of Beirut and
    Jaffa 163
Guy of Montefeltro 160

Hadrian V, pope 3, 50, 55, 57, 69 n.
Haifa 73
Hamah (Hamath), 112 n.
Hayton 92, 98, 171, 172, 183, 195, 201,
    202–3, 212–14, 215, 224; *Flos historiarum
    Terre Orientis* 195, 201, 202–3, 214,
    215
Heidelberger, F. 5, 8, 242
Henry VII of Luxembourg, emperor 104,
    188, 196, 210 n., 240
Henry II, king of Cyprus and
    Jerusalem 78, 98, 101, 131, 154, 155–6,
    163, 164, 165 n., 193, 194, 195, 203, 212,
    213, 215, 240, 248; plan for crusade
    204, 212, 213
Henry of Ghent 113
Hetoum II, king of Lesser Armenia 78,
    79, 162, 163, 195, 211, 214
Hillgarth, J. N. 9, 104, 189, 198, 208
Holy Land, conquest by the Mongols
    (1300) 162–6; in public opinion 166–75
Holy Land, loss of 70–3, 74–139, 150,
    258–60; explanations of 114–34, 150;
    impact of 258–60; in public
    opinion 112–34; reaction to 74–139; *see
    also* Acre, fall of
Holy places 146, 171, 174, 189; of
    Jerusalem 163
Holy Roman Empire 104, 146, 147
Holy Scriptures 31, 34, 36, 114, 115, 116,
    118, 122, 123, 126, 127, 129
Holy Sepulchre, *see* Jerusalem
Holy Trinity, language school at
    Miramar 106
Homs 162, 164
Honorius IV, pope 8, 51, 63–6, 71, 107
Hospitallers 11, 18, 47, 50, 99, 132, 140,
    168, 192, 194, 195, 254, 259, 260;
    blamed for the fall of Acre 130–1;
    conquest of Rhodes by 187, 222–4, 228,
    230–1; crusade planning 103, 109, 136,
    137, 202, 204, 208, 249; in Cyprus 77,
    153–6; enforcing the blockade of
    Egypt 193, 231–3; expedition to Syria
    (1300) 163–4; navy of 82, 109, 204;
    passagium particulare of 18, 193, 194, 204,
    219–38, 262, 265, 271; and the property
    of the Temple 244–5; proposals for
    reform 154–5, 227; proposals for union
    with Templars 75, 76, 109, 137, 197; *see
    also* Military Orders
Hostiensis 100–1
Housley, N. 3, 5–7, 8, 263
Hugh III, king of Cyprus 37, 59
Hugh, bishop of Byblos 68

Hugh Revel, master general of the
    Hospital 37
Hulagu, il-khan of Persia 43, 44, 169
Humbert of Romans 3, 19, 22–3, 25, 27,
    28–35, 36, 37, 48, 49, 75, 82, 91, 92, 97,
    120–1, 134, 226; *De praedicatione
    Crucis* 28, 34; *Opus tripartitum* 3–4, 25,
    28–35, 91
Hungary 58, 61, 65, 75, 86, 106, 117, 153,
    209, 210

Ibelins, family of 194
Iberian peninsula 228
Ibn Furat 112 n.
Ido Poilechien 61
India 126
Innocent III, pope 21, 36, 39, 41, 48, 76,
    81, 150, 158, 240
Innocent IV, pope 27, 39, 41, 100–1,
    264
Innocent V, pope 3, 50, 52–5, 57, 158
Ionian sea 231
Iorga, N. 4
Ireland 46
Isol le Pisan (Zolus Bofeti) 164, 174
Italy 46, 160, 161, 209, 210, 240, 253, 263,
    265; eschatological expectations in 175;
    Italian communes in the Latin
    Kingdom of Jerusalem 40–1, 96, 122,
    132; Italian crusaders 7, 64, 69–70, 132;
    maritime cities of 40–1, 135, 192;
    popular movements in 236; Templar
    property in 244; tenths in 66; *see also*
    crusade, in Italy

Jacques de Plany, Templar knight 244
Jacques de Vitry 24, 41, 94, 95, 131,
    206 n.; *Historia Orientalis* 94
Jaffa 16
James I, king of Aragon 18, 36, 37–8, 45
James II, king of Aragon 67, 89, 109, 137,
    176, 187, 203, 208, 233, 238, 240, 241,
    253; and the crusade against
    Byzantium 177, 178, 185; and the
    crusade against Granada 190, 192, 207,
    217, 228–9, 250; and the Holy
    Land 151–2, 188–93, 228–9, 261, 267;
    relations with the sultans of Egypt 83,
    167, 189–92; sends aid to Acre 69
James II, king of Majorca 66, 106, 190, 240
James Colonna 161
James Doria 77, 79, 82
James of Molay, master-general of the
    Temple 99, 154, 155, 163, 164, 183,
    197, 199, 201, 202, 212, 256, 257; plan
    for crusade 201–2
James of Verona 112
Jativa 106
Jerusalem 18, 20, 43, 44, 52, 73, 74, 88,
    90, 95, 106, 111, 116 n., 119, 120, 123,

124, 134n., 140, 158, 163, 166, 170, 177, 200, 219, 226, 250, 252, 265, 267; celestial 116n., 122; conquest of by the Mongols (1300) 163, 167–8, 169; Dome of the Rock 163n.; Golgotha 120; Holy Sepulchre 4, 49, 56, 119, 166, 167, 171, 179, 200, 237; Mount of Olives 119
Jerusalem, kingdom of 2, 17, 18, 19, 20, 37, 40–1, 49, 51, 52, 59, 60, 61, 62, 66, 67, 68, 69, 70–1, 75, 93, 95, 108, 110, 122; criticism of 94–6; crown of 37, 59, 190, 191, 198, 210–11; Italian communes in 40–1, 96, 122; recommendations for maintenance after recovery 99–100, 110, 136, 153
Jews 28, 30, 119, 129, 136, 236
Joachim of Fiore 49, 116n., 121, 122–4; *Expositio magni Abbatis Joachim in Apocalipsim* 123; Joachimite expectations 142–3, 170; *Super Hieremiam Prophetam* 123
Joan of Navarre, queen of France 198
John XXI, pope 27n., 51, 54, 55, 59, 158
John XXII, pope 81, 204, 249, 264
John, duke of Brabant 236
John, Italian friar 112–13
John of Antioch 163
John of Brienne, king of Jerusalem 60
John of Brittany 171, 196
John of Chalon-Arlay 237
John of Grailly 37, 67, 69, 122
John of Ibelin, lord of Arsuf 61
John of Ipres 195n.
John of Joinville 130, 180–1; *Life of Saint Louis* 130, 180–1
John of Monte Corbino 88
John of Paris, chronicler 165, 255
John of Parma, prophecy of 170
John of Villiers, master-general of the Hospital 79, 113n., 115, 174
John Pecham, archbishop of Canterbury 138
Joseph of Cancy, a Hospitaller 63

Kaffa 231
Kaiserheim 167
Koran 124, 127
Kreuxbrueder 235
Kurds 212

Lambeth 136
Laodicea (Lycaonia) 106
La Spezia 145
Last Days 118–21, 122–3, 130, 134, 139, 142, 170, 265; *see also* eschatological expectations
Last Emperor, prophecy of 119–20, 170
Lateran, third council of 80
Lateran, fourth council of 21, 36, 39, 50, 76, 80

Latin Kingdom, *see* Jerusalem, kingdom of
Laurent, M. H. 3, 90
Lausanne 45
Lebanon 203
Leo IV, king of Lesser Armenia 196, 214
Lerida 171
Leros 223
Lesbos 231
Lesser Armenia 82, 95, 98, 185, 189, 193, 215, 228, 256; aid from the West 77, 78–9, 150, 156–7, 195–7, 201, 212, 219, 221; alliances of with the Mongols of Persia 89, 162, 163, 164, 166, 167–8; attempts at union with the church of 87, 105; as stopping point of a general crusade 99, 103, 106, 207, 212–13, 252, 256
Levanto 145
Liège 20; St Jacques of Liège 117, 134
Limassol 223
livre tournois, *see* monetary system
Lombardy 69, 210
London 20; New Temple 136; St Paul's 86
Louis IX, king of France 20, 28, 36, 37, 51, 52, 76, 113, 134, 199; aid to the Latin Kingdom 19; crusade to Egypt 17, 213; crusade to Tunis 8, 15, 35, 50, 59, 258, 259, 267; as the ideal of crusading king 127, 147, 181–2; relations with the Mongols of Persia 43–4
Louis X of Navarre, king of France 211, 220, 242, 255
Louis of Evreux 211
Low Countries 236
Lübeck 234
Luceta, abbess 68
Lucia of Antioch 60
Lucia of Gouvain 61
Lupus de Cuscia 126
Lusignans 59
Luttrell, A. 154, 223 n.
Lyons 141, 182, 183, 199
Lyons, conference of 193
Lyons, first council of 39, 50, 80
Lyons, provincial council of 137
Lyons, second council of 7, 10, 11, 12, 13, 16, 61, 75, 76, 94, 98; as beginning of new era in crusade planning and making 90–1, 93, 139, 200, 202, 247, 248, 258; crusade planning at 18–19, 21–2, 36; decisions of 36–44, 49, 50; Mongol embassy at 89; prohibition of commerce with Moslems 39–41, 80, 82, 83, 178; tenth of 39, 46, 47, 51–2, 54–5, 56–8, 65, 153, 157; union of the churches at 158–9; see also *Constitutiones pro zelo Fidei*; Gregory X

Index

Macedonia 184
Magdeburg 47
Majorca 27, 106, 171, 243, 251
al-Malik an-Nasir, sultan of Egypt 190
Malta 202
Mamluks 38, 49, 50, 78, 88, 94, 96–9,
  140, 163, 169, 172, 196, 203, 215, 216,
  230, 261; army of 96, 216; *see also*
  Egypt
Manuel Zaccaria, admiral 77
Maraclea 164
Marco Polo 112
Mardin 172
Margaret of Beaumont 60
Margaret of France 66
Maria of Anjou 159
Maria of Antioch 37, 59, 211
Maria of Lusignan 187
Marino Sanudo Torsello 92, 98, 101, 171,
  177, 183, 187, 194 n., 201, 203, 204, 205,
  212, 213, 215, 218, 225; *Conditiones Terrae
  Sanctae* 178, 204, 205, 218; *Liber
  Secretorum Fedelium Crucis* 178, 204, 205
Marj-al-Suffar 172
Marseilles 40, 41, 43, 224
Martin IV, pope 51, 58–66, 67, 71, 107,
  158–9, 264
Matthew of Clermont, marshal of the
  Hospital 122
Matthew Paris 117, 254
Mellorus de Ravendel 108
Memoria Terrae Sanctae 215
Menko, Premonstratensian abbot of
  Bloemhof (Floridus Hortus) 118
Menteshe 232
Merlin, prophecy of 117, 119–21
Messina 121
Michael VIII Palaeologus, Byzantine
  emperor 42–3, 158, 159, 177
Michael IX Palaeologus, Byzantine
  emperor 87, 179
Michaud, J. F. 2
Milano, council of 137 n., 138
Military Orders 17, 23, 37, 60, 79, 92, 95,
  109, 110, 140, 153–6, 161, 162–5, 168,
  193, 194, 196, 197, 198, 199, 200, 202,
  208, 241, 243, 244, 245, 246, 259, 260;
  criticism of 76, 97, 130, 131, 153,
  220–1, 243, 259; in Cyprus 153–6; fleets
  of 154; proposals for union of 75,
  109–10, 135, 137, 138, 154–5, 197, 198,
  200, 202, 210, 211, 217, 241; *see also*
  Hospitallers, Templars
Minorca 251
Miramar 27, 107
missions, to the Saracens 25–8, 101,
  124–5, 250; to the Mongols 87–95, 103,
  214; schools for missionaries 106–7,
  246–7
Modon 231

Mohammad 30, 33, 34, 102, 125, 126, 127,
  134
Mollatt, M. 17
monetary system 271–2
Mongols 30, 50, 75, 109, 117, 126;
  attempts at conversion 87, 88, 90, 105;
  conquests of the Holy Land by 89, 108,
  120, 152, 162–75, 190, 236, 263, 265;
  diplomatic relations with the
  West 43–4, 87–8, 89–90, 189, 214–15;
  in crusade planning 82, 88–9, 99, 203,
  204, 212, 213–16, 260; relations with
  Lesser Armenia 162–4, 195–6
'Monk of Lido' 40
Montpellier, conference of 190, 192, 207
Morea 231
Morocco 103, 228, 251
Moslems 6, 21, 64, 68, 79, 86, 90, 95, 127,
  131, 132, 148, 154, 169, 172, 173, 232,
  243, 251; apocalyptic conversion
  of 119–20; attempts at conversion
  of 25–9, 105; justifications for war
  against 29–33, 70; missions to 25–8,
  200; praise of 96–7, 122, 125–6;
  prohibitions against trade with 39–40,
  202, 203; trade with 40–1, 95, 151, 202,
  203
Mossul 125
Mount Athos 185
Mulai, emir 163
Münster 235 n.
Murcia 106
Myconos 204

Naples 144, 180, 185, 190, 233
Narbonne 224
Narjon of Toucy, admiral 60
Navarre 53
Naxos 204, 231
Negroponte 184, 231
Nephim 69
Neymerich Dusay 191
Niccolò of Poggibonsi 163 n.
Nicholas III, pope 3, 51, 55–8, 158
Nicholas IV, pope 12, 52, 64, 91, 93, 94,
  106, 107, 108, 109, 111, 113, 132, 139,
  140, 149, 151, 154, 176, 178, 189, 193,
  218; and the crusade against
  Byzantium 159; and the crusade against
  Sicily 71, 84, 137, 141; and the Holy
  Land crusade 66–7, 69, 71–2, 74–91,
  258, 259; reaction to the fall of
  Acre 74–6, 77; and redemption of
  crusading vows 153, 263
Nicholas Boccasini, *see* Benedict XI
Nicholas de Hanapes, patriarch of
  Jerusalem 67, 122, 127, 131
Nicholas de Salix, Franciscan prior of the
  Holy Land 145
Nicholas Falcon 195 n.

Nicholas Tiepolo 69
Nicosia 77, 174
Nile 73, 228
Nola 232
North Africa 106, 217
Norway 86
Norwich 47
Nubia, *see* Ethiopia
Nuremburg 37

Odo des Pins, master-general of the Hospital 155
Olim tam in generali, papal bull 83
Oljaitu, il-khan of Persia 214, 215
Orkham, Turkoman emir 232
Orontes 99
Orvieto, treaty of 60–1
Oschin, king of Lesser Armenia 197
Otho of Grandison 71, 79
Otto Visconti, archbishop of Milan 138
Ottokar, king of Bohemia 23, 46
Ottokar von Steiermark 127, 133
Ottoman Turks 158
Oxford 246

Palermo 156, 177
Palestrina, fortress of 160
Papacy 7–8, 103, 104, 105, 109, 115, 119, 124, 132–3, 140, 145, 161, 162, 176, 202, 207, 214, 264, 267–8; criticism of 94, 96, 115, 132, 170, 226, 254; and the European crusades 5–6, 58, 61–2, 63–7, 84, 85, 140–2, 147–62, 228–9, 251–3, 261, 263; fleets of 77–9, 151; and the Holy Land crusade 20–2, 36–67, 69–91, 147–62, 181–99, 230, 246–57, 268; Jubilee of 1300 162, 169–71; missionary activity 87–90
Paris 27, 65, 106, 113, 246, 255, 256; Île-de-la-Cité 255; Île-des-Javiaux 256; university of 27, 102, 113
Parma 69 n., 70
'Particular' crusade (*passagium particulare*) 16–19, 20, 22, 23, 32–3, 37–8, 49, 50, 52, 53, 97–8, 103, 109, 201, 202–5, 212, 262; aims of 16–19, 97–8, 109, 202–5; forces of 97–8, 109, 201, 202, 203, 204, 212, 225, 228; Genoese of 1300 165–6; naval bases of 98, 201, 202, 204; papal-Hospitaller of 1309 18, 193, 204, 212, 219–38; *see also* crusade planning; *de recuperatione Terrae Sanctae* treatises; general crusade
Pastoraux 11, 117, 236, 256; of 1251 117, 236; of 1320 236, 256
Pèlerinage de Charlemagne 48
Pera 231
Persia 43–4, 87, 98, 162, 163, 169, 171, 172, 173, 188, 189, 190, 191, 195, 214, 215

Perth, council of 47
Peter I, king of Cyprus 213
Peter III, king of Aragon 55, 62, 65
Peter Colonna 161
Peter John Olivi 142
Peter of Morone, *see* Celestine V
Peter of Peredo 227
Peter of Pleine Chassagne, bishop of Rodez and patriarch of Jerusalem 225, 227, 230
Peter of Zittau 240
Pheraclos, castle of 223
Philermos, castle of 223
Philip I Augustus, king of France 271
Philip III, king of France 19, 21, 37, 45, 46, 52, 53, 54, 55, 56, 57, 58, 63, 65, 66, 85
Philip IV, the Fair, king of France 7–8, 12, 13, 27, 66, 88, 104, 105, 112, 124, 155, 184, 204 n., 209 n., 230, 256; and Benedict XI 176; and Boniface VIII 133, 148, 149, 157, 159–60, 161–2, 263; and Clement V 182–3, 187, 193, 196, 199, 220–1, 241–3; and the Council of Vienne 240, 244, 247, 251, 253; and the crusade against Byzantium 186, 187, 217; and the crusade of 1309 222, 226–7, 237; and the Holy Land crusade 63–4, 84–5, 120, 137, 147, 161, 162, 182–3, 198–9, 222, 226–7, 230, 242, 255, 266–7; image of 145, 147, 181–2, 241, 246, 248, 253, 254, 261; and the Military Orders 197–9, 210, 220–1, 240–2, 244, 262; will of 25 n.
Philip V, king of France 211, 220, 255, 262
Philip of Courtenay, titular emperor of Constantinople 60
Philip of Flanders 16
Philip of Taranto 158, 178, 186, 256
Picardy 117, 236
Pierre of Aminnes 19
Pierre Dubois 92, 101, 104, 146, 147, 159, 183, 200, 202, 206, 208–12, 217, 248; *De Recuperatione Terrae Sanctae* 147, 201; *Oppinio cujusdam* 202, 210, 211, 212; *Pro facto Terre Sancte* 210–12; *Summaria brevis* 159, 209
Pietro Gradenigo, doge of Venice 163 n., 173, 184
Pisa 40, 43, 77, 83, 105, 224, 225; bank of Clarenti 157; consul of Acre 131
Poitiers 188, 195, 197, 199, 214, 220
Poland 30, 65, 75, 153, 234
Portugal 243
Portus Palorum 99
Power, E. 208
Prawer, J. 35
Premonstratensians 117, 195
prices 271–2

Processus adversus Fredericum et Siculos rebellos 174
Provence 45, 65, 66, 253
Prussia 30, 106
Ptolemy of Lucca 234, 235, 237
Purcell, M. 3, 8

Qalawun, sultan of Egypt 72–3, 97, 179

Rabban (Mar) Suama 88
Ragusa 178
Rahbatal-Shām (Meyadim) 215
Ramadan 172
Ramon Lull 26, 75, 76, 90, 97, 98, 100, 101, 102–7, 108n., 109, 110, 111, 120, 138, 143–4, 171, 173, 183, 190, 192, 197, 198, 200, 202, 203, 206, 207, 217, 223, 229, 247, 248, 249; *Blanquerna* 102, 106; *Epistola pro recuperatione Terrae Sanctae* 102, 110; *Liber de acquisitione Terrae Sanctae* 107, 202, 207, 217; *Liber contra Antichristum* 102; *Liber de fine* 107, 183, 198, 202, 207, 217; *Petitio conversione infidelium* (1295) 144; *Petitio pro conversione infidelium* (1294) 143, 144; *Tractatus de modo convertendi infidelis* 102, 107, 110, 198, 207
Ramon of Peñafort 27, 106
Raoul of Brienne, count of Eu 211
Rashid ad-Din 172
Raymond Gaucelin, troubadour 53
Raymond of Pins, papal nuncio 230
Red Sea 98
Regensburg 118
Reggio 85
Regnans in excelsis, papal bull 221, 239, 240, 243
Reims, council of 136
Rhine 236
Rhodes 98, 141, 187, 203, 219, 222, 223, 224, 228, 230, 231, 232, 244, 261; conquest of 199, 219, 222–4
Riant, P. 4, 94, 121
Richard, J. 10
Richard I, 'the Lion Heart', king of England 35
Richard Mepham, dean of Lincoln 47
Ricoldo of Monte Croce 83, 97, 114, 122, 124–8; *Epistolae* 114, 124–8; *Improbatio Alcorani* 124–5 n.
Riley-Smith, J. 5, 10
Robert, English knight 227
Robert, king of Naples 191, 232
Roger Bacon 25, 26–7, 142 n.
Roger de Flor 179
Roger Lauria 165, 166, 189, 202
Roger of San Severino, count of Marsico 59, 61
Roger of Thodino 77, 79
Rogeronus 202

Röhricht, R. 2–3
Romagna 65
Roman Empire 42, 100, 101, 248
Romania, empire of 159, 179, 228
Rome 45, 53, 56, 84, 106, 107, 139, 144, 158, 169, 171, 174, 195, 240, 246; St John Lateran 240; Vatican palace 57
Rosetta, island of 205, 207
Roux of Sully 69
Ruad, island of 98, 191, 203, 243; conquest of 164–5, 193 n., 194, 201
Rudolph of Habsburg, king of Germany 23, 42, 45, 46, 53, 56, 72, 267; and the Holy Land 53, 56, 87, 267

St Basil, order of 132
St Blasius, annals of 234
St Dominic 127
St Florian, monastery of 237
St Francis of Assisi 127
St Jacques, Benedictine monastery in Liège 117, 134
St James, order of 47
St John, order of, *see* Hospitallers
St Martin, abbey of 271
St Peter 51, 52, 119, 170, 180
St Simeon (Soldinum) 99
St Thomas 215
Safed 15, 96 n., 179
Safi 103
Saladin 243
Salamanca 246, 247
Salimbene de Adam 20, 49
Salvator noster in, papal bull 21–2, 76
Salzburg 118
Sancho of Aragon 177, 179, 223
Saracens, *see* Moslems
Sardinia 61, 65, 152, 159, 192, 251
Saxons 30
Scandelore (Alaya) 78
Schöpp, N. 3
Scotland 46, 47, 65, 71, 86, 152, 187, 188
Sempad, king of Lesser Armenia 156
Senlis 177
Sens 21; council of 136, 137, 145
Serbia 185
Setton, K. M. 3, 7–8
Si mentes fidelium, papal bull 48
Siberry, J. E. 4
Sicily 18, 36, 43, 45, 46, 60, 105, 180, 183, 185, 187, 190, 201, 251; crusade against 52, 58, 62–7, 69, 70, 71, 84, 108, 133, 137, 140, 141–2, 151, 153, 156–9, 160, 161, 166, 174, 189, 281, 263, 267, 268; Sicilian Vespers 61, 62, 63, 107, 141
Sidon 73
Silesia 234
Simon of Beaulieu, bishop of Palestrina 149

Simon of Brie, cardinal priest of St
  Cecilia 45, 48, 58
Sis, council of 195
Skyros 231
Slavs 23
Smyrna (Ismir) 231, 236
Sodom 127, 129
Spain 103, 106, 110n., 207, 209, 229, 251,
  252, 253
Spinola, family of 165
Statin 37
Stephen of Blois 16
Sternfeld, R. 3
Stickel, E. 35, 134
Strayer, J. R. 8, 266
Sweden 65
Syria 1, 18, 95, 98, 146, 154, 155, 164, 166,
  179, 213, 214, 215, 231, 262, 264; in
  crusade planning 99, 101, 103, 150, 203,
  205, 213, 249; Mongol invasions 89,
  140, 162–3, 164, 168, 171, 172, 174

Tagliacozzo, battle of 46
Tarsus 195, 213
Taurus 106
taxation, of the church for crusades 36,
  46–7, 56–7, 242–3, 246, 250–1, 253;
  opposition to 46–7, 57, 246, 250–1; *see
  also* crusading tithes
Tedisio Doria 77
'Templar of Tyre' 113n., 171
Templars 18, 23, 47, 49, 67, 75, 76, 77, 78,
  79, 82, 103, 109, 130, 131, 132, 136, 137,
  140, 153, 154, 155, 161, 163, 164, 165,
  168, 191, 192, 193, 194, 197, 199, 201,
  202, 207, 208, 210, 211, 220, 221, 239,
  240–6, 254, 259, 262; property
  of 241–2, 244–5, 250–1, 254, 260; trial
  of 112, 165, 221, 240; *see also* Military
  Orders
Temple, order of, *see* Templars
Teutonic Knights 103, 106, 122, 131, 161,
  224
Thadeo of Naples 83, 114, 115, 120,
  121–4, 128; *Hystoria de desolatione civitatis
  Acconensis*, 114, 121–4
Thessalonica 185
Thessaly 185
Thibaud Gaudin, master-general of the
  Temple 79
Thibaut of Chepoy 184, 185, 186
Thier, L. 5, 8
Thomas Agni of Lentino, patriarch of
  Jerusalem 19, 50
Thomas de la Moore 254
Thomas of Tolentino 79
Throop, P. A. 3–4, 7, 16, 35, 134
Tiel 234
at-Tina 73
Tinos 204

tithes, *see* crusading tithes
Tommaso Ugi of Siena 214
Toron 15
Tortosa 73, 164, 243
Tournai 118
Tours 21
*Tractatus dudum habitus ultra mare per
  magistrum et conventum Hospitalis ...*
  219
trade embargo 39–40, 50, 138, 202, 268;
  through maritime blockade of
  Egypt 79–83, 91, 97, 103, 109, 110, 150,
  151, 154, 166, 177, 178, 193, 201, 202,
  203, 204, 205, 206, 207, 212, 219, 224,
  231–3, 254, 258, 259, 260–1; *see also*
  commerce
Tripoli 17, 41, 67–9, 71, 73, 99, 103, 109,
  126, 165, 166, 203, 226; Cistercian
  monastery in 69; convent of Poor
  Clares 68; fall of 67–9, 71, 73, 92, 95,
  113, 117, 119, 122, 125, 126, 127, 170,
  191
Tripoli (in Barbery) 103, 106
Tripoli, vision of 68–9, 117–19, 134
Tunis 8, 15, 20, 35, 44, 50, 54, 59, 65,
  101, 106, 205, 207, 211, 258, 259,
  267
Turcopoles 17
Turkey 101, 205
Turks 179, 230, 232–3
Tuscany 69, 210
Tyre 40, 41, 73, 80, 146

Uppsala 134n.
Urban II, pope 18
Urban IV, pope 48, 107, 158

Valencia 65, 106
Venice 46, 59, 73, 84, 99, 105, 106, 156,
  173, 193, 202, 206, 210, 230; aid to
  Outremer 69; communities in Cyprus
  and Lesser Armenia 77; communities
  in Outremer 40–1; criticism of 95;
  and the crusade against Byzantium 61,
  180, 184, 186; diplomatic relations with
  Egypt 178–9; and the Hospitaller
  conquest of Rhodes 231–3; navy
  43, 204, 205, 224; trade with the
  Moslems 41, 81, 83, 178, 225, 261
Verum 118
Via ad Terram Sanctam 215
Vienne 141
Vienne, council of 11, 12, 13, 107, 186,
  203, 207, 217, 238, 239–57, 268; *naciones*
  at 253, 260; tenth of 242–3, 253
Vignolo de Vignoli, Genoese pirate 223,
  232
Villehardouin, dynasty of 231
Viterbo, treaty of 58
Vox in excelso, papal bull 242–3

Wales 46, 86, 153
Walter of Guisborough 153
warrior-king (*bellator rex*) 109–10, 202, 207, 210–11, 217
Westphalia 234
Weichard of Polhaim 118
Wieruszowski, H. 104
William Adam 206 n., 215
William Durant 'the younger', bishop of Mende 203, 217, 249, 250, 251; *Informatio brevis ...* 249–50; *Tractatus de modo generalis concilii celebrandi* 249
William Le Maire, bishop of Angers 250–1
William Sanudo 204
William II of Agen, patriarch of Jerusalem 18

William of Beaujeu, master general of the Temple 37, 38, 82, 122, 127
William of Nangis 62, 132
William of Nogaret 92, 104, 146, 148, 200, 215, 217, 243, 248, 249, 250, 256; plan for crusade 249–50
William of Plaisians 220, 241
William of Roussillon 19
William of Tripoli 22, 25–6, 35, 91, 97; *Tractatus de Statu Saracenorum* 25–6, 91
William of Tyre 48, 206 n.
William of Villaret, master-general of the Hospital 154, 155, 195

Zaccariae, family of 204